AF707177

Winding Dali's Clock: The Construction of a Fuzzy Temporal-GIS for Archaeology

Christopher Thomas Green

BAR International Series 2234
2011

Published in 2016 by
BAR Publishing, Oxford

BAR International Series 2234

Winding Dali's Clock: The Construction of a Fuzzy Temporal-GIS for Archaeology

ISBN 978 1 4073 0796 1

BAR Publishing is the trading name of British Archaeological Reports (Oxford) Ltd. British Archaeological Reports was first incorporated in 1974 to publish the BAR Series, International and British. In 1992 Hadrian Books Ltd became part of the BAR group. This volume was originally published by Archaeopress in conjunction with British Archaeological Reports (Oxford) Ltd / Hadrian Books Ltd, the Series principal publisher, in 2011. This present volume is published by BAR Publishing, 2016.

Printed in England

BAR
PUBLISHING

BAR titles are available from:

BAR Publishing
122 Banbury Rd, Oxford, OX2 7BP, UK
EMAIL info@barpublishing.com
PHONE +44 (0)1865 310431
FAX +44 (0)1865 316916
www.barpublishing.com

FOREWORD

Archaeology is fundamentally concerned with both space and time: dates, chronologies, stratigraphy, plans and maps are all routinely used by archaeologists in their work. To aid in their analysis of this material, the use of Geographic Information Systems (GIS) by archaeologists has become widespread. However, GIS are conventionally ignorant of time. Thus, if archaeologists are to achieve the fullest potential in the application of GIS to their studies, GIS are needed that properly take into account time as well as space.

A GIS capable of dealing with temporal data is referred to as a temporal-GIS (TGIS), and commercial TGIS systems currently exist. However, these are locked into a model of modern clock time. Archaeological time does not sit well within that model, being altogether fuzzier and less precise. Nor are commercial TGIS able to address the questions that archaeologists ask of their spatio-temporal data. Thus, a TGIS is needed that deals with the types of time that we encounter as archaeologists, lest we end up shaping our data and questions to the inherent capabilities of non-archaeological TGIS.

The creation of that new TGIS is the subject of this book: a fuzzy TGIS built specifically for the study of archaeological data that also takes into account recent developments in the theory of temporality within the discipline. The new TGIS needs to be flexible and powerful, yet to ensure that it is actually used it must remain within the software horizons of GIS-literate archaeologists.

The new TGIS has been applied to two case studies, one in prehistoric Derbyshire and one in Roman Northamptonshire, producing informative and interesting new results. It is hoped that others will fruitfully use the TGIS and that, as a result, new forms of spatio-temporal analysis might come to be applied to archaeological studies.

* * * * *

Thanks go to Mark Gillings and Jeremy Taylor for their excellent advice, support and supervision during the completion of the PhD around which this book is based. Further thanks go to Matthew Beamish and University of Leicester Archaeological Services for access to and assistance with the Willington data. Thank you again to Jeremy Taylor for access to and assistance with the Roman Northamptonshire data. Thank you also to David Wheatley and David Mattingly, my PhD examiners, for their insightful comments and for giving me a passing grade. Finally, thank you to my friends, colleagues, and especially my family for listening when needed.

* * * * *

Please note the following:-

All Ordnance Survey data used in the figures presented in this book were obtained under their OpenData license.

All URLs given in this book were correct as of January 2011.

Tables referred to in the text appear in their own section of the book between the final chapter and the bibliography.

The software described in this book can be downloaded from either http://arcscripts.esri.com/details.asp?dbid=16280 or http://www.zen26819.zen.co.uk/TGIS.zip

CONTENTS

Foreword 1

List of figures 3

List of tables 4

Introduction 5

Conceptions of time in archaeology 9

What temporal questions do archaeologists ask? 22

Temporal GIS 39

The creation of the TGIS 57

Radiocarbon dating at Willington, Derbyshire 70

Pottery in Roman Northamptonshire 115

Analysis and conclusions 137

Tables 145

Bibliography 152

List of Figures

1.1 - C14View
3.1 - Harris matrix
4.1 - Linear, branching, and cyclical times
4.2 - TimeMap output
5.1 - Graphs of different probability models implemented in the TGIS
5.2 - TGIS user interface
5.3 - Flowchart showing TGIS functionality
6.1 - OxCal outputs for fallen trees and associated features
6.2 - OxCal outputs for fire clearance of the floodplain
6.3 - OxCal outputs for Burnt Mound I and associated contexts
6.4 - OxCal outputs for Burnt Mound II
6.5 - OxCal outputs for alluviation
6.6 - OxCal outputs for environmental column
6.7 - OxCal outputs for ceramic sequence
6.8 - Map of phases associated with radiocarbon dates
6.9 - Graph comparing probabilities of conventional and Bayesian dates
6.10 - Graph comparing probabilities of different phases
6.11 - Map of Bayesian dates 11810-4000 cal BC
6.12 - Map of calibrated dates 11820-4000 cal BC
6.13 - Map of Bayesian dates 4000-3500 cal BC
6.14 - Map of calibrated dates 4000-3500 cal BC
6.15 - Map of Bayesian dates 3500-3000 cal BC
6.16 - Map of calibrated dates 3500-3000 cal BC
6.17 - Map of Bayesian dates 3000-2500 cal BC
6.18 - Map of calibrated dates 3000-2500 cal BC
6.19 - Map of Bayesian dates 2500-1500 cal BC
6.20 - Map of calibrated dates 2500-1500 cal BC
6.21 - Map of Bayesian dates 1500-921 cal BC
6.22 - Map of calibrated dates 1500-910 cal BC
6.23 - Maps of tree-throw / usage dates in c.250 year time-slices
6.24 - Maps of fire clearance dates from in c.250 year time-slices
6.25 - Maps of Burnt Mound I dates from in c.250 year time-slices
6.26 - Map of Bayesian dates 2340-1840 cal BC
6.27 - Maps of Burnt Mound II dates from in c.250 year time-slices
6.28 - Map of Bayesian dates 1290-1000 cal BC
6.29 - Map of inverse distance weighted interpolation 3750-3500 cal BC
6.30 - Map of inverse distance weighted interpolation 3500-3000 cal BC
6.31 - Map of inverse distance weighted interpolation 2500-1500 cal BC
6.32 - Map of inverse distance weighted interpolation 3366-2750 cal BC
6.33 - Map of inverse distance weighted interpolation 2750-2250 cal BC
6.34 - Map of inverse distance weighted interpolation 2250-2000 cal BC
6.35 - Map of inverse distance weighted interpolation 2000-1614 cal BC
7.1 - Map of site locations
7.2 - Ceramic fabric phasing
7.3 - Map of Taylor 1996 trend surface for Early phase
7.4 - Map of Taylor 1996 trend surface for Early / Mid phase
7.5 - Map of Taylor 1996 trend surface for Middle phase
7.6 - Map of Taylor 1996 trend surface for Mid / Late phase
7.7 - Map of Taylor 1996 trend surface for Late phase
7.8 - Graph comparing temporal profiles for finewares against ceramic assemblage for the villa at Sivier
7.9 - Graph comparing effect of deposition rates on probability profile
7.10 - Map of TGIS trend surface for Early phase
7.11 - Map of TGIS trend surface for Early / Mid phase
7.12 - Map of TGIS trend surface for Middle phase
7.13 - Map of TGIS trend surface for Mid / Late phase
7.14 - Map of TGIS trend surface for Late phase
7.15 - Comparison of Taylor 1996 and TGIS trend surfaces for each phase
7.16 - Graph showing 25% rate of change phasing determined by TGIS
7.17 - Map of TGIS trend surface 110-10 BC

7.18 - Map of TGIS trend surface 10 BC - AD 90
7.19 - Map of TGIS trend surface AD 90-115
7.20 - Map of TGIS trend surface AD 115-240
7.21 - Map of TGIS trend surface AD 240-265
7.22 - Map of TGIS trend surface AD 265-415
7.23 - Map of TGIS trend surface AD 415-500
7.24 - Map of zones for sub-regional analysis
7.25 - Graph comparing probability profiles for sub-regions
7.26 - Graph comparing probability profiles for sub-regions as percentage
7.27 - Map of location of Southwick
7.28 - Graph comparing probability profile for Southwick against overall profile as percentage

LIST OF TABLES

3.1 - Summary of dating techniques not directly implemented in the TGIS
6.1 - Probabilities of tree-throw / usage dates in c.250 year time-slices
6.2 - Probabilities of fire clearance dates in c.250 year time-slices
6.3 - Probabilities of Burnt Mound I dates in c.250 year time-slices
6.4 - Probabilities of Burnt Mound II dates in c.250 year time-slices
7.1 - List of pottery fabrics and their respective dates
7.2 - List of pottery phases and their respective dates

CHAPTER ONE: INTRODUCTION

> *Through my blue fingers, pink grains are falling, haphazard, random, a disorganized stream of silicone that seems pregnant with the possibility of every conceivable shape... but this is illusion. Things have their shape in time, not space alone. Some marble blocks have statues within them, embedded in their future.* (Moore and Gibbons 1986: Chapter 4, pp. 24)

Within direct human experience, everything in the universe exists in both time and space: no object can exist only in space or time in isolation from the other. Whenever we study an object or process in space or time alone, we are inevitably dealing with a simplification: a representation of essential reality. In many cases, this may be a justifiable or adequate situation, depending upon the specific questions which we intend to study. However, when a more complex or specific question arises, the restricted aspatial or atemporal model may then prove incapable of approaching that new hypothesis with sufficient detail or rigour. Thus, when we fix a specific value (or small subset of values) for time or space in our studies, we enter into an intractable compromise that fundamentally limits the scope and complexity of our analyses.

The use of Geographic Information Systems / Science (GIS) is an example of this problem. GIS are computer systems created for the storage, manipulation, study and display of spatial data. Their usage has become widespread in a variety of different fields, including archaeology. Yet conventional GIS take no account of time: temporality may be expressed only as a single fixed state, or as a series of aggregated period layers at best. Time is what separates geography from geometry, and yet temporal detail is foreign to these systems. Timeless space limits conceptualisation and representation: this is a problem for many different users of GIS, but particularly so for archaeologists. Time and chronology are the cornerstones of our discipline: daily, we deal with temporal data, whether radiocarbon dates, pottery typologies or stratigraphic diagrams. Yet none of these types of chronological information are accessible in their fullest detail when GIS are applied to our data.

The archaeological study of space whilst simultaneously ignoring or freezing time can often prove fruitful, but falls short of achieving the full potential of true spatio-temporal analysis. As a result, the integration of time into GIS should provide archaeologists with the ability to pose and answer far more detailed and precise spatial and temporal questions than facilitated by current implementations of GIS technology. As one would expect in a research field as dynamic as spatial technology, the fact that atemporal GIS is incomplete has been noted by many previous researchers, and geographers have produced a number of different studies and software packages that attempt to negotiate and address this inadequacy. These are referred to in the geographic information science literature as temporal-GIS (TGIS), a commercial example of which would be Discovery Software's STEMgis (see Chapter 4). However, these conventional TGIS come with new problems of their own when we, as archaeologists, attempt to appropriate and engage with them.

Geographers' TGIS are locked down into modern clock time, which provides the rigid framework within which data are structured and analysed. Yet the types of temporal data yielded by archaeology rarely fit comfortably into the model provided by the precisely quantified temporality of the present day. Archaeological time is generally imprecise, dates usually being expressed as ranges, often with complex variation in their internal probabilities. In essence, it is much fuzzier than clock time, distorted from its physical basis like the melting clocks of Salvador Dalí. Some archaeological dates might fit into a more precise time-frame, but even these will usually possess a greater degree of uncertainty in their interpretation. For example, a dendrochronological date obtained for a ship timber might be as accurate as a particular season in a particular year, but this date only applies to the felling of the tree. If the timber was aged after felling or salvaged from an old ship and reused, uncertainty creeps back into our temporal interpretation.

Furthermore, in a trend that has accelerated over recent years, archaeologists have theorised at great length about the nature of our temporalities. Many of these temporal theories are, again, incompatible with conventional TGIS. The multiple scales of the *Annales* paradigm or non-linear dynamics do not sit easily with the types of time to which the developers of TGIS envision the application of their software. In essence, geographers' TGIS are designed to deal with the demands of public utilities, city councils, traffic management, etc. These applications may expect to deal with immense volumes of data, but this is data recorded according to a single, precise schema. Archaeological dates and theoretical temporalities do not possess such a monolithic, singular conception of time at their core.

Therefore, if we are not catered for by commercial TGIS developers, we are left with two potential solutions. We can compromise, bending our data and the questions we might ask of it to the inherent capabilities of the commercial TGIS available (a form of technological determinism). Alternatively, we must look to our own solutions. Of course, other archaeologists have considered these questions and developed various solutions (e.g. Castleford 1992, Johnson 1999, Daly and Lock 1999). Yet these previous projects have never fully gotten to grips with the issues discussed above. TGIS applications created by archaeologists have tended to be either very case-specific, or to deal with tangential issues (see Chapter 4). No archaeological or conventional TGIS has ever been created that is able to deal with the inherent uncertainties (or fuzziness) of archaeological dates; nor has a system been created that takes into account alternative archaeological models of time or provides the

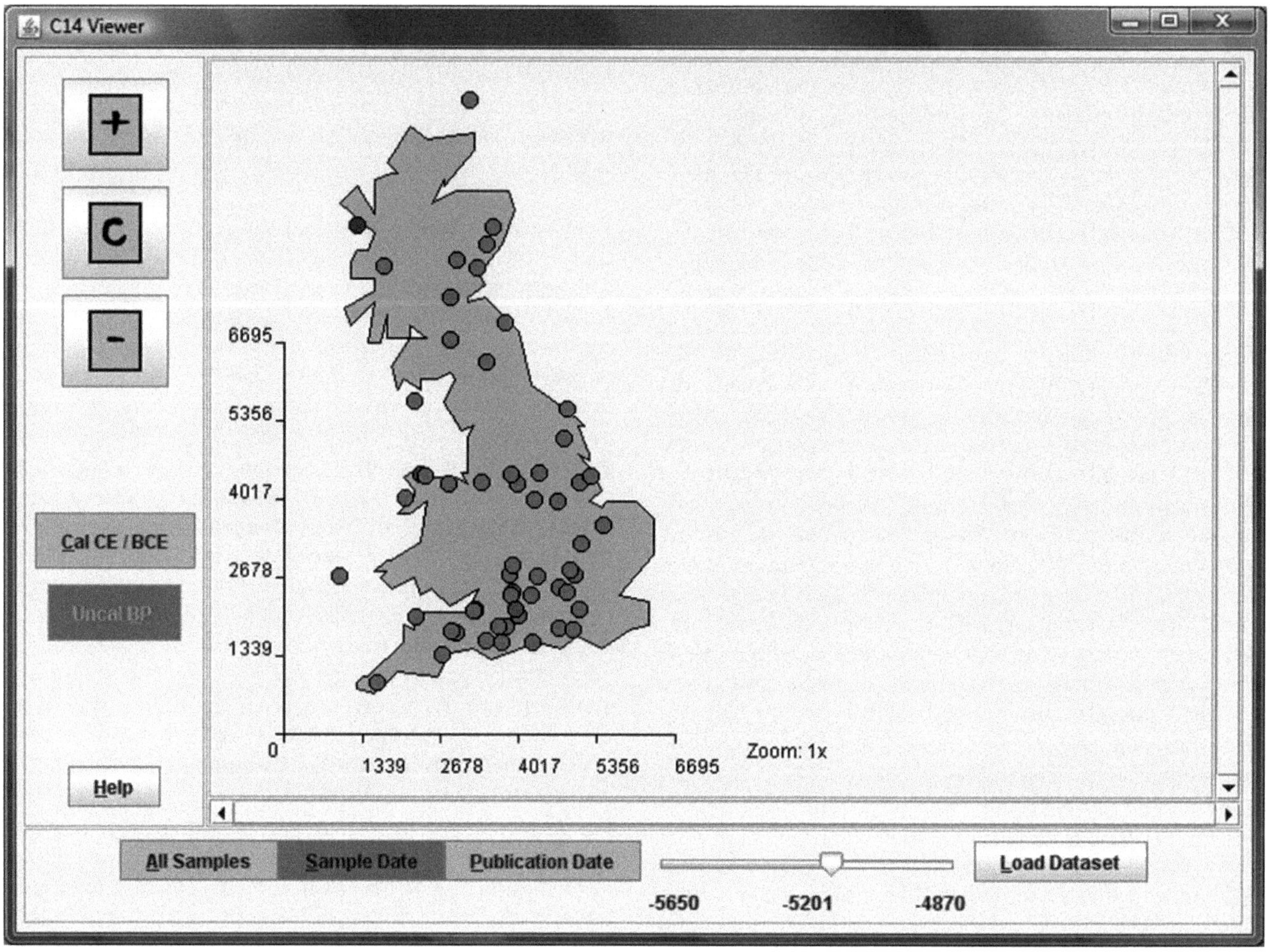

FIG 1.1 - C14VIEW, A SIMPLE JAVA-BASED TGIS FOR THE VISUALISATION OF RADIOCARBON EVIDENCE. THE CIRCLES ARE INDIVIDUAL RADIOCARBON DATES, ORIGINALLY COLOURED ACCORDING TO PROBABILITY, RISING FROM YELLOW, THROUGH GREEN, THEN BACK DOWN TO BLUE.

capacity to undertake anything approaching true spatio-temporal analysis.

Therefore, the central aim of this project is to create a fuzzy temporal-GIS for the specific study of archaeological data that takes into account recent developments in the understanding of the nature of archaeological time. Furthermore, in order to encourage the widespread uptake of TGIS and awareness of the shortcomings of conventional GIS, this new software is based around an existing GIS package that is widely used by archaeologists. To explore its functionality and illustrate how such a system can critically enhance and enrich existing studies it is applied to two case studies, in order to test what fresh information and conclusions the new software might unlock or reveal.

This study builds upon the foundations laid down in my MSc dissertation, for which I created a simple TGIS (C14View) for the visualisation of radiocarbon dates (Green 2002). This program displayed the spatial location of radiocarbon dates: as the user moved a slider bar through the temporal span of the current data, the probabilities of the displayed dates would be calculated on the fly and displayed on the map using colour coding (Fig 1.1). The probability was calculated on the assumption that it obeyed a normal distribution and, as such, was only statistically correct for uncalibrated dates. It did provide a simple calibration procedure, but the calculation of this was much less precise than that provided by conventional calibration software such as OxCal. The software also produced pop-ups giving more information about the probability of single or multiple dates.

C14View was an interesting experiment, but was created primarily with GIS considerations rather than archaeological issues in mind. As such, the solution was inevitably imperfect. Firstly, the software was a bespoke solution written in Java and, as such, lacked the powerful spatial analysis tools available in a commercial GIS software environment. Thus, it was also unable to display any geographic background beyond a single green polygon. Secondly, as noted, C14View was not able to make accurate calculations of calibrated radiocarbon probabilities, or any calculations regarding other types of archaeological date. Finally, the user was only able to select a single year of interest for analysis, rather than the more usual date ranges used in archaeology. In essence, C14View was not truly a TGIS, but rather a tool for visualisation. The roots of this current work can be seen here, however, particularly in its emphasis on probability as the key issue of concern for archaeological TGIS.

Moving on to look in more detail at the content and structure of this book, Chapter 2 will discuss the theoretical basis of archaeological time. Many archaeological models of time have taken inspiration from anthropology, but in this chapter we shall see how the theories of that discipline are themselves often problematic. Central to the discussion is a careful examination of the important interrelationship between linear time (B-series) and time as tense (A-series). This is expanded to consider archaeological models of time that include concepts such as multi-scalar time, such as the three scales of Braudel or the multiple Chaos Theory inspired scales of non-linear dynamics. The nature of time as experienced by past persons is also an important and growing strand of archaeological thought, perhaps most cogently expressed in Gosden's System of Reference. The final important archaeological model of time discussed is the biographical approach advocated recently by Gavin Lucas. As the discussion will illustrate, whilst at first very different and potentially conflicting in their positions, all of the above have one particular theme in common: the desire to move beyond conventional, totalising chronology. Drawing upon this, any TGIS that wishes to carry any relevance in future archaeology needs to be created with an awareness of this underlying imperative.

Chapter 3 will consider time in everyday archaeology and the application of the theoretical temporalities discussed in Chapter 2 to real data. Put simply, rather than focus on what archaeologists are saying about time, the emphasis of this chapter is on what they are actually doing with it. There are many kinds of date available to archaeologists, both absolute and relative. Absolute dates include those obtained through scientific analysis, such as radiocarbon or thermoluminescence, and those of directly datable objects such as coins. Relative dates are those obtained by reference to absolute dating schemes, such as pottery typologies. All of these different types of evidence have their own integral uncertainties, advantages and difficulties. Furthermore, relative dating is extended through stratigraphy to take in the whole archaeological site. Again, a TGIS that is to be truly useful to archaeologists needs to at least achieve a degree of rapprochement within this essential temporal complexity.

Chapter 4 will work through previous work on TGIS by both geographers and archaeologists. As noted above, geographers' TGIS are wedded to modern clock time, but they have undertaken several important studies into the representation of time in GIS and databases, the nature of temporal query languages, and in regards to the visual language of animation. Archaeological work on TGIS carries more relevance, but mostly as critique rather than in practice. Practical projects have tended towards subjects with little wider applicability, such as the study of changes in lake levels in the past. The only more widely applicable archaeological TGIS is TimeMap (Johnson 1999). That software has excellent capabilities in regards to animation and interpolation of variables between known snapshots, but has largely concentrated of late on the less relevant issue of web delivery of archaeological data. That is an important area of research, but arguably not one of central concern to archaeological TGIS.

Chapter 5 will discuss methods for dealing with uncertainty and the construction of the new archaeological TGIS. It is the contention of this book that the ability to deal with temporal uncertainty should be the first and most important aspect of any archaeological TGIS, due to the fuzzy nature of archaeological time discussed in Chapters 2 and 3. There have been previous attempts to deal with this uncertainty using GIS, but they have all relied upon simplification of the inherent complexity of our dating methodologies. The new TGIS constructed for this project will finally give full rein to this fuzziness. It has been programmed to function within ArcGIS, a widely available commercial GIS. As a result, the TGIS should be readily available to many GIS-literate archaeologists with the hope of fostering an increased desire for TGIS within that community. As noted above, GIS without time is incomplete, so the cultivation of these metaphorical roots is of great importance.

Chapters 6 and 7 will consist of the two case studies. The first will focus upon a dataset consisting of a suite of radiocarbon dates obtained during the excavation under the aegis of PPG16[1] of a quarry site at Willington in Derbyshire. These dates were taken to help answer questions about Neolithic use of naturally-fallen trees, later fire clearance of the woodland during the Late Neolithic and Bronze Age, and the chronological associations of two burnt mounds revealed during the excavations. The site is a complex palimpsest and the excavators encountered some difficulty in extracting spatial patterns from their results. The application of the new TGIS methodologies to this data not only enables more of this past spatial detail to be recovered, but raises interesting conclusions regarding residuality and also the construction of sampling strategies.

The second case study will consider a dataset consisting of a large corpus of Roman pottery recovered through field survey in Northamptonshire. This contrasts well with the Willington case study, being of greater spatial extent and consisting of an alternative form of dating method: typology. The material had previously been subject to trend surface analysis to attempt to discover any underlying, dominant trends in pottery supply as they changed over time (Taylor 1996). However, this analysis had been somewhat restricted due to the necessary use of overlapping periods in its phasing, based as it was upon the categories within the typological scheme. The probabilistic nature of the TGIS removes the need to

[1] Planning Policy Guidance 16, which governed rescue archaeology in the UK from 1990 until 2010.

restrict analysis to these problematic categories and also enables the construction of a new phasing, based upon the rate of change of pottery supply across the region. Furthermore, interesting conclusions are also reached when examining the individual probability curves for sub-regions within the overall region.

Finally, Chapter 8 will analyse the results of the project and its successes. A key element of this is to consider how the different archaeological models of time discussed in Chapter 2 manifested themselves within the TGIS, and how the TGIS could be applied to future analysis undertaken within these new paradigms. We shall also see how future versions of the TGIS might build upon the important groundwork laid down in this book.

Summarising the discussion to this point, archaeology is fundamentally about time. GIS has become widespread in archaeology, but is currently ignorant of time. If we are to fully realise the full potential of spatial analysis, we need to integrate this important temporality, whilst paying due regard to its specifically archaeological uniqueness. The project discussed in this book attempted to undertake this essential task.

CHAPTER TWO: CONCEPTIONS OF TIME IN ARCHAEOLOGY

> *Once upon a time, time itself was the hidden thread of any story, so that people concentrated on the events which unfolded in sequence, rather than the sequence itself. Over the last two centuries, time has become not just a medium but a message, and is now a central concern in many disciplines from astrophysics to anthropology. Strangely, archaeology is not in the forefront of thought about time, although time has always been a central concern.* (Gosden 1994: 1)

Breaking with the convention that all discourses on time begin with a statement about the fundamental mystery of their subject (Murray 1999b: 1), I begin by noting that all archaeologists ought to agree that time is of central importance to our discipline (Ramenofsky 1998: 1). To borrow a metaphor from Braudel (1980: 47), time sticks to the archaeologist's thinking like soil to a gardener's spade. However, Lucas has suggested that perhaps it is because time is so important to archaeology that we tend to take it for granted and fail to appreciate fully how the ways we understand time affect our practice as archaeologists (2005: 1).

As will be seen, the key term is diversity when considering the approaches that have been taken to time in archaeology, some complementary and others less so. However, discourse on time in archaeology has been surprisingly slow to develop (Lucas 2005: 31). Shanks and Tilley noted that no other discipline can boast the temporal perspective on humanity that archaeology has, but that nevertheless no distinctive temporal theories had been produced by archaeologists. They believed that all theories of social change used by archaeologists had been derived from other subjects (1987b: 137). For the time being at least, this remains the case. Puzzlement about time is ancient, it being the crucial element in all human activities. In order to understand it we need concepts which can capture temporality and change, more than we need improved dating methods (Gosden 1994: 2, 7) A desire to understand time is a fundamental human attribute, in order to conquer time and attempt to evade its grasp. Further, how we understand time is how we understand change (McGlade 1987: 22-23).

As such, it is of fundamental importance for archaeologists to get a firm handle on time in order to understand change in the past. In the context of this project, an understanding of archaeological conceptions of time was key to the production of useful software based on sound archaeological theory. The aim was not to produce a synthesis or new archaeological theory of time, but rather to ensure that the software was able to deal with the kinds of conceptions and questions that archaeologists would like to consider. Therefore, I shall not be delving into the works of the primary thinkers referred to by archaeologists in any great detail, as the purpose of this chapter is to consider how archaeologists have used that material, not to consider whether their interpretations are correct.

As noted, if late to develop, there are now a multiplicity of approaches to time in archaeology which must be taken into account, themselves drawing inspiration from many centuries of thinking principally deriving from outside of our discipline. Discovering which of these hold greatest potential was the key task here, as this helped shape the functionality and capabilities of the temporal-GIS produced.

Some of the most widely cited discussions of time in archaeology derive from the discipline of anthropology. Ingold argued that time (along with landscape) was the essential point of topical contact between archaeology and anthropology (1993: 152). Archaeology has a tendency to borrow concepts from anthropology (Gosden 1994: 192), and anthropological models were used in the original construction of time in archaeology: the ethnographic present being used to create a model of prehistoric time (Lucas 2005: 122; Gosden 1994: 4). Further, anthropological models hold the potential to provide fresh inspiration. Thus, it is to anthropology that we turn first.

Time in anthropology

Durkheim's social time and metaphysics

The anthropology of time may be traced back to Durkheim's 1915 work, *The Elementary Forms of the Religious Life*. As we shall see, Durkheim's conception of time has influenced archaeology and, as such, requires close examination. Central to Durkheim's thesis was his belief that time was derived from social life via a circular relationship: the collective representation of time of a society is both derived **from** society and dictates **to** society. Gell stated that this rejected the naïve realist assumption that time merely is what it is and raised the possibility that collective representations of time actually create time as a phenomenon perceived by human beings. Durkheim's idea was that time exists for us because we are social beings, coming to this concept due to the similarly all-encompassing nature of time and society (Gell 1992: 3-5).

There is an important lesson here and one that has been taken on board by anthropologists and lately by archaeologists (Murray 1999b: 5). Experience of time permeates our lives and is integral to human existence, but the way that we conceptualise that experience varies with culture, historical era, context, and an individual's position in society. Thus, the meaning and values given to time are inherently context dependent (Adam 1994: 503). Lucas illustrated this point by noting how phrases such as "last week", "at Christmas", and "in 2006" are all very culturally specific. For example, the phrase "half five" can mean both 4:30 (in Icelandic or German) and 5:30 (in English) depending on context (i.e. if one is speaking with a German or to an Englishman). This expresses a **slight** difference in the perception of time (2005: 61).

This suggests that a contrast exists between modern, measured clock time and socially derived time. The creation of global time highlighted the diversity of socially derived forms of time in existence, and the dissonance between these is the cause of profound thought and anxiety. People of European descent forget that our conception of time and space as a neutral framework was created over a period of centuries as an aid to empire building (Gosden 1994: 2, 15). As such, archaeologists should be careful not to make assumptions about the past based too closely on modern clock time. After all, as Adam has noted, modern western life does not take place exclusively in modern clock time: we talk about good and bad times, opening and closing times, right and wrong times. Even for modern western humans, time is not exhausted by measured clock time (1994: 508).

Shanks and Tilley stated that different temporal orientations shape history (1987b: 127). By way of example, prior to the 1850s Europeans lived in a world less than 6,000 years old, created by their god in 4004 BC. This contrasts sharply with the deep sense of time possessed by Indians, where the age of some Bodhisattvas was 10^{32} years - far older than the age we now know the universe to be (Agrawal *et al.* 1999: 30-31). Discourse in archaeology has raised problems regarding bringing together measured time and socially derived time, questioning in turn how far abstract chronologies derived from radiometric dating can take us in the investigation of social time (Gosden 1994: 7). This is a key question, but no entirely satisfactory solutions have as yet been offered. We shall return to this issue later.

As important as Durkheim's recognition of the social basis of time perception was, his work was also unfortunately responsible for the rise of metaphysical cultural relativism. This has rather dominated anthropological work on time ever since and also crept into archaeological explanation. Gell believed that this fatal defect arose as a result of Durkheim's attempt to promote sociological analysis to the level of metaphysics, by identifying collective representations of time with Kantian categories, a class of general concepts (time, space, number, etc.) believed by philosophers to underlie discursive thought (Gell 1992: 5).

Gell summarised Durkheim's lines of argument as follows:

1. The objective world cannot be experienced except via the categories.
2. Time is one such category.
3. Can we think of time except in terms of periods? (Durkheim's answer: No)
4. Are conventional time periodisations socially-derived? (Durkheim's answer: Yes)
5. Therefore, we cannot have any experience of the objective world except in the light of socially-derived periods of duration which constitute the category time. (And ditto for the other categories)
6. Therefore, all experience of the objective world is socially derived.

Refuting this, Gell noted that there is no reason to say no to 3, that the empirical case for 4 is underwhelming as many time periodisations have straightforward ecological origins (i.e. Easter, Christmas, etc.) and, finally, that the reasoning behind 5 is invalid as, firstly, we often refer to events by non-periodic (i.e. ordinal) schemes and, secondly, most periodic schemes are based on astronomical phenomena (1992: 11-13). Accordingly, the theoretical basis of much work that followed on from Durkheim is flawed (1992: 14):

> *By claiming that sociological analysis constituted an independent route towards the strictly metaphysical goals of rationalist philosophizing, Durkheim opens a door through which all manner of demons are able to crowd in.*

We are dealing here with chronotypes - essentially kinds of time, like morning, evening and afternoon, or linear and cyclical. The metaphysical yearning seen by Gell finds perhaps its best illustration in this concept of the chronotype and the promotion of models revolving around structuralist binary oppositions. For example, the ultimate goal of Lévi-Strauss was to understand the human mind itself. He divided societies into those that were hot (i.e. recognised history) and those that were cold (i.e. did not recognise history as we would know it, but had a more mythical take on their past). He saw the human mind in cold societies as operating through a series of conceptual oppositions: left / right, night / day, black / white, past / present, mythic time / present time etc. He referred to the variety of cultural forms using this simple oppositional structure and used it to explain transformations. He suggested that cold societies used rituals to overcome the paradox between mythic time and present time. Leach took this further, claiming that pre-modern / non-Western societies regarded all time in terms of opposites (Gosden 1994: 47-48; Lucas 2005: 62-63). Lévi-Strauss's concept of hot / cold societies has been rightly criticised on the grounds that it denies many societies a sense of history (Lucas 2005: 63). However, Shanks and Tilley noted the interesting implications of his work in the relation of social temporality to the rhythm and nature of social change. The temporality of tradition is short, thinly layered over the authority of a timeless mythical past. This results in a compression of chronology, with tradition thus operating on a different scale to modern history (1987b: 130).

In the wake of Lévi-Strauss and Leach, the use of chronotypes and binary oppositions has become widespread. An archaeological example of these is the work of Douglass Bailey on the Bulgarian Chalcolithic / Neolithic, in which he concluded that such societies were driven by cyclical and linear chronotypes, sometimes in opposition, sometimes combined (1993: 218). This work suffers from over-simplification in its use of these opposed binary chronotypes (Lucas 2005: 82-83). Lucas believed that the use of oppositions may be a useful way of conceptualising a subject, but that they should not be overplayed and may become more harmful than useful, as in the case of chronotypes. Demolishing the supposed opposition between linear and cyclical time, Lucas wrote that the distinction between them was spurious, due to the interdependence of each in the perception of the other. We cannot measure linear time without cyclical units, whether regular (e.g. hours) or irregular (e.g. cigarettes smoked on a journey) (2005: 93).

Changes within cycles have only one direction: people do not get younger or become reborn after death; only events, not time, may repeat (Adam 1994: 510, 520). Accordingly, the simplistic binary oppositions and chronotypes inspired by Durkheim's followers should be questioned rather than accepted uncritically. Essentially, linear and cyclical time are two sides of the same coin, so how can there be tension between them? Shanks and Tilley also wrote about cyclical time, where past, present and future project onto one another, emphasising continuity (1987b: 131), but this chronotype was not presented as opposed to linear time and, thus, presents less conceptual difficulty.

Gell believed that, although cultural relativism is essential to anthropology (otherwise, what would be the point in doing anthropology at all, if there were no differences between cultures?), problems arise when anthropologists (and, by extension, archaeologists) attempt to use cultural relativism to make grand metaphysical statements about the nature of being (1992: 54-56). There are multiple times across and within societies (of different people and groups, etc.), acknowledged by Lucas as a positive element of Durkheim's work (2005: 66). However, we must be aware of the limits beyond which we cannot push this realisation, as Gell pithily recorded (1992: 315):

> *There is no fairyland where people experience time in a way that is markedly unlike the way in which we do ourselves, where there is no past, present and future, where time stands still, or chases its own tail, or swings back and forth like a pendulum. All these possibilities have been seriously touted in the literature... ...but they are all travesties, engendered in the process of scholarly reflection... There are only other clocks, other schedules to keep abreast of, other frustrating delays, happy anticipations, unexpected turns of events and long stretches of grinding monotony.*

People are thus in error when they attempt to make distinctions between different types of time, when they are really discussing distinctions between different processes (Gell 1992: 315). For example, a harvest may be a cyclical event that takes place every year, but it still exists primarily in linear time and will never be exactly the same from one year to the next. The process has cyclical elements but the time in which it takes place remains essentially linear. This is an important lesson: it is an attractive, but dangerous, trap to attempt to turn the anthropology (or archaeology) of time into metaphysics. In summary, perception of time is socially influenced, but distinct from real universal time (where real time may be understood as that of astronomical bodies). Ingold conceived of the relationship between temporal perception and real time as a strip, with time along one side and thought along the other. If cut, this strip would encompass chunks of thought along one side and chunks of time along the other (1993: 158-159). I find this an attractive analogy.

Gell and the A/B series

In his *Time in Anthropology* (1992), Gell attempted to achieve a reconciliation between universal time and relative time through combining the phenomenology of Husserl (which we shall consider below) and the philosophy of McTaggart (Lucas 2005: 66). Gell took his application of McTaggart's thinking from Mellor (1981), stating that he used it to enable him to get the basic metaphysics of time clear in order to examine phenomenological approaches cogently (Gell 1992: 150). McTaggart, whilst attempting to prove the unreality of time, distinguished between two kinds of time: the A-series and the B-series. The A-series differentiates between events according to criteria of pastness, presentness and futurity; in other words, time as tense. The B-series differentiates between events according to criteria of before and after; in other words, a tenseless, ordinal time; time as linear chronology. Philosophers may generally be divided into two camps that hold for the primacy of either the A- or B-series over the other as capturing the essence of real time (Gell 1992: 151; Lucas 2005: 21; Mellor 1981: 4).

Gell stated that many see the the A-series as essential to the idea of change, as it is difficult to see how change could be accommodated within the B-series. However, the only way to make these tensed criteria of past, present and future stick to events is to introduce a B-series of dates at which events possess these A-series characteristics (1992: 151-152). As Mellor noted, although dates are permanent, tenses are not (1981: 7). Further, pastness, presentness and futurity are fleeting attributes, gained and lost by events through the passage of time (Gell 1992: 158). Mellor argued that if one event affects another, it must be earlier and, thus, that causal order fixes temporal order: the B-series is thus defined by causation and change is defined as variation through time (1981: 8). Therefore, it **is** possible to accommodate change within the B-series. In addition, Mellor noted that the B-series can be defined directly from observation without reference to the A-series at all. The converse is not true, as defining pastness, presentness and futurity is only possible by reference to dates (1981: 28).

Thus, Gell summarised Mellor's conclusion as being that B-series time is real world time, but that all our actions in the world arise from choices made according to tensed A-series beliefs. Although real time may be summarised down to the fact that events have dates, this is not sufficient to explain human perception of time and of their position in time. We must make a distinction between real temporal facts and the way in which people think and communicate. The underlying B-series nature of time is an unattainable limit, accessible only through our A-series cognitive resources (Gell 1992: 156, 159-160). B-series time is a universal process on which subjective A-series time provides a commentary (Bradley 2002: 7): in essence, B-series time is time and A-series time is tense (Gell 1992: 166).

So, what can we take from this? We must always retain some hold on real or scientific B-series time, as this provides a common measure by which different A-series perceptual systems can be compared (Bradley 2002: 7). As such, it is clear that a B-series conception of time ought to be the base layer of our model of time in archaeology. On top of such a conception, we can build models of temporal perception which will likely hold a more A-series character. But how can we model these A-series and B-series aspects of time? For that, we must return to older and existing conceptions of time in archaeology.

Time in archaeology

Detailed discussion of the concept of time in archaeology is relatively recent, with early discussions tending to focus on the construction of chronologies (Lucas 2005: 28). However, awareness within archaeology is growing of the fact that archaeological time not only orders history, but has a fundamental importance in the creation of archaeological phenomena (Murray 1999a: 2). As a result, Lucas has challenged us to find new ways of thinking about the archaeological record (2005: 37). We shall look in this section at the various ways in which archaeologists have attempted to do this. Firstly, we shall look at the progression of B-series time in archaeology from time perspectivism, through *Annales* approaches to non-linear dynamics. Then we shall consider the phenomenological A-series approaches of Thomas and Gosden, drawing on the thinking of Heidegger, Husserl and Bourdieu. Finally, we shall look at why it may be necessary to move beyond chronology and examine biographical / narrative methods for achieving this.

Time perspectivism and evolutionism

An early attempt at examining the concept of time in archaeology was Geoff Bailey, from the early 1980s. Coming from the background of economic prehistory, Bailey defined two contrasting assumptions about the effects of time-scale on archaeological interpretation:

1. Behavioural determinism - the belief that patterns of behaviour seen in the material record can only be explained in terms of contemporary parallels.
2. Time perspectivism - the belief that different time-scales bring into focus different features of behaviour, thus requiring different explanatory principles.

In the second, explanations appropriate to contemporary behaviour may become less so over longer time-scales (1981: 103): behaviour at any point in time being influenced by processes operating at scales from the neuropsychological (milliseconds) up to the evolutionary (millions of years), with many more in between (Bailey 1987: 7). In the first, the effect of time can effectively be ignored. However, although the approaches may appear mutually exclusive, they need not be. Indeed, Bailey noted that some element of behavioural determinism is inescapable in archaeological interpretation (1981: 103), much like cultural relativism in anthropology (see above).

However, Bailey feared that an over-reliance on modern day parallels would result in archaeology becoming derivative of the disciplines from which the parallels were taken, which are limited by being inevitably based on short-term observations. Archaeological dates generally tend to have a fairly coarse resolution (see Chapter 3), this coarseness accentuated by the margins of error of radiometric dating. As such, Bailey saw a special role for archaeologists in studying processes over long durations, taking inspiration from Braudel's *longue durée*; he saw the key temporal question of archaeology as being a decision as to what time-scale is most appropriate to the study of a phenomenon (Bailey 1981: 104-106, 109-110, 112). Although he acknowledged the existence of different conceptions of time, like Gell (see above), he noted that different times are really just ways of describing different sorts of processes (Bailey 1983: 168).

Bailey elucidated his preference for a time perspectivist approach, suggesting that it gave archaeology a sense of autonomy and purpose and that it offered the possibility of placing checks on human notions of our own historical importance by providing an alternative perspective. He suggested that archaeology covers a much greater time span, at a greater range of timescales than any other discipline concerned with humanity and that we are best placed to explore the interactions between the various processes that result in human history (Bailey 1987: 17). Lucas noted that, as a result, Bailey concentrated on long-term processes (Lucas 2005: 29).

Bailey's arguments came under considerable critique from post-processual archaeologists (Lucas 2005: 29). For example, Shanks and Tilley questioned what constitutes long-term rather than short-term since even environmental phenomena can operate over short timescales - for example, the tsunami of Boxing Day 2004. They accepted that different temporal scales apply to different processes, but believed that this implied a contradiction between abstract chronological time and saturated substantial time as specific to process. Thus Bailey needed to explain how '*...scale and process or logic has anything to do with temporality.*' (Shanks and Tilley 1987b: 122) They also questioned why Bailey felt the need to peg out archaeology's independence as a discipline, stating that he was proposing a disabling theoretical fragmentation. Whilst this particular criticism of Bailey is fair, since fortifying the independence of archaeology does not seem a useful project, Shanks and Tilley mainly saw Bailey giving primacy to nature over history: they believed that his history dismissed human experience as belonging to another scale, subject to the law and will of nature (1987b: 123-124).

Essentially, time perspectivism gives no account to human agency in the past, the past being defined by environmental process and thus becoming an evolutionary model.[1] This is a separation of past from present and, as Joyce and Preucel wrote, the relationship of archaeologist to the events described is fixed as retrospective and removed (2002: 35). Shanks and Tilley noted that evolutionary theories of social change are thoroughly embedded in archaeology, justified by stating that they are ideally suited to deal with long durations. Such theories suggest that history resides in a basic set of processes: yet processes are always different from each other and no such basic set exists (1987b: 138-139, 141). Further, Trigger noted that evolutionary theories tend towards over-generalisation and triviality (1978: 43). Also, as McGuire noted, cultural evolution appears to make the path of history inevitable (1992: 155).

Squair took the criticism of Bailey's time perspectivism to a whole new level of polemic (Murray 1999a: 13):

> *The explanatory perspicacity of the archaeologist has been attributed to the apparently enviable temporal position in which she is located. The archaeologist resides in the future of some particular past's present, and enjoys the privilege of retrospect. It is this ability to anticipate the future perfectly, because it too has already occurred, that endows the archaeological observer with omniscience.* (Squair 1994: 102)

Squair argued that taking such a time perspectivist position on the past as perfected negates the lived experience of the past and elides effective human agency (1994: 104-106). Murray mounted a defence of Bailey, seeing a certain amount of theoretical "Nimby"-ism in the vigour of the post-processual assault. He noted that although time perspectivism **can** deny a role to the social in archaeological explanation, this was not the same as saying that it **must** (1999a: 14-15).

Much of the criticism of Bailey failed to develop ideas beyond critique (Lucas 2005: 30) and Shanks and Tilley failed to demonstrate how existing social theory could be developed to deal with the kind of problems outlined by Bailey (Murray 1999a: 13). Furthermore, if viewing the past as perfected is untenable, then so is the converse. A past that is co-extensive with the present and constructed in our own image is a passive past, with the actions of past people merely serving our present needs, removing their independence. We may only be able to understand the past through the prism of the present, but the erasure of all difference between past and present is unnecessary (Moore 1995: 53). Knowledge of the past must be a complex result of past and present (McGuire 1992: 215). To conclude, the time perspectivism of Bailey teaches us that we must think about appropriate timescales for analysis and we must acknowledge the effect of processes of long duration. However, we must not concentrate exclusively on the long duration to the exclusion of all other timescales, as down this path lie the perhaps rather limiting perspectives of evolutionism and environmental determinism.

Braudel and the Annales *school*

Clearly, Bailey took inspiration for his treatment of long-term processes from Braudel's concept of the *longue durée* (Bailey 1981: 106). The concepts formulated by Braudel and the *Annales* school bear further examination as they have had some significant influence on archaeological conceptions of time.[2] In Braudel's schema, processes are divided into those that operate over the short, medium and long terms (Braudel 1980: 3):

> *The first is an inquiry into a history that is almost changeless, the history of man in relation to his surroundings. It is a history which unfolds slowly and is slow to alter, often repeating itself and working itself out in cycles which are endlessly renewed. ... this facet of history, ...exists almost out of time and tells the story of man's contact with the inanimate...*
>
> *Over and above this unaltering history, there is a history of gentle rhythms, of groups and groupings, which one might readily have called social history...*
>
> *Lastly comes the third part, concerned with traditional history, history, so to speak, on the scale not so much of man in general as of men in particular. ...the history of events: a surface disturbance, the waves stirred up by the powerful movement of tides. A history of short, sharp, nervous vibrations.*

Lucas summarised these three scales: the short term deals with events or individuals (the focus of traditional history), the medium term with social or structural history

[1] Such as the scientific model of social evolution proposed by Friedman (1982).

[2] From the more evangelical: Bintliff 1991a; to the more sceptical: Knapp 1992a.

(such as persistent social or economic forms of organisation), and the long-term with very slow-moving processes (such as environmental processes). These three scales are intertwined and affect the course of each other (2005: 15).

Bintliff saw the medium and long terms as structures which are largely beyond the direct perception of past persons, noting that, as a result, Braudel's system was termed structural history. He believed that its reconciliation between the general and the particular made structural history a landmark of infinite potential (1991b: 7-8). However, Thomas criticised Braudel's schema for its lack of links between the three layers of history and the lack of any interplay between longer term structures and human agency (1996: 35), making individuals '*...the dupes of long-term social processes...*' (Gosden 1994: 134). Bintliff saw a way out of this difficulty in the use of problem history, by which he meant approaching history with a problem defined at the outset (why did X do Y?). He felt that one could study the long-term by looking at the decisive, yet fleeting, moments that reveal the social and intellectual structures of society (1991b: 14-15): to use a literary example, think how the (failed) revolution in *Les Miserables* brings (violently) to the surface the underlying tensions of society in post-Revolutionary France. Further, long-term change may not have been noticeable to past persons, but that does not mean that it is not useful to researchers investigating past change over long periods (Llobera 2006: 160).

However, the lack of links between layers remains a difficulty with *Annales* approaches. Indeed, Braudel himself paid little mind to any scale other than the long duration (Bintliff 1991b: 9; Thomas 1996: 36). The long duration has also proven of particular interest to archaeologists (such as Bailey) due to its emphasis on slow systemic change. Bradley noted that this facilitated archaeological work, due to our regular difficulty with measuring rapid change. He saw an analogy between the chronological resolution of prehistory and the nature of ritual time (1991: 217). Ritual time falls firmly within Braudel's medium term (as it is instituted and maintained by corporate bodies and elites) with habitual time tending towards the apparently timeless (to the person in the past) long-term. Indeed, Gosden claimed that the formalisation of ritual and its often conservative nature helped to take social formations out of time and give them an appearance of being natural (1994: 123). These are chronotypes, but interdependent rather than opposed. Essentially, ritual time is made to carry the appearance of the long-term, though remains firmly medium term in origin. Thus, perhaps the study of the medium to long duration can be a valid methodology for the study of prehistoric ritual. However, long-term change is generally described in evolutionary or ecological terms, Braudel's *longue durée* tending to privilege biological and ecological factors over the social (Gosden 1994: 8, 135): this incipient determinism is problematic (McGlade 1999: 146).

Knapp saw Braudel's scales as different aspects along a continuum (1992b: 10), making it questionable that Braudel's structure is too idealised. Bintliff feared that numberless structures might lead to the dissolution of any methodological framework (1991b: 16). However, this is slightly hysterical. Another difficulty noted by Knapp is that an archaeological *longue durée* is likely to be much longer than the kind of durations with which Braudel dealt under the same nomenclature (1992b: 13). There is a need for careful consideration of temporal scale at the level of research design and methodology (Smith 1992a: 31). *Annales* approaches encourage us to work on a large scale (Sherratt 1992: 139-140), but they do not necessarily take archaeologists much further intellectually than we have already gone (Bulliet 1992a: 131-132). In essence, the hierarchies proposed by Braudel are defective due to their insufficient coherence and the fact that they could not be effectively integrated, a weakness that is even more severe for archaeology (Fletcher 1992: 36) Arguably, therefore, Braudel's ideas provide inspiration but few answers that stand up to rigorous examination, although he did set new thinking into motion.[1] Indeed, Braudel himself only saw his scheme as a means of explanation rather than a model of time *per se* (Harding *et al.* 2006: 90).

Non-linear dynamics and multiple temporalities

> *Evolution gives rise to a nested hierarchy of spatio-temporal structures, from microbes to Gaia, which continue to evolve.* (Allen 1997: 54)

A possibly more nuanced manner of dealing with varying timescales without becoming trapped in idealised schema and environmental determinism may be found in non-linear dynamics. Einstein's theory of relativity caused a major shift in the western world's conceptions of time: '*Einstein replaced the single Newtonian clock with as many clocks as we like; these clocks tell different times, but they are all correct...*' (Gosden 1994: 5) Whilst Ramenofsky stated that non-linear time is only relevant on the scale of the universe (1998: 77), that does not mean that it cannot provide inspiration for new models of time. Material culture and monuments can change their forms at radically different rates, sometimes very rapidly and sometimes emphasising continuity (Bradley 2002: 11). Douglass Bailey concluded his work (discussed above) with a plea that reconstructions of prehistoric society address prehistoric perspectives of time which were multiple, interacting and conflicting (1993: 219). This discourse over the continuity / discontinuity of change arose out of the ideas of punctuated equilibrium in evolutionary theory and catastrophe theory, with ideas concerning non-linear forms of change being directly borrowed from chaos theory (Gosden 1994: 5). Such non-linear evolutionary models inspired the idea of history as a process of long-term stability punctuated by periods of rapid transformation, rather than as a linear process (Lucas 2005: 16-17).

[1] Continuing to this day: see Harding 2005.

In this sense, non-linear dynamics has much in common with Braudel's concept of multiple scales, but it attempts to provide a more mathematical approach to explaining the relationship between these scales, a key weakness of the *Annales* paradigm (as noted above). Both approaches agree that history involves different rates of change and that much discontinuity in history may result from the conjunction of different temporalities. Thus, it may be argued that archaeology should refocus from a consideration of change to consideration of rate of change, and even the changing rate of change. Non-linear models see social structures as complex systems exhibiting two key characteristics: all systems have a certain inherent instability, and this instability may collapse or transform the system when amplified. This instability arises from human agency and other idiosyncratic behaviour (Lucas 2005: 17). McGlade elucidated the fundamental point: non-equilibrium behaviour as an intrinsic property of social systems can provide self-organisation, and thus is the instigator of qualitative restructuring of such systems as they evolve (1999: 149). These ideas provide a much more nuanced take on multiple temporalities and their inter-relationships than Braudel's overly-regimented schema, though are dependent upon a very systemic view of society and social formations.

As Lucas noted, this brings to mind the old concept of the palimpsest, which originally meant a manuscript which had been erased in order to be written on again. In archaeological terms, palimpsest refers to '*...traces of multiple, overlapping activities over variable periods of time and the variable erasing of earlier traces.*', much untidier than a simple layering (2005: 37). A development of this model recorded by Lucas was the English Historic Landscape Characterisation project, which suggested that the present-day landscape is not so much a palimpsest of fragmented, fossilised landscapes of the past, but rather a historical process incorporating multiple temporalities which carry different resonances in the present day. This gives a much richer image (2005: 41-43) and brings a flavour of non-linear dynamics to landscape archaeology.

In this vein, Ingold adopted what he called a dwelling perspective for his consideration of the temporality of the landscape, according to which the landscape is constituted as an enduring record of the lives and works of people who dwelt in it in the past, telling (or rather being) a story to both archaeologist and aboriginal. Telling this story guides your listener into the world. Perceiving the landscape becomes an act of remembrance accomplished through perceptual engagement. Ingold coined the term taskscape to encompass the entire assemblage of tasks (and their temporality) which those dwelling in the landscape took part in and which are engraved into the landscape as a result. However, the landscape is in a perpetual state of construction which never reaches completion. He saw the temporality of the taskscape as a pattern of resonances, embracing the totality of rhythmic phenomena. As in music, any rhythm may provide the rhythm for any of the others. This rhythmic pattern of human activities rests nested within the wider rhythms of the world, from the animal kingdom up to the extremely slow rhythms of geological time (1993: 152-164).

Ingold saw archaeologists as studying the meaning of the landscape not by interpreting its many layers of representation, but by probing ever more deeply into them. Every feature could be a clue, a key to meaning rather than a vehicle for it. This discovery procedure is what distinguishes the dwelling perspective. Since the process of dwelling is fundamentally temporal, understanding landscape through the dwelling perspective must start with a recognition of its temporality (1993: 172). The musical metaphor worked by Ingold explains the rhythmic and resonant nature of past activities as they are refracted in the landscape of the present.

McGlade and Van der Leeuw noted that non-linear dynamics can provide human agency with more influence on history than Braudel's schema, but that it should not be considered the primary motivating force. Agency can account for the sudden abrupt changes which cannot be accounted for by long-term systems. Conventional chronology's emphasis on continuity softens out these sudden transitions and they end up underplayed or ignored in explanation (1997: 4-5). As McGlade separately noted, these fluctuations dismissed as noise contain a subtle structure which mathematicians call chaos. Evolved structures (dissipative structures) rely on the action of these fluctuations, which are damped when below a critical threshold and thus have little systemic effect. When they move beyond this threshold, however, they are amplified and generate new macroscopic order. These fluctuations arise from processes such as reproduction, cooperation and competition on the individual and community levels (McGlade 1999: 149-150). Essentially, this is a highly structured form of disorder. However, this is a redefinition of evolution, not a dismissal. Evolution of society over the long-term is seen, thus, as a history of discontinuity in social space, punctuated by a sequence of phase changes caused by both conscious and unconscious actions. These discontinuities are thresholds of change, where human agency or idiosyncratic behaviour assumes paramount significance in the formation of social structure (McGlade and Van der Leeuw 1997: 8, 11).

Another model of the past which explicitly acknowledges the existence of multiple temporalities is Foucault's concept of genealogy. This attempts to deal with a history of the fundamental givens of human thought, recorded by Thomas as the human body, emotions, instincts, etc. It does this by attempting to write histories that trace these entities back to their emergence, producing directed rather than totalised accounts. Genealogy concerns itself with discontinuity, rupture, and reversal, rather than progressive evolution. Things change over time, but they may disappear and form transient formations with logics of their own. Multiple paths may be traced through the past, each bringing into focus forgotten voices buried within conventional history (Thomas 1996: 38-39). This can, of course, also be applied to more prosaic human institutions (Harding 2005: 97) and is a very useful way to think about multiple temporalities: no particular history is best and different models can reveal to us different aspects of the past, genealogy charting '*...the contingent, complicated and motley character of past events.*' (Gosden 1994: 140)

Multi-temporality more generally can also reflect the different time periods in a thing's life. Lucas quoted Olivier's example of an Iron Age burial which had three associated periods: the deceased's lifetime, the time between death and burial, and the burial event itself. Each period is associated with different items: items used and repaired during lifetime; items gilded after death; food offerings placed into the grave at burial (Lucas 2005: 38). That is a rather specific case, but such multi-temporality applies to nearly all archaeological finds. For example, a pot might be associated with several life phases: gathering of raw materials, manufacture (with various sub-phases), use (with phases of repair and disuse perhaps), perhaps a period as an antique, and finally destruction (and disposal) or burial as a grave good. How do you date such an object and which date is most valid? Is it the thermoluminescence date which tells us when the pot was made or, perhaps, the radiocarbon date on burnt residues which tells us when the pot was used? The answer: all of these temporalities are equally valid and should be considered.

Time may be both plural and cultural, or singular and absolute, depending on context: maintaining a simplistic opposition between clock time and social time ignores their deep interpenetration and is unlikely to be of aid in moving archaeology onto firm conceptual ground (Murray 1999a: 2-3). Both *Annales* and non-linear dynamics approaches enable archaeologists to consider continuity as well as change, enabling multiple histories and devaluing monolithic universal history, such as evolution. Such multi-temporal approaches enable the analytical capture of some of the non-spatial nature of time, leading to a much richer representation of it (Lucas 2005: 18, 25).

As we have seen, excepting their systemic base, non-linear dynamics have fewer conceptual difficulties than *Annales* approaches, as they provide a clearer relationship between their varied scales of analysis and are more flexible in terms of their multiple timescales. Whereas we are limited to three distinct timescales within Braudel's schema, with non-linear dynamics we can adjust our scale to fit our data. As such, some sense of the multiple temporalities of non-linear dynamics would appear to have the most potential for elucidating the B-series, real, aspect of time, perhaps holding the key to a coherent explanation for the relationship between structure and events (McGlade 1999: 154). However, how can we model the A-series aspect of time, that of temporal perception? For that, we must consider the ideas of Heidegger, Husserl and Bourdieu as appropriated by archaeologists.

Heidegger and being in the world

Time is not an abstract entity, but a quality of human involvement with the world; a crucial feature of everyday life. There is a contradiction between measured time and experienced time: experienced time is a complex mixture of past, present and future and defies measurement (Gosden 1994: 1-2, 187). Even in the modern western world, experienced time is not linear or quantitative: Adam made it clear that if it were, any reference to good or bad times etc. would be meaningless (1994: 509). Practical daily involvement with the world is fundamentally temporal. This experienced time is not reflected in the kind of B-series models discussed above: they aid in structuring the past, but cannot help us to elucidate time as experienced by people in the past. As such, Gosden believed that time and space needed to be rehumanised in the face of abstract chronologies (1994: 6, 15). The thinking of Heidegger provides a possible route into this elucidation and rehumanisation.

Thomas's work - *Time, Culture and Identity* - was an attempt to create an explicitly Heideggerian archaeology (1996: 2). Heidegger's project was an attempt to reassess the nature of human being in the world (Gosden 1994: 42). We are not in the world in a purely spatial sense, like an object in a room. In truth, we are in the world in the way people are in love or in the theatre, with an immediate and active involvement in it. This model of being does not rely on meaning, ideas, or truth as the correct perception of reality, but instead looks at how we and our surroundings are interwoven (Gosden 1994: 42, 44). This is connected to material culture:[1] whilst some modes of involvement are preserved as habits or traditions, others hold a physical presence. How we engage with the world is immanent in our relationships to things, objects being bound up with human projects as part of the web of relations in which we are embedded (Thomas 1996: 20). Clearly, this provides archaeology's route into the consideration of being in the world. Thomas noted that things can serve as witnesses to a human past (1996: 81) and what goes for things, also goes for places (Thomas 1996: Chapter 4).

Heidegger coined the term *Dasein* to refer to the being of creatures who are self aware in such a way that their existence matters to them: this is a practice, a way of being, rather than a thing (Thomas 1996: 39-40). *Dasein* is self-interpreting: it wants to know the hows and whys of existence. It is a state of involvement, which is quite distinct from possessing a view or theory of the world (Gosden 1994: 110-111). But where does time come into this? Heidegger contends that we are fundamentally temporal: time is written into our constitution in our possession of a past, a present, and a future. The time of the world, public time, and clock time are abstracted from

[1] See Thomas 1996: Chapter 3.

this inherent temporality (Thomas 1996: 40). This is partly an argument for the B-series being derived from the A-series, which we have already seen to be problematic and difficult to sustain. However, *Dasein* does have the potential to provide a method for approaching experienced A-series time.

Dasein cares. This care is deeply bound up with *Dasein*'s temporality: understanding / coping is planning for the future; falling into habit is absorption in the present; mood and feelings are concerned with the past and *Dasein*'s continuity through time (Thomas 1996: 43). The main emphasis of these three ecstasies is towards the future, with the present serving as a point of oscillation between past possibilities and future potentials (Gosden 1994: 112). *Dasein*'s past hands down possibilities to its present. However, not all aspects of this past contribute equally and this past is indeterminate (like the future): our past is necessarily reconstructed from fragments, juggled and played off against each other to create a coherent story (Thomas 1996: 45, 52-54). This is a metaphor which ought to read as very familiar to archaeologists.

Heidegger made a distinction between this personal time of *Dasein* and public time, where public time is large and open and perpetuated by institutions which exceed the lifespan of *Dasein* (Gosden 1994: 112). Public time is infinite and external to any one *Dasein* and, as such, it does not care for it. Therefore, public time can be quantitative. Thomas argued here that public time is derived from experienced time (1996: 44). In contrast, Gosden argued that while public time approaches the longer duration, we must consider it in tandem with personal time. The weakness of Heidegger's work was that he never gave context to *Dasein*, avoiding making empirical analyses or considering the nature of historical change and the processes of history. His disdain for public time also meant that he did not attempt to analyse it. He also refused to deal with human relations (1994: 8, 45, 114, 124). As such, Heidegger encouraged us to consider human involvement with the world and to emphasise the personal nature of experienced time, but he did not provide us with a useful mechanism for taking such an approach. So, where might such a mechanism be found?

Husserl and phenomenology

As mentioned above, Gell sought an answer to this question in the theory of Husserl (Gell 1992: Chapter 23). Gosden's work - *Social Being and Time* - also sought solutions from the same source (1994: Chapter 5). Husserl's ideas seem more useful in practice than Heidegger's, as the most careful and intricate account of subjective time available to us, even today (Gell 1992: 222). Husserl popularised the term phenomenology, although the concept was not his creation. Phenomenology is, essentially, the study of the nature of the appearance of phenomena as they are perceived by the senses and understood by the mind, and raises issues regarding perception and perception-derived knowledge (Gosden 1994: 102).

Husserl's theory regarding time can be boiled down to two basic concepts: retention and protention. The origins of these concepts can be seen in the explanation of Husserl's teacher, Brentano, as to how we hear a five second note played on an oboe as a continuous tone: we hear only the current tone, but this is enriched by our short-term memory of the earlier part of the note. The recent past is fed forward and matched against present input: if a match is found, there is a perception of a temporally continuous time object (the enduring tone). Husserl introduced distinctions that enable us to explain how we can distinguish this short-term feed forward from the after-the-event memory of the note. Brentano's explanation results in a fragmented series of now moments: Husserl's concept of retentions (as opposed to the reproductions of longer-term memory) was designed to overcome this problem, being (in contrast to perception and memory) what we have of the temporally removed parts of experiences from within the now moment (Gell 1992: 222-223). This could be likened to the difference in a computer between the RAM (accessed and used in the active processing of data) and the hard disk (used for the permanent / semi-permanent storage of data).

Retentions give events depth or echo, resulting in a flow of time as a combination of succession and retention. This can be seen as mediating between A-series tense and B-series sequence (Lucas 2005: 23-24). As Gell explained, retention and protention are the horizons of a temporally extended present, providing it with thickness and temporal spread. To return to the metaphor of the oboe, the beginning of the tone is not remembered as such, but rather the tone prolongs itself within a single, constantly modifying present moment. Retentions are part of the current consciousness of the present and become distorted and diminished as they are pushed back from our current awareness. In counterpart, we also have protentions: perspective views on the future phases of current events as they emerge from the near future, likewise distinguished from reproductions of a fantasised future present. They are not anticipations of future presents, but projections of the evolution of the current present which may be disappointed or fulfilled (1992: 223-224, 227). In this way, as a person's present moves forwards, varied and multiple perceptions of his/her past and future come into being, influence each other and then fade from consciousness at differing rates.

Retention is not just embodied, but also present in material things, the physical properties of which trigger anticipation. Thus, a structure of recursiveness is derived from the interaction of the human and the material. (Gosden 1994: 124). As with Heidegger's ideas, this link to material culture provides a route by which archaeology can attempt to utilise Husserl's paradigm. Lucas suggested that Husserl's model could be combined with non-linear models of archaeological change, with different retention lengths attached to different events like echoes. As such, the echo of past events at any point in time will have different effects on the present, according to the length of said echo. Thus, important events in the far past would have greater impact than some recent events (2005: 26). In this way, retention and protention can serve as useful metaphors for the construction of archaeological models of change.

Gosden assessed the strength of Husserl's theories as being the way that they gave people an active role in the constitution of their relationship with time and the material world. For Gosden, the weakness lay in his tendency towards transcendental idealism; in other words, viewing the world as constructed solely through acts of intention, in his search for a secure foundation for all thought (with the resultant forlorn hope of aiming for universality), and in his concentration on perception over acts in the world (which robs acts of their context). Gosden believed that Husserl gave us '*...the deep constituting power of human communities in their creation of time...*', but that Heidegger was needed to place human agents in the world (1994: 106-107). Husserl and Heidegger thus give us a strong theoretical base for constructing models of the A-series perception of time, but it is a very personal time. How do we extract such personal temporal perceptions from archaeological evidence?

Bourdieu, practice, and Gosden's System of Reference

Gosden sought further inspiration from the work of Bourdieu on practice (Gosden 1994: Chapter 5). Bourdieu coined the concept of *habitus*: systems of structures predisposed to function as such, thus being principles that generate and organise practice and that can be objectively adapted to their outcomes without needing conscious final plans or complete mastery of their operations. This is built up from experience, not consciously mastered, representing the sedimentation of past practice which then guides future practice unconsciously. Gosden saw *habitus* as reconciling subjective and objective approaches by explaining how past experience becomes embodied and guides human agency. This has parallels with Heidegger's theories, but Bourdieu is more practical (although both lack any real consideration of material culture) (Gosden 1994: 117, 121).

Gosden combined all of these ideas into his System of Reference, in which a structure of retention and protention creates habit and in which human acts form a chain or network of actions, each implicitly referring to many others. Thought and unthought actions continually blend in human action, unthought habit acting as the prime motivator and giving life a general forward direction, but aided by human agency when problems occur. Gosden saw public time as being constituted from the mass of unconscious times and believed that it drew its power from the structure of habitual times, repetition breeding habit and public time continually becoming naturalized, with conscious action fading into unconscious. In essence, public times gain their authority through appearing as natural as possible - material culture facilitates this. Gosden went on to define three forms of time: harmonious, disjoint and concatenating (1994: 124-126, 137, 188-189; Daly 2003: 21) - I think of these through the musical metaphor of the differing styles of Mozart, John Cage and Carl Orff. To take this musical metaphor slightly further, however, I suggest that perhaps time can be all three of these at once, much like music can encapsulate harmony, disharmony and a sense of growing momentum in a single piece (or even all at the same time). Gosden also, much like Braudel, defined three levels on which time operated within its infinite depth: the time of our lives from birth to death; public times of consciously manipulated symbols and meanings; and the larger group resources of habit and recursiveness, which are solidly material. These three levels interpenetrate each other in complex ways; for example, individual life is grounded in habit (1994: 137).

Social formations contain combinations of different times created through practice, in turn creating a temporal structure or style, by which Gosden wished to characterise them. Material culture is key to this process: we only possess such a great depth of social relations because we use material objects in the creation of these relations in fundamental ways; yet we only possess such a depth of material involvements due to the way in which we can combine socially to alter our world. Gosden linked this closely to power relations, due to the relationship between public time and habit. He concluded that the consideration of the long-term was the province of archaeology (1994: 187, 190, 194), much like Bailey (see above). I believe that Gosden's System of Reference is the closest that archaeologists have come to successfully capturing the essence of A-series temporal perception.

However, Bradley sounded a note of caution when he stated that the recovery of past conceptions of time by archaeologists might only be possible in the most exceptional circumstances. He noted that both Gosden and Thomas operated in rather abstract terms, Gosden losing track of the intricacies of the archaeological record and Thomas lacking links between his theoretical and practical discussions (Bradley 2002: 14). Further, McGlade saw a gender bias in Gosden's opposition of social (female) and public (male) time (1999: 145). Finally, there is little evidence of archaeologists engaging as of yet (in print at least) with the science of temporal perception, as opposed to the philosophy. Nevertheless, the System of Reference carries great appeal, but to approach it successfully would require a strong B-series base. Beyond our B-series taking inspiration from the concept or echo / resonance, it would be difficult to model the A-series along Gosden's guidelines without first ensuring that our B-series is theoretically and practically stalwart.

Beyond chronology

A central theme which can be extracted from the foregoing discussion is the desire to move beyond the conventional archaeological conception of time as chronology. This was the main thesis of Lucas's work - *Time in Archaeology* (2005). Bradley noted that the conception of time as chronology is relatively recent (1991: 209), but what do we mean when we use this familiar term? Lucas defined chronology as the science of computing dates. Different chronologies may be used in different contexts, but the key distinction is between those that are absolute and those that are relative. Absolute chronologies are based on a time framework that is independent from the dates being studied and are usually expressed through a calendrical system. Relative chronologies are constructed around the relationships between the objects being studied. Primary relative chronologies are built up directly from the data, secondary ones (such as the Three Age System) from primary chronologies. Relative chronologies are often tied into absolute chronologies through historical evidence or scientific dating (Lucas 2005: 3-5):

Relative chronologies:	Absolute chronologies:
Primary:	*Historical*:
Stratigraphy	Associative
Seriation	Intrinsic
Typology	
Secondary:	*Scientific*:
Periodisation	Radiocarbon
	Dendrochronology
	etc.

Chronology is fundamental to the way archaeology is practised (Lucas 2005: 114), so what is the problem? Why should we not simply construct a temporal-GIS to deal with this, a chronological-GIS perhaps (C-GIS)? There is a parallel between the way in which people in the past constructed their cosmos through daily practice and the way in which archaeologists construct the present through the practice of archaeology (Lopiparo 2002: 84). Further, archaeology exists within our modern conceptions of time and space and, thus, all our findings take on meaning as a function of that framework, not independently of it (Leone 1978: 25). As such, McGlade recorded that absolute chronologies subordinate the intrinsic times of the material culture being examined. He referred to an unpublished Spanish PhD thesis by Gonzalez-Marcén, which demonstrated that different typologies constructed respectively around male and female grave goods resulted in radically different chronological ordering (McGlade 1999: 143). This is the crux of the problem with chronology - its thorough embedding in modern concepts of time. Modern western people tend to see time as a road, fading into haze behind us and into fog before us. In this three-spaced continuum, we move down the road, the fog parting and the haze filling in, with a brief moment of clarity forming the present (McGuire 1992: 215). However, archaeological time does not necessarily share this directedness and its moments of clarity exist at many different times, not just in the present. Also, archaeologists are not only interested in determining matters of fact and discussing chronological relationships (Trigger 1978: 25), but ought to be considering much wider issues. Further, chronology also lends itself towards evolutionist models of change, due to the two concepts' shared directional and linear temporal construction (Lucas 2005: 13).

Ramenofsky stated that, in archaeology, the concepts of time and chronology have become blurred and conflated. She advocated the assessment of chronologies for reliability and validity, but noted how easy it is to confuse resolution, accuracy and validity in such assessments. She concluded that time and chronology are separate and that chronology is simply a tool to divide time appropriately (1998: 74-75, 82-83). Squair suggested that chronology has monopolised time, subsuming alternative temporalities and forming an unconditional context, to be referenced but not subject to critique (1994: 95-96). Squair - and his predecessors Shanks and Tilley (1987a: Chapter 5) - saw the way out of this problem in Durkheimian social time, but as we saw above, there are difficulties with such an approach.

We have also considered several other methods designed to move beyond chronology, in B-series non-linear approaches to temporal ordering and in A-series approaches to temporal perception. Lucas stated that these conceptions of time do not abandon chronology *per se*, but rather use a different conception of time. These multi-temporal conceptions of time suggest that chronological frameworks need radical modification. In essence then, although archaeology is dependent on chronology, chronology should not be the only conception of time used in archaeology. Lucas noted how its linear and uniform nature has tended to influence our models of change (i.e. evolutionism). Non-linear approaches rely on chronology, but escape from its temporal conception (2005: 18-19, 38, 114-115). As archaeologists move beyond conventional chronology, so must their tools; as such, any C-GIS would risk becoming obsolete before it even came into existence.

Narratives, biographies and the metaphor of memory

> *Narratives are structured in time and impose forms of order on disordered reality.* (Gosden 1994: 54)

The final alternative archaeological conception of time that we shall consider is the construction of narratives and biographies, as cautiously advocated by Lucas (2005: Chapter 4). In respect of the relationship between artefacts and time, Gell outlined how it is a '*category-mistake*' to assign dates to an object. Only events have dates, objects have histories. These histories include dated events, but we only think objects have dates because we naturally associated them most strongly with the events surrounding their creation (1992: 28). Thus, narratives or biographies may be key to a fuller understanding of an artefact's temporalities. These biographies may be complex: the mobility of objects allows complex biographies to develop, as objects enter into a wide variety of relationships throughout their active lives (Daly 2003: 39). This has much in common with the ideas of resonance / echo seen in the form of the landscape and in Gosden's System of Reference. It may also be linked to the event-based recording of archaeology inherent in the CIDOC-CRM ontology (see Chapter 3; Binding *et al.* 2008).

Holtorf's life history of a potsherd from a 5th/6th century BC Sicilian site is a interesting example of an artefact biography. He wrote a long narrative of the discovery, excavation and analysis of the sherd, emphasising the ways in which the material properties and identity of a thing are renegotiated in different social circumstances. He noted that the sherd acquired its identity through a series of specific archaeological formation processes. Holtorf concluded that while the modern (i.e. post-discovery) biographies of ancient artefacts are not as exciting as their lives in the past, their past lives as understood by us are a direct consequence of their modern biographies (2002: 49, 54, 62-63).

Changes in the character of material culture may have arisen from deliberate attempts to emphasise or downplay tradition and Bradley saw in this the metaphorical embodiment of archaeological sequence in narratives or plots (2002: 12). There is an affinity between the writing of narratives as sequentially ordered stories and the overweening archaeological convention of time as linear chronology (Pluciennik 1999) and, further, it is only convention that drives people to write narratives in the past tense: there is no reason not to write them in the present or future voice (Joyce 2002a: 54-55). Bias is also a particular difficulty with the writing of narratives. McGlade proposed non-linear dynamics as a solution to escaping the biases intrinsic to these processes (1999: 159) and Joyce noted how experimental writers in archaeology propose the rewriting of narratives for multiple respondents. She stated that such a process possessed the power to create new knowledge through consciously re-framing our understanding at different scales and in different space-time relationships (2002c: 143). Whether this proves to be the case is difficult to see as of yet. I suggest that the biggest issue is probably the logistical impossibility in writing such multiple biographies - such as that of a Roman jar in Lucas's work (2005: Chapter 4) - for more than a tiny subset of archaeological material.

Related to these themes is the conception of time through the metaphor of memory. Lucas illustrated this idea evocatively in his account of the vivid connection to the past felt by an Icelandic archaeologist when he came face to face with the farmer who owned the land on which he had been excavating a grave. Both ancient skeleton and modern farmer possessed a curious gap in their front teeth. For the archaeologist, the distinction between modern and ancient man must suddenly have blurred, giving him an unusual sense of personal acquaintance with the skeleton. Lucas suggested that it was as if he was remembering who that skeleton had once been, the memory resurrected after centuries of oblivion, in a special case of what archaeologists try to do every day: remember the past (1997: 8). He worked this idea into the conception of archaeology as a process of recovery from amnesia. Amnesiacs attempt to produce total recall through the collection of fragments of their past. Archaeology uses the same method, but Lucas stated that archaeologists do not ever expect to achieve total recall. However, archaeological amnesia is so complete that the metaphor is far more appropriate than mere forgetfulness (2005: 133-134). These concepts point towards the study of conceptions of the past in the past, which we shall consider further in the following chapter.

Conclusions

> *Is it possible to re-think the nature of time, the nature of temporalization that archaeology creates and sustains, or is archaeology, in fact, defined by this temporality as much as it defines it? Fittingly, perhaps, only time will tell.* (Lucas 2005: 136)

As outlined at the beginning, my aims in the construction of this chapter were to review and evaluate conceptions of time in archaeology. Gell stated that there is no need to be in awe of time, as it is no more mysterious than the rest of our experience of the world (1992: 314). However, if not mysterious, we have seen how time is problematic. The past is inseparable from the event of archaeology, constituting a past being choosing a future (Shanks and Tilley 1987b: 41). Thus, we must approach how we consider time with care, as the construction of the past in the present is a political act (McGuire 1992: 217).

As noted at the outset, archaeology generally takes its conceptions from other disciplines. Sherratt suggested that archaeology will have come of age when non-archaeologists organise symposia and write essays on what they can learn from us, rather than vice versa (1992: 140-141). This is not yet the case. Archaeology takes many of its concepts from anthropology. Durkheim provided us with the important observation that perception of time is, to some extent, socially constructed. However, in their desire to build grand metaphysical statements, Durkheim and his followers pushed this conception further than it could stand. Gell provided us with McTaggart's division of time into the A-series and the B-series. The B-series was argued to be the basis of time as it is in essence and the A-series argued to be the basis of human perception of time. The tension between A-series tense and B-series time has the potential to be incorporated into effective models of time in archaeology.

Moving on to conceptions of time in archaeology, we considered Bailey's advocacy of taking long-term approaches to time in archaeology, but I noted the pervasive spectre of evolutionism which all too easily infected such approaches. This is problematic because change in human societies does not necessarily always follow the environmentally-determined progressive pattern seen in conventional Darwinian evolution (although, naturally, evolution can provide a useful model for the development of some processes). The influence of the *Annales* school and specifically Braudel on archaeological conceptions of time was then considered. I concluded that this paradigm encourages us to examine time over different timescales, but that it was too regimented and lacked the linkages necessary between its various scales of history. Some of the concepts inspired by non-linear dynamics were suggested as a better route into considering multiple temporalities, providing stronger links between the various levels on which history operated and much greater flexibility in the definition of said levels, but tending towards excessive structuralism. Foucault's concept of genealogies was seen as a useful way to think about non-linear temporal structure.

We then considered how archaeologists have attempted to approach temporal perception: Thomas's work on Heidegger's ideas surrounding embodiment in the world, and Gosden's work on Husserl's phenomenological approach and Bourdieu's ideas regarding practice. I concluded that Gosden's resultant System of Reference was intriguing and potentially useful, but noted how these various conceptions of temporal perception were somewhat problematic to realise fully in practice, as we shall consider further in the next chapter.

We noted that chronology is fundamental to the way archaeology is practised. However, it can also be significantly problematic, due to its inherently linear and regimented structure. Other approaches, such as those outlined above, provide a more nuanced way to approach archaeological conceptions of time. It seems unlikely that archaeology will ever abandon chronology entirely, but we need to rethink the way we use it if we are to avoid becoming trapped once more by purely evolutionist models. Consideration of artefact biographies provided the important observation that artefacts do not in themselves possess intrinsic dates, but are associated with events which possess dates. Construction of biographies or narratives for all archaeological objects was noted to be an impossible project, but such constructions do invite us to think further upon the multiple temporalities inherent in objects.

Thus, there are many different ways in which archaeologists have conceived of time. Some stand up to validation better than others, but all provide different inspirations to take forward into the future. The key themes are the need to move beyond monolithic chronology and to take a more fluid stance which acknowledges multiple temporalities and non-linear models of change. Consideration of temporal perception is potentially very enlightening, but requires a solid foundation in B-series time: archaeological consideration of A-series temporal perception is perhaps something best left for the future, once we have a stronger handle on B-series temporal structure. We shall come to see how this body of theory on archaeological time influenced the TGIS produced for this project in Chapter 8. Before we go on to consider this, we need to look carefully not at the ways in which archaeologists theorise about time, but at the ways in which they deal with time and temporal questions in their everyday practice. As I hope to demonstrate in the next chapter, despite radically different theoretical agendas (as discussed above), the basic chronological questions asked by archaeologists and the techniques used are remarkably consistent across the discipline.

CHAPTER THREE: WHAT TEMPORAL QUESTIONS DO ARCHAEOLOGISTS ASK?

Temporality in everyday archaeology

Moving on from considering how archaeologists **think** about time, it is imperative that we consider how archaeologists actually **deal** with time in their daily practice. The actual temporal questions that archaeologists ask when considering chronological questions are rather more mundane than the kind of questions we might imagine on the basis of the review offered in the last chapter, but they are the foundation of all archaeology. Fundamentally, and arguably independent of the particular theoretical framework espoused, temporal enquiry in archaeology may be summarised down to two principal queries (where X and Y are archaeological entities, be they artefacts, contexts, sites, etc.):

- How old is X (or Y)?
- What is the temporal relationship between X and Y?

Stratigraphy provides the key to answering the second question, providing a floating, relative chronology which then needs anchoring using material which may be aged. This is the first stage of archaeological interpretation, upon which **all** further models and interpretations are built (Barker 1993: 207). For the first major section of this chapter, we will begin by considering the ways in which archaeologists approach the first question before moving on to the second.

Dating - how old is X (or Y)?

In the early days of the discipline, dating questions could only be answered through the construction of chronologies that stretched back to historically dateable events. This involved a heavy element of estimation and relied upon a fundamentally evolutionist model of change. The advent of radiocarbon dating and other scientific methods forced a reappraisal of these old chronologies. Today, scientific dating has become essential to validate archaeological interpretation (Aitken 1990: 1). However, despite the improved accuracy of these methods, the placing of archaeological evidence within an absolute dating framework remains fuzzy and it is often the case that a handful of dates must stand as proxies for a whole site or even a whole artefact type. Furthermore, absolute dates are often only accurate to within several decades: it is not possible to make temporal choices of a similar precision to our spatial choices, as the same degree of control in measurement is not possible (Lucas 2005: 32-33). This fundamental uncertainty of archaeological dating techniques provides a particular problem for any software that seeks to manage and explore temporal data, but it is also something that computational methods have great potential to assist with (see Chapter 5). All of the above is potentially compounded by the large variety of different dating methods available to archaeologists. As such, it is necessary to consider this menagerie of techniques and their varied applications and uncertainties.

Dates may be relative (i.e. produced by reference to another dating scheme) or absolute (i.e. based on an independently measurable quantity). Most scientific methods provide absolute dates, but this is not always the case. Further, dating techniques require the presence of some form of clock in the material being dated and some archaeologically relevant event to start that clock. Some reset events are more relevant to archaeological questions than others (e.g. compare the thermoluminescence reset caused by kiln firing against the radiocarbon reset caused by the death of organic material - sometimes many years before past human usage). Furthermore, scientific dates can be expensive to obtain and may be destructive to the material being dated: as a result, the archaeological questions to be answered by the dating technique should be weighed up and considered carefully. Aitken suggested a hierarchy of reasons for obtaining a scientific date (1990: 2):

1. Dates that have a worldwide significance;
2. Dates that establish or strengthen a region's basic chronological framework;
3. Dates that help resolve doubts about a site's relation to its region's basic chronological framework;
4. Dates that help test a new technique or assess the performance of an old one.

This view seems rather too attached to what we might characterise as the tyranny of chronology, but consideration **is** necessary of the balance between the benefits of obtaining a date against the cost (both budgetary and destructive). Dates are also obtainable using less expensive methods, such as those obtained from typologies and from coinage (or other historically datable material). We will now consider the main dating techniques used in UK archaeology in more detail (and other techniques very briefly).

Typology:

Type:	Relative
Datable materials:	Manufactured artefacts, primarily pottery
Uncertainty:[1]	Provides an age range, usually to the nearest 25 years at best
Timespan:	From the time when humans began making artefacts
Calibration:	None

[1] Barker 1993: 196

Gräslund wrote that the definition of typology is not always clear, being used to denote both an evolutionistic gradation of types and yet **also** the results of such a procedure or indeed any classification however derived: typology is used to mean almost all conceivable analyses of similarity or classification (1987: 5). Herein, I shall take typology to mean specifically the use of chronologically ordered material as dating evidence. There are several issues that should be considered when approaching the use of typologies as dating evidence. Firstly, is the question as to why finds are studied separately from features in archaeology. The pioneers of typology were highly influenced by Darwinian evolutionism, as seen in the perceived progression from stone through bronze to iron and, in effect, evolutionary typology became a paradigm for understanding prehistory. Lucas has stated that this evolutionist basis collapsed in the wake of a new concept of culture, but noted that the legacy of studying finds out of context remains today (2001: 74-75, 78, 79).

Although the association of two finds in a single context should not be read as evidence of their contemporaneity, the greater the number of associated finds the more reliance can be placed on those associations as being meaningful: also, the more material on which a typology is based the more valid it can be considered (Barker 1993: 197, 228-229). As such, greater consideration of context in the construction of typologies might aid their accuracy. Typologies can be constructed either by attribute association (defining types by recurrent attributes) or by object clustering (defining types by grouping objects of overall similarity) (Lucas 2001: 97). Typologies should not be sequenced on the basis of appearance alone: variation in types is not always logical and varies regionally; for example, Anglo-Saxon pottery in eastern England tends to be finer than (later) 12th century pottery from the same region. When constructing a typology through attribute association, every characteristic of sherds should be taken into account (Barker 1993: 195-196).

Lucas has argued that attribute association was too monothetic in its conception of a type, but noted that object clustering is also problematic due to its dependence on a key to inform us where to split the clusters and due to the use of different clustering algorithms producing different results. In any event, object clustering is merely attribute association reached by another route. We divide objects in the way that we do due to our tendency to associate the durability of objects with their essential characteristics; in other words, their material being, and their impermanence with qualities such as meaning. He also questioned why we consider some objects to be the same (such as Roman and modern cutlery) when they might have had very different existences, and also why we consider some objects to be different (such as ceramic, wooden and metal bowls) when they may have had the same function (2001: 95-98, 103). These are clear difficulties with the conventional construction of typologies.

Furthermore, typologies should never be considered as perfected. As new evidence is discovered or comparisons to other dating techniques - such as AMS radiocarbon dates (Vellanoweth 2001) - made, typologies may be subject to revision (to which they ought always to remain open). This is an aspect which could be approached fruitfully using computerised approaches, as dynamic updating of data is a key strength of information technologies. Taking into account all of the above, the use of typologies as dating evidence should be approached critically, with careful consideration of the quality, applicability and accuracy of the typology used and an openness to future change. There is also a possibility that some of the alternative models of time discussed in the previous chapter may come to improve the ways in which typologies are constructed and used. Objects moving through contexts carry traces of earlier times: in essence, typological variability is a shorthand for continuity and change in design traditions. As such, it is necessary to temporalise this variability and differentiation between objects (Lucas 2001: 105). We can see in this some sense of non-linear dynamics and of biography. Related to typology is seriation and event based recording, which will be discussed later.

Coins and other historical dates:

Type:	Absolute
Datable materials:	Coins and other historically datable artefacts (e.g. inscriptions etc.)
Uncertainty:	Down to a single year
Timespan:	From the invention of coinage etc.
Calibration:	None

Coins and other specifically datable artefacts give the most precise archaeological dates. However, the major issue with the use of coin dates is that they only provide a *terminus post quem*[1] for their context of discovery. This is in common with **all** datable material. All too often a coin may be said to date a context, when in fact all that it does is state that the context formed after the date on which the coin was minted. Further, the date of minting the coin does not provide a *terminus ante quem*[2] for any layer beneath the context in which the coin was found. If there are many datable objects in a context, it is the latest that provides the *terminus post quem* for the context (assuming that it was not introduced by post-depositional processes): the older material may have been redeposited in antiquity. However, it may be possible to improve the dating of a context through estimation of deposition dates based upon an analysis of condition (although this is complicated by wear after burial), or economic conditions (Barker 1993: 196-197, 205-206, 226) - more on which later. Therefore, historically derived dates are in some

[1] The latest datable object in an archaeological context only provides the date **after** which the context was formed (Barker 1993: 224).

[2] If a context is sealed or cut by another **securely dated** context then it must date to **before** the date of the cutting / sealing context (Barker 1993: 224).

ways easier to deal with than dates derived by other methods, but they should still be approached critically. As we saw in the last chapter, any datable material will only ever provide a date for a single event (e.g. the minting of a coin or the throwing of a pot) and this event should not be conflated with the event of deposition.

Radiocarbon:

Type:	Absolute (relative if calibrated)
Datable materials:[1]	Wood, plants, seeds, bone, antler, ivory, teeth, tooth enamel, shells, slag; (also possibilities with stalagmitic calcite, pottery, baked earth, rusty iron,[2] desert varnish,[3] bog bodies[4])
Uncertainty:[5]	±1% (AMS); ±0.25% (conventional); plus calibration error
Timespan:[6]	250 to c.75,000 BP
Calibration:[7]	Dendrochronological

Radiocarbon dating is based upon the spontaneous decay of radioactive carbon-14 (^{14}C). The time elapsed since the death of a living organism can be calculated from the concentration ratio of ^{14}C to $^{12}C/^{13}C$, but it should be noted that the clock starts at the time of formation; in other words, the inner rings of a tree will date to earlier than the outer parts.[8] AMS dating may date smaller samples than conventional radiocarbon dating, although with slightly greater uncertainty. The major difficulty with radiocarbon dating is the need for calibration (due to variation in the past of atmospheric ^{14}C). Radiocarbon dating produces ages in radiocarbon years which need converting to calendar years if the dates are to be compared with non-radiocarbon dates and if rates of change are to be considered accurately. Calibration is achieved through the radiocarbon dating of dendrochronological material (see below), revealing an overall long-term trend superimposed by short-term fluctuations. The difference between radiocarbon years and calendar years thus varies over time, sometimes being very significant (c.800 years during the 3rd millennium BC) and sometimes less so (c.250 years between 800-400 BC). Calibration curves have been constructed in North America and in Europe and these agree to within a few years. Variation is present, however, between the northern and southern hemispheres (Aitken 1990: 4, 45, 56-58, 61, 67, 77, 84, 92-93).

The European dendrochronological curve has been extended back to 9,908 cal BP and a less detailed calibration produced which approaches 50,000 years (Bailey 2006: 717), but there is no universally accepted method for Pleistocene calibration (Dolukhanov *et al.* 2001: 700). Reference to associated thermoluminescence dates (for example, of burnt flint, chert and silex) can also be used to suggest the offset in uncalibrated radiocarbon dates (Richter *et al.* 2000: 76, 86). Beyond the limits of calibration, correction needs to be made for initial errors in the calculation of the ^{14}C half-life, and conversion to calendar date can be confusing due to the irregular form of the calibration curve (especially in the translation of error bands). Problematically, due to the wiggliness of the calibration curve, one radiocarbon age may correspond to several calendar dates (Aitken 1990: 93, 98-99, 108). Problems are also caused by environmental factors such as the marine reservoir effect (a systematic age difference between terrestrial and marine animals), which can be transferred to humans if they consumed large amounts of marine protein in their diet (Yoneda *et al.* 2002).

Furthermore, obtaining only a single date for a context is problematic due to concerns over contamination, intrusive material and errors. Thus, it is ideal to obtain more than one date for each dated context (Aitken 1990: 95-96), and careful statistical analysis of large data samples may be used to assess the congruence between dates (Dolukhanov *et al.* 2001: 700). If it is proved that a series of dates are contemporary, it may be justifiable to average the dates and to quote tighter error limits (Dolukhanov *et al.* 2001: 711), although this should be supported statistically to avoid accusations of fudging the evidence (Aitken 1990: 96-97). It is desirable generally to err on the side of caution and to quote the error margins of a radiocarbon date at $\pm 2\sigma$ (i.e. 95% confidence) (Aitken 1990: 107). Further discussion of the nuances of working with radiocarbon probabilities, and in particular Bayesian methods for doing so, will follow in Chapter 5.

Archaeologists often appear reluctant to explain why they accept some radiocarbon dates and reject others. Pettitt *et al.* questioned whether our working databases of radiocarbon dates should be whittled down to a more manageable rump of verified dates, being more critical and placing less reliance on intuitive responses to validity (2003: 1685). There are certainly a growing number of examples of contradictions between radiocarbon dates and their stratigraphic associations (Bailey 2006: 718), but we should not unwittingly place greater emphasis on constructed relative chronologies over absolute dates just because they disagree with our assumptions about past change: either aberrant dates have moved post-deposition or our assumptions are wrong (Pettitt *et al.* 2003: 1687).

Pettitt *et al.* derived a scoring system for assessing the reliability of radiocarbon dates based on the following areas of potential error (2003: 1687-1689):

Chronometry:

1. Contamination by older / younger carbon and the measurement of irrelevant carbon fractions;

[1] Aitken 1990: 3 - Fig 1.1
[2] Cook *et al.* 2003
[3] Dragovich 2000
[4] Van der Plicht *et al.* 2004
[5] Aitken 1990: 76-77, 84, 92-93
[6] Aitken 1990: 4 - Fig 1.2, 84; Taylor 1997: 69
[7] Aitken 1990: 44
[8] For more detail see Aitken 1990: Chapter 3.

2. Dating of different carbon fractions (e.g. bone has greater potential for contamination than other substances);
3. Accuracy;
4. Sample materials and radiocarbon measurement;
5. Sample measurement and reporting.

Interpretation:

1. Certainty of association of dated sample with human activity (e.g. was artefact made by humans);
2. Relevance of dated sample to specific archaeological entity of concern (e.g. does the event of interest relate to the radiocarbon zero point);
3. Quantity and nature of dates for archaeological horizon;
4. Sample materials and stratigraphic issues (e.g. small AMS samples may be very stratigraphically mobile).

On a positive note, comparison and quantification exercises comparing the results and methods of different radiocarbon laboratories demonstrated concordant results for the vast majority of participating organisations (Scott *et al.* 2004). However, radiocarbon dates should always be assessed for reliability according to the above criteria.

Luminescence:

Type:	Absolute
Datable materials:[1]	Unburnt sediment, burnt flint and stone, pottery, baked earth, stalagmitic calcite; (also possibilities with volcanic material, slag)
Uncertainty:[2]	±5-10%
Timespan:[3]	Present to 100,000+ BP (30,000-50,000 BP for pottery)
Calibration:	None

Luminescence dating is divided into two types: thermoluminescence (TL), which is generally used to date pottery and other baked clays, and optically stimulated luminescence (OSL), which is generally used to date sediments.[4] TL / OSL has a wider applicable age range than radiocarbon dating, but with generally poorer precision. Over the past millennium, the accuracy of the two techniques is comparable (and for TL / OSL better over the past three centuries) and TL / OSL should be preferred outside of the limits of radiocarbon calibration due to giving an age in calendar years (Aitken 1990: 141). Uncertainty in calculating a TL / OSL date arises from uncertainty over the level of background radiation present in the soil (which varies according to wetness and homogeneity of the soil matrix, but which may be reduced by averaging dates from different samples within the same context) and from systematic error in laboratory measurement (Aitken 1990: 152-154, 164; Feathers 2002). However, at present there is probably little likelihood that the error limits of TL / OSL dates will be much improved. Estimation of the accuracy of TL / OSL dates may be tedious and some aspects of uncertainty are unquantifiable. Experienced TL / OSL laboratories err towards caution in giving error limits. When comparing TL / OSL dates with other types of date the overall error bands should be used. When comparing TL / OSL dates prepared in the same laboratory conditions, however, the systematic error may be ignored (Aitken 1990: 164-165).

Dendrochronology:

Type:	Absolute
Datable materials:[5]	Wood
Uncertainty:[6]	Sometimes better than the nearest year
Timespan:[7]	Present to c.10,000 BP
Calibration:	None

As well as its use in radiocarbon calibration, dendrochronology is also used to directly date ancient wood to a far greater level of accuracy than radiocarbon dating. Through counting back tree rings and matching patterns in sequences between different samples, it has been possible to build up various tree ring chronologies that stretch back far into the past (Aitken 1990: 36-38).[8] The date of felling is the nearest that dendrochronology can get to the date at which a piece of wood was used in the past. For this to be possible, the sapwood must be intact and unfortunately this part of the tree is particularly susceptible to decay. If sapwood is present, the felling date can be determined sometimes to less than a year; if not, the accuracy of the date obtained may be taken to be within a decade or two. Obtaining a date requires a sample of greater than 100 tree rings in width, although combining samples from a site to produce a site master sequence improves the accuracy of this process. Another general requirement for obtaining a successful date is that the species of tree dated matches that of the master sequence, although this is not always necessary (Aitken 1990: 38-39, 46-47). Procedures are needed to deal with the rare occasions when dendrochronological dates are challenged and also to deal with dating clusters (Dean 1997: 54), but the major difficulty with dendrochronology is uncertainty over the re-use or ageing of timber. The stunning accuracy of some dendrochronological dates makes it sorely tempting for

[1] Aitken 1990: 3 - Fig 1.1
[2] Aitken 1990: 164-165; Aitken 1997: 211
[3] Aitken 1990: 4 - Fig 1.2, 166
[4] For more detail see Aitken 1990: Chapter 6.
[5] Aitken 1990: 3 - Fig 1.1
[6] Aitken 1990: 46
[7] Aitken 1990: 4 - Fig 1.2; Dean 1997: 46, 52-53
[8] For more detail see Aitken 1990: Chapter 2.

archaeologists to conflate them with dates of use or deposition, but this should not be done: timber was often allowed to age before use or recycled and, as noted, dendrochronology may only at best date a tree's felling. Seasonality of a site or deposit can also be assessed using other techniques (which do not provide a calendar date).[1]

Other methods:

There are many other scientific dating techniques available to archaeologists as well as those reviewed above (summarised in Table 3.1, page 146), but these are less commonly used or are generally applied more often to Palaeolithic archaeology outside of the UK. They also have a tendency to follow similar statistical behaviours to one or another of the above techniques (generally TL). In the context of this project, this would make it relatively straightforward to introduce these techniques into the software at a later date.

Summary:

The above review of dating techniques illustrates the care that should always be exercised in the assignment of a date to an archaeological context, the huge variation in dating techniques available to archaeologists and, thus, the complexities of combining different sources of dating evidence. Generally, a dated item may only provide a *terminus post quem* to the context in which it was discovered. Statistical measures may help in the assignment of a date to a context if more than one datable item is available. The different uncertainties and accuracies of different dating techniques distinctly complicate this process. Discovering a fruitful computational method to approach this necessary combinatory operation was a key challenge of this project.

My software attempts to facilitate the incorporation and combination of the following dating techniques: typology, coinage / historical dates / dendrochronology, radiocarbon, and dates that follow a normal distribution such as thermoluminescence. A summary of their varied uncertainties (as explicated above) is as follows:

***Dating technique*:**	***Uncertainty*:**
Typology:	Age range, to nearest 25 years at best, for date of manufacture.
Coinage / historical:	Down to 1 year for date of manufacture.
Dendrochronology:	Less than 1 year (at best: often closer to ±10-20 years) for date of felling.
Radiocarbon:	Usually around ±1% uncalibrated, compounded by calibration.
Thermoluminescence:	±5-10%, following statistical uncertainty.

How can we reconcile these multiple uncertainties (both between and within different dating techniques)? To return to an example from the previous chapter, think of a hypothetical pot: it may be dated typologically to AD 50-100, produces a TL date of AD 100±160 and has food residues which have been AMS radiocarbon dated to AD 110-130 (calibrated). How would we reconcile these three different dates? They each relate to slightly different things: the typological date tells us when other similar pots elsewhere were deposited, yet it might in fact be a later pot which harked back stylistically to an earlier era or to a curated artefact. The TL date tells us definitively the period during which the pot was fired, yet the period bracketed by its uncertainty is too broad to be of much use (this example is slightly false of course: it would be a waste of money to obtain a TL date for a pot which could be so closely dated typologically). The radiocarbon date tells us when the pot was used in food production. Yet the pot would appear to have been old by that time if the typological date is to be trusted. Does this cast doubt on the radiocarbon date or the typological date, or is this a true representation of the pot's life history? Furthermore, none of these dates may tell us when the pot was deposited in the ground, the key event for archaeological interpretation. Only in conjunction with other dating evidence found in the same context may that date be ascertained to a reasonable degree of accuracy. This adds a whole other level of combinatory complexity, providing an excellent justification for investment in event based recording systems such as CIDOC-CRM (Binding *et al.* 2008).

Intuitive dating:

There is a final dating method that does not appear to have been discussed in detail by archaeologists: that is intuitive dating, where an archaeologist assigns a date to a layer, context or object using his/her intuitive knowledge. To a greater or lesser extent, this is done daily in archaeology - think of the field archaeologist who digs up a sherd of pottery and decides that it is Roman based on his/her previous experience. Often, dates assigned by intuition in the field will be confirmed or revised once artefacts reach the hands of specialists. A particularly clear example of intuitive dating is that of Boudiccan fire deposits. Burnt layers in areas sacked by Boudicca during her rebellion of AD 60/61 if found by archaeologists could conventionally only be dated using radiocarbon dating - with an accuracy of, say, 50-100 years at best - or by the dating of associated assemblages. However, in certain circumstances such deposits can be tentatively given a more closely defined date to the Boudiccan destruction. For example, excavation of the Claudian quay at Regis House in London showed a phase of building destroyed by a fire which may have dated to the razing of the city by Boudicca; the succeeding Neronian quay was dendrochronologically dated to AD 63, suggesting that reconstruction was rapid and that the quay in-fill comprised a sealed pre-Boudiccan

[1] For example, Higham and Horn 2000: 439 or Mannino *et al.* 2003

assemblage (Brigham *et al.* 1996: 35-36). The excavations at Number 1 Poultry showed burnt deposits that were also tentatively dated to the destruction (Burch *et al.* 1997: 131).

Millett used material associated with the Boudiccan destruction horizon in Colchester in an attempt to establish the limits within which other factors determining pottery supply, usage and deposition might obscure underlying chronological structures and render assemblages unreliable for precise dating. Pottery from Boudiccan deposits is unusual archaeologically in that it was still in use at the time of deposition, rather than the more usual waste material discarded at the end of its use-life. Millett concluded that Samian pottery from true Boudiccan deposits ought to appear newer than other Samian deposits; for example, he suggested that older material from the first Colchester potters' shop might more likely date from controlled fires used to clear up the old fort in preparation for the founding of the town than from Boudiccan destruction levels (1987: 93-94, 104, 106-107). Thus, with careful consideration of associated assemblages, burnt layers dating to the mid-1st century in the relevant geographic area can be suggested to date (more closely) to AD 60/61. This must be done with prudence and the resultant date should only be taken as a possibility, not a proven case, yet this example illustrates how intuitive dating is a valid and useful archaeological method. However, it adds another level of complexity to the uncertainty of dating and to the task of integrating different sources of dating.

Stratigraphy and formation processes - what is the temporal relationship between X and Y?

Moving beyond the question of discovering how old an archaeological artefact or context might be, we must now consider how archaeologists approach their other fundamental temporal question: discovering the temporal relationship between two archaeological contexts (or other entities). On the face of it, the answer is straightforward: stratigraphy. However, the issues surrounding stratigraphy are not straightforward. Firstly, the concepts of *terminus post quem* and *terminus ante quem* must be rigorously applied when considering the relative dating of contexts (Barker 1993: 224) and, secondly, a system must be employed to record the stratigraphic relationships between the different contexts.

Archaeology originally took inspiration from the geological concept of stratigraphy, a cyclical process of deposition and denudation, and transformation. However, unlike geological stratification, archaeological stratification is a by-product of human activity, as there is no evidence that humans deliberately made sites with archaeology in mind (Harris 1989: 4, 19-20). As such, Harris redrafted the axioms of geology to make them more relevant to archaeology (Harris 1989: 30-32, 34):

The Law of Superposition: In a series of stratigraphic units as originally created, the upper units are younger and the lower are older, for each must have been deposited on, or created by the removal of, a pre-existing layer.

The Law of Original Horizontality: Layers deposited in unconsolidated form will tend towards the horizontal. Layers found with tilted surfaces were originally deposited as such, or are conformal to the contours of a pre-existing basin of deposition.

The Law of Original Continuity: Stratigraphic units as originally created will be bounded by basins of deposition, or thin down to feather-edges. If any edge of a deposit is exposed in a vertical view, part of its original extent must have been removed by excavation or erosion, and continuity must be sought or absence explained.

The Law of Stratigraphical Succession: Stratigraphic units take their place in the stratigraphic sequence from their position between the undermost of the units which lie above it and the uppermost of the units which lie below it and with which the unit has a physical contact, all other superpositional relationships being redundant.

The limited extent of archaeological strata and their interruption by upstanding strata (such as walls) results in archaeological stratigraphy being multilinear. In 1973, Harris used his laws to capture this multilinearity in his Harris matrix. The matrix is a sheet of paper covered in rectangular boxes in which the stratigraphic relationships on a site may be recorded and exhibited (1989: 33-34), topologically (Wheatley and Gillings 2002: 41).

The matrix recognises three possible relationships between stratigraphic units (Harris 1989: 36; Lucas 2001: 56):

- No direct physical relationship (unlinked);
- Superposition; in other words, earlier / later than (linked with a single line);
- Correlated as separate parts of a once whole deposit; in other words, equivalent to (linked via a double line).

Indirect temporal relationships are then discovered through grouping the matrix into broad phases, either on the basis of major horizon breaks in the sequence or through artefact association (Lucas 2001: 56). The matrix is designed to be used during excavation, so that the excavator possesses a full record of a site's stratigraphic sequence by the time that the dig is concluded. The principles of archaeological stratification are not affected by the age of a site. Further, stratification is formed by processes of deposition and degradation: thus, an appreciation of the difference between cyclical and linear processes is necessary for an archaeologist to understand and interpret stratigraphy correctly. Unlike geological stratification, it is uncommon for archaeological stratification to become inverted, as archaeological deposits do not tend to be solidified in the same way as the rocks with which geologists work. Another major difference between archaeological and geological stratification is the importance in archaeology of interfaces as well as strata. Interfaces may be the surfaces of extant strata or the surfaces of strata that have been removed. Further, upstanding features (such as walls) affect the nature of the strata either side of them (Harris 1989: 36, 40-24, 44-45, 48, 54).

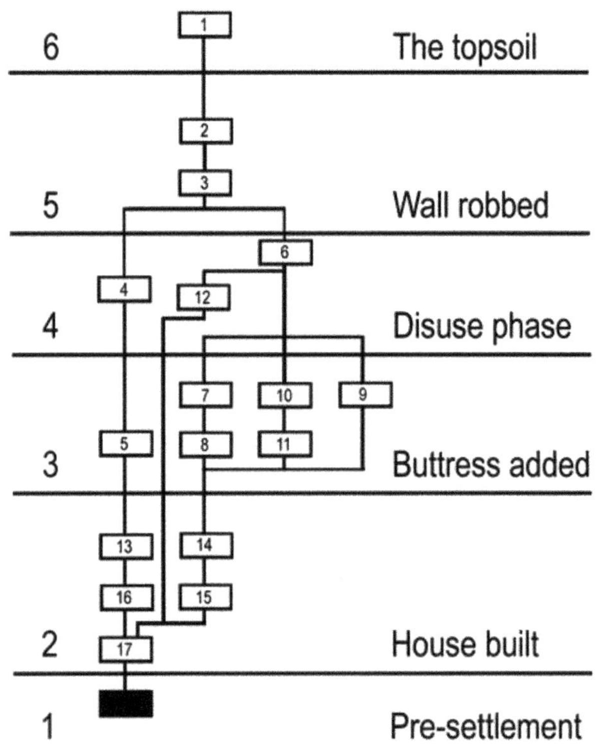

FIG 3.1 - AN EXAMPLE OF A HARRIS MATRIX (ADAPTED FROM BARKER 1993: 232).

The Harris matrix liberated stratigraphy from the section (Lucas 2001: 57) and its use has been widely adopted in archaeology (albeit largely in the UK: e.g. it is far less visible in US archaeology), from the highly refined version used by the Museum of London archaeological unit (Spence 1993; Lucas 2001: 58), through Polish (Kobyliński 1993) and Mexican (Hammond 1993) archaeology, and even for the archaeology of standing buildings (Davies 1993). From a computational point of view, the matrix has been combined with AutoCAD for the reconstruction of computerised composite plans and 3D models (Alvey 1993). However, despite its popularity the system is not beyond criticism. Harris himself saw the major criticisms of his matrix as, essentially, geological elitism and defended it on those grounds (Brown and Harris 1993: 7-9), but his defence fails to tackle problems with his system that are very real. For example, Lucas wrote that it best illustrates the conception of archaeological fieldwork as the recovery of an archaeological record: no longer a physical site, it becomes an abstract data structure (2001: 58). Also, there are a growing number of examples of radiocarbon dates that show individually dated artefacts found in close association within the same stratigraphic layer yet separated by thousands of years, or (conversely) successive stratigraphic layers with indistinguishable radiocarbon dates (Bailey 2006: 718).

In this vein, Barham questioned the focus on stratigraphy over formation processes, insofar as little attention has been paid as to whether recorded stratigraphic data are meaningful. He suggested that the Harris system does not cope well with sites which have undergone considerable post-depositional transformation: on such sites, the use of soil micromorphological techniques for the study of formation processes is essential if full interpretation of archaeological context is to be achieved (1995: 145, 152-153). Formation processes were first fully considered by Michael Schiffer: they are processes, both cultural and non-cultural, that have acted (and continue to act) upon the buried remains of the past, producing the archaeological record that we discover today. These processes mediate between past activity of interest and its surviving traces. Schiffer favoured the transformation view that the archaeological record is a transformed or distorted view of a behavioural system. He held that past behaviour cannot be read directly from archaeology, but that the distortions introduced by transformation can be rectified through appropriate analytical and inferential tools. Also, transformation can also create patterns of its own. Through ethnoarchaeology and experimental work, Schiffer formulated experimental laws and empirical generalisations, which could then be used to rule out formation processes as the cause of specific traits in the archaeological record (1987: 7, 10-11, 21-23).

Nevertheless, there can be no doubt that formation processes do have a large effect on the constitution of the archaeological record. A crucial assumption of the Harris system is that a single context represents a single bounded event (Lucas 2001: 157), yet a single act of deposition may produce multiple deposits and multiple processes may act in the formation of a single deposit (Schiffer 1987: 265-266). Harris's system is dependent upon the visual identification of context boundaries, but scientific techniques may reveal boundaries invisible to the naked eye. Further, the Harris system is predicated upon the assumption that stratigraphy at the time of excavation closely matches stratigraphy at the time of formation. However, Barham wrote that stratigraphic units could be better considered as sedimentary storage systems that inherit properties at deposition, but which are transformed over time. In this regard, he made four specific criticisms of the Harris matrix (1995: 152-156):

- The failure to separate pre- and post-depositional attributes during recording;
- The blending of primary and secondary properties when defining contexts;
- The failure to define criteria by which human modified stratigraphies can be separated from natural stratigraphy;
- The failure to record sedimentological structures on-site.

These problems may result in the recording of two contexts as separate due to post-depositional change in their visual appearance when they were, in fact, the same context at the time of deposition, or in difficulties in integrating any sedimentological structures discovered post-excavation into the perfected matrix (Barham 1995: 155-156; Lucas 2001: 159). Barham thus suggested that sedimentological methods needed integration into the Harris system, enhancing but not replacing it (Barham 1995: 161). In addition, the Harris matrix also does not cope well with interfaces that are too small to excavate separately (Lucas 2001: 155; also see Matthews *et al.* 1996).

Due to its ignorance of the formation of individual deposits, employing the Harris system results in the dominance of the formation process of the site over those of the individual contexts. Although the site itself acquires a richer temporality, the individual contexts become homogeneous and equivalent moments of no specific duration: their only temporality lies in their place within the site sequence. It is possible that the discreteness of a context relates to its degree of openness (i.e. how well sealed it is). This concept may help bridge the distinction between form and content (Lucas 2001: 158, 160-161). Another difficulty with conventional stratigraphy is that it rarely tells us anything about temporal span (Gräslund 1987: 10): the Harris matrix needs supplementing to provide any sense of the duration or longevity of a context. Harris suggested that this could be achieved by lengthening or shortening the individual boxes on the diagram, but this would only illustrate the temporality of production, not of use: for example, a hearth overlying a floor would be shown as later in the matrix (as it was constructed later), but the floor may have been used for as long (or even longer) than the hearth. This longevity is, however, difficult to measure and must generally be estimated at best. Again, consideration of formation processes may help in this (Lucas 2001: 161; Lucas 2005: 39-40).

In this regard, Lucas suggested using the temporality of the matrix to produce a relative measure of longevity, which could then be calibrated. An adapted matrix of this type would not require a neat phasing for a site, but would display a multi-layered temporality, emphasising both continuity and change (2001: 162). This was attempted separately at Çatalhöyük and by Carver. Here we begin to see a sense of the practical application of some of the ideas considered in the last chapter. The idea is not to replace the Harris system, but to complement it and to provide a more complex model of the site: bringing to the fore the event-character of units and conjoining with the temporalities of succession and duration. The coarse phasing of a site once the matrix has been constructed is little different in the final synthesis from the days prior to Harris: the system loses any desired multilinearity. By considering formation processes and the longevity of events, this multi-temporality may be better maintained (Lucas 2001: 162, 179), illustrating the differing rates of change and temporalities on a site (Lucas 2005: 41). I suggest that such a system might also aid in the dating of individual contexts by aiding in the explanation and thus integration of different types of dating evidence.

In his stratigraphic system, Carver attempted to illustrate duration as well as topology. Horizontal lines show the position of layers, relative to each other with later layers above earlier layers. Vertical lines show features, such as pits, ditches, post-holes etc. The arrows at the end of each vertical line show the limits of their possible existence, their life. However, the ordering is not unequivocal, as it is not possible to relate each layer to every other and there will be ambiguity of position where different episodes of deposition were not physically in contact. As a result, the more ambiguous the stratification of the site, the more ambiguous Carver's stratigraphic diagram (Carver 1987: 132-134). However, this uncertainty and lack of clarity is the reason why systems like Carver's have never threatened the hegemony of the Harris matrix: the matrix gives an illusion of perfectedness that complements the concept of preservation by record. System's like Carver's, which come closer to revealing the intrinsic uncertainties of archaeology, sit less comfortably within this ingrained mindset.

Carver saw finds as providing the pre-eminent route into interpretation. For the next stage of his system, he entered quantification of the pottery in each layer into a table which displayed typology against stratigraphy. This construction is a seriation table, effectively a statement of the variation through time of pottery populations. The axes of the table are adjusted to find the best fit of stratigraphic order to typological order. When analysing the table, several conclusions can be drawn: the presence of many different types of pottery in a layer shows that a deposit is secondary and contains much residual material; a vertical edge on the curve shows rapid development; a jagged edge shows a secondary dump. The full potential of this system has not been fully developed, but it is most useful for showing which layers and features could be contemporary and which could be primary. Seriation makes typology more representative, but does not aid its uncertainty. Through the combination of stratigraphy and seriation, an ordered account of events may be produced, but this is a floating chronology which needs anchoring using dating evidence (1987: 133-134). Carver's system possesses a richer temporality than the Harris matrix, but has not been widely adopted and has not achieved its full potential.

Carver's system has some similarities to the CIDOC-CRM ontology that is being increasingly adopted for archaeological recording. This ontology is conceived of as the semantic glue used to link disparate sources and types of information for cultural heritage material. English Heritage's CRM-EH adaptation of this system is based upon the archaeological concept of the context, modelled as a place from which constituent material has been removed by a series of archaeological events. Its entities include mappings of stratigraphic relationships, phasings, finds, and environmental samples. In essence,

CRM-EH and CIDOC-CRM are event based models: events which would once have been recorded only implicitly in archaeological databases are explicitly recorded (or recreated) under these new ontologies (Binding *et al.* 2008: 281-282; 284). Perhaps it can be hoped that a growing awareness amongst archaeologists of the event structures of excavation might also lead to a greater acknowledgement of the event structures of past objects and contexts themselves, as represented in the biographical approaches discussed in Chapter 2.

Theoretical fieldwork

At Çatalhöyük, Hodder attempted to develop post-processual methodologies for archaeological fieldwork, experimenting with new ways of recording and working in the field. Çatalhöyük consists of a pair of mounds in central Turkey, dug by Hodder from 1993 onwards (1996c: 3, 6). The site is largely Neolithic, with radiocarbon dates for its 20m sequence from 6400 to 5600 BC (2000a: 3). Hodder wished to explore the potential of deep, contextualised methods of description. His aim was to define typologies and terminologies on the basis of both general and local information, acknowledging their sensitivity to context and their capability of redefinition dependent on perspective. He saw this contextuality as leading to multivocality, interactivity and reflexivity (1996c: 6). Hodder also wished to explore stratigraphy as social process: at Çatalhöyük, wall alignments are built on top of previous walls wherever possible. There was very little change in house layout, possibly due to previous walls providing the best available foundations for later walls, and yet there were changes in the use of space through time. Walls were not always completely rebuilt, buildings became more or less complex internally, and houses might become open areas for refuse or vice versa (1996a: 363-364). It is unclear as yet whether his new methodologies aided him in exploring these phenomena and questions.

Hodder's main theme at Çatalhöyük was non-dichotomous thinking; in other words, the breaking down and questioning of boundaries. He saw four ways into this: reflexivity, relationality or contextuality, interactivity, and multivocality. This was manifest through various strategies, including flexible sampling strategies, keeping finds in context, strong links between laboratory and fieldwork staff, an integrated and fluid computer database including diaries, video recording of excavation, reflexivity and self-criticism, anthropologists examining archaeologists, web publication including hypertext links to primary evidence in narratives, virtual reconstruction, and teams from around the world using their own techniques (Hodder 2000a: 6-9).

These are largely laudable aims, but how did they work out in practice? Recording the reactions of the British team,[1] Farid wrote that the uncertainty could be disconcerting, frustrating and disempowering, but also a challenge. The project attracted significant interest and funding, and thus it was much easier to experiment with sampling strategies than on most sites. However, the commitment to avoiding interpretation prior to excavation meant that the teams ended up sampling everything with little element of choice. This strategy also resulted in digging the site using gridded baulks to enable micro-sampling. Ironically, this fragmented the site, contradicting the overarching aim of breaking down boundaries. Examination of finds in context also proved very difficult to accomplish, with artefacts still being removed and sent off to different labs for analysis by material type. Interaction between field and laboratory staff was problematic due to time constraints on peoples' availability and on the length of time which different techniques needed to produce results (Farid 2000: 19, 21-24).

The interactive database was limited by resource, time and process constraints, meaning that immediate recording was not always possible. The maintenance of diaries was found to be very useful, especially by new and external researchers as they provided informative background in addition to the dry and coded conventional recording forms.[2] Some advocated the sole use of narrative recording, but even the best professionals need reminding of all the categories of information which ought to be recorded and it would be hard to effectively query narrative data. However, Farid did note that the fluidity in the written record resulted in large differences in recording between seasons, requiring the constant revision of previous seasons' data. Over time, this process might become unsustainable (2000: 24-25, 27).

Hamilton suggested that elements of the work at Çatalhöyük might eventually be built into a postprocessual methodology for archaeology (2000: 121). There is, however, more scope for experimentation on research excavations. Farid noted that time constraints and the need to minimise information loss mean that rescue / contract archaeologists have to use pro-forma recording sheets and pre-planned methodologies (2000: 28). The methods employed at Çatalhöyük do not supercede conventional excavation or recording: rather they should be seen as augmentations or enhancements, some more successful than others. Narrative recording encourages a plurality of interpretations, but I can see little evidence of a plurality of temporalities despite considerable fluidity in terms of site practice. The basic temporal questions of archaeology seem to be answered in the same old ways.

Bender *et al.* also aimed to recognise plurality of interpretations during fieldwork in their work at the Bronze Age settlement of Leskernick. They wished to explore the relationship between archaeology as discourse on the past and contemporary practice. Excavation reports usually eliminate the processes of

[1] See Tringham and Stevanovic 2000 for the American view.

[2] Including Harris matrices (Leibhammer 2000: 134).

reconstruction and interpretation that take place as a dig is in progress. They suggested that the interpretative debate that runs through excavation needs highlighting to offset the spurious perfectedness of archaeological reporting. To achieve this, Bender *et al.* presented the excavation of a stone row at Leskernick as a process. Thus, they wrote a biography of the dig, a series of narrative diary entries describing what took place each day, their various impressions of the processes (wild horses chewing at flags, a heavy female presence, the aesthetics of the trench furniture), and their initial interpretations and thought processes (1997: 147-151, 158-160). However, the final results of the excavation were presented in a conventional manner, with sections and plans and context entries.

The work of Bender *et al.* has clear parallels with that at Çatalhöyük: recognising the plurality of interpretations and recording thought processes through the life of a dig. Again, however, the fundamental temporal questions of archaeology appear to be answered in the conventional way. It is revealing that the diaries do not answer this query definitively. When writing a narrative, any author must have a clear idea in his/her head as to what elements their reader will wish to know about, as the recording of everything that took place during the day's work would be impractical. Yet when I come to answer specific questions from their narrative, they have neglected to tell me what I want to know. This is a problem: there can be little correction or addition with the benefit of hindsight to archaeological recording, as the material recorded is destroyed forever in the process of excavation.

Theories of time in practice

Having considered the ways in which archaeologists deal with temporal questions in day to day practice, the integration between this archaeology at the coal face and the different conceptions of time reviewed in the previous chapter requires consideration. After all, what good does theorising do if it does not inform practice? Lucas noted how much more difficult it was to think of ways in which post-processualism had affected archaeological fieldwork (at least until the work referred to above), quite in contrast to the far-reaching practical effects of the New Archaeology (2001: 2, 14). Similarly, the ideas discussed in the previous chapter have had little influence on archaeology in the field as of yet, and some may prove too difficult to realise fully in practice.

Multiple temporalities

We shall begin by looking more closely at the influence of the *Annales* school on archaeology. The effects of *Annales* thinking seem to have been mostly limited to field survey and broad synthetic themes, as areas which are particularly conducive to study over long timescales. In the introduction to his volume on *Annales* approaches to archaeology (Bintliff 1991a), Bintliff analysed his Boeotia survey within an *Annales* framework. He recognised the effects of medium term processes, such as famine and economic decline, historically in inscriptions about specific events and archaeologically in the depopulation of towns and in evidence of severe erosion (1991b: 19, 26). However, there is little detail given in his case study and little sense of the longer term rhythms of the *longue durée.*

Performing a similar, yet more detailed, analysis of his Biferno valley survey, Barker dismissed the potential of archaeology to study short-term events on the grounds of chronological imprecision (incorrectly: prehistoric material may be fuzzily dated, but many events are still visible, such as the making of a pot, the construction of a house, etc.), but saw archaeological evidence for medium term processes (albeit over timescales Braudel himself would have considered to be more the domain of the *longue durée*). In this, he saw trends in settlement that he believed to form a good example of the interplay between the various levels of history in Braudel's model. He also noted how the valley's environment was key to the course of prehistoric settlement, with the earliest farmers selecting the best locations to practice their system of agriculture. This conclusion illustrates the perceived effects of the *longue durée.* Moving beyond the prehistoric, Barker recognised the effects of medium term processes in social and economic trends represented by the archaeology, but concluded that the primary aspect which had determined the course of the past was the constraints and opportunities provided by the valley as an agricultural resource (i.e. by the *longue durée).* He saw the valley environment as shaping the nature of human relationships between each other and with the '*outside world*' (1991: 39, 45-46, 51, 53-54). This clearly illustrates the incipient environmental determinism which can all too easily infect *Annales* approaches to archaeology.

Jones (1991) considered why the expansion of the Roman Empire stopped where and when it did. He felt that a full understanding of this problem could only be achieved through consideration of the decisions and actions of Roman emperors and generals, the political, social and economic structures operating around the frontiers, and the background of the natural environment (i.e. Braudel's three scales of history). Jones felt that the common process in frontier studies of attempting to discover specific historically recorded events from archaeological evidence does not help us to answer the big questions, such as the choice between fortifying frontiers or attempting total conquest. Examining the differences between Hadrian's Wall and the Antonine Wall, which appear to be strategically the same, Jones saw a more fruitful approach in considering the differences between the two regions through the association of patterns in the archaeological record with settlement, production and exchange systems. Over the *longue durée*, structures such as climate, topography and subsistence prove little different between the regions of the two Walls. Jones thus concluded that the frontiers were defined by a balance between will and resources, with the people living between the two walls lacking the socio-economic structure for integration into the empire; by the time that this situation had changed, the Imperial appetite for conquest had evaporated. The *Annales* element of Jones's study is in his choice of the medium term as the appropriate scale of analysis for his evidence (despite the title of his paper implying that he was taking a longer term approach). The environmental factors of the *longue durée* remain essentially constant throughout the period and so few answers can be found there. Jones's one attempt to bring the *longue durée* into his work does not really come to fruition: he noted that the Tyne-Solway line was selected for topographical reasons (1991: 93-104), but this is not the influence of the *longue durée* as it is not a question of change, simply one of context.

The papers in Knapp's *Archaeology, Annales and Ethnohistory* (1992a) tended to focus mainly on the use of *Annales* approaches as a tool to link historical and archaeological evidence.[1] For example, Knapp himself sought to redress the balance of the dominance of textual sources in the socio-historical and politico-economic analysis of Bronze Age societies in the Levant. Documents tend to deal with elites, archaeology with more humble folk. His solution to this problem was through improved chronologies placed within an *Annales* framework. Better chronologies provide a better understanding of the relationships between internal production, interregional exchange and international relations. Knapp applied the structural analysis of La Roy Ladurie: beginning with an empirically evident structure (Egyptian imperialism in the Levant) and moving back in time to consider at least one traumatic event that helped to establish the structure, then finally attempting to place that event within pre-existing structure. This is reminiscent of Bintliff's problem history (see Chapter 2) and refocuses analysis back onto the event in order to assess in depth its significance as an agent of change or stasis, uncovering pattern rather than defining narrative history. Knapp wrote that his analysis aided in consideration of the Egyptian influence on the Levant, but also demanded careful analysis on the causal events for that phenomenon. Archaeologically, this Egyptian presence is constant and intensifying; historically, the Levant is seen as key to Egyptian trade with more distant lands (1992c: 84-85). However, Knapp's paper is far more effective in its integration of different types of data than in examining questions about change. Writing in a similar vein, Moreland quite successfully demonstrated the complex relationship between structure and agency that is absent from Braudel's schema (1992: 125).

As a final example of the practical application of *Annales* approaches in archaeology, Smith applied a Braudelian model to his study of change in Postclassic central Mexico. Much changed during that period: demographically, politically, economically, the emergence of the city state, and warfare. This is recorded both archaeologically and in native historical chronicles, but the potential of combining these two sources of data to construct models of change had not been fully realised. Smith believed that they had been juxtaposed prematurely before either had been sufficiently analysed on its own terms, particularly in the assignment of historical dates to archaeological phases. The accuracy of the archaeological chronologies could not match the fast-paced action of the historical sources: this lends itself to the archaeological examination of medium- and long-term change, but further refinement would be required to consider shorter-term processes. In some areas, archaeology and history disagree (e.g. the founding of Tenochtitlan; the extent of the Toltec empire; evidence for late Postclassic warfare). In some areas they agree more closely (e.g. demographic change). In many areas, no comparison is possible due to the different temporal rhythms involved in different processes (e.g. warfare may have occurred too quickly to register a strong archaeological signature). Smith concluded that insights gained from Braudel had aided in his attempts to correlate archaeology and history, particularly over the *longue durée* in the archaeological consideration of demographic change and the historical consideration of the rise of urbanism. He also believed that an *Annales* approach could help relate chronology building to research goals and fieldwork results (1992b: 51-53, 65, 68-69).

These examples illustrate how the influence of *Annales* approaches on archaeology has been rather limited in scope and in ambition: it has mostly been seen in field survey and broad regional syntheses and mostly restricted to a rather simplistic tool for aiding description and chronology building. The Bintliff volume in particular struggles to live up to its promise to explain key *Annales* concepts, and fails to take into account both much of Braudel's thinking and particularly also that of other *Annalistes*. The Knapp volume is a stronger, more critical work, finding the most useful application of *Annales* thought in helping to bridge the gap between historical and archaeological sources (Delano Smith 1992: 540-541). This could be seen more as a fresh archaeological take on *Annales*, as Braudel's original model was weak in its bridging of the layers of history. By contrast, although superficially similar theoretically to *Annales*, the influence of non-linear dynamics on archaeology has been quite different. Firstly, it has been much less widespread, probably due to most archaeologists' inherent (and possibly justified) fear of anything too scientistic. Beyond that, however, non-linear dynamics contrasts to *Annales* insofar as studies showing its influence tend to be much more focussed in subject, both thematically and spatio-temporally.

For example, Mithen saw the study of the mechanics of the extinction of mammoths as particularly conducive to mathematical modelling due to three factors: a clearly defined problem (what level of hunting would cause extinction); a modern analogy in elephant populations (though I find this assumption unconvincing, due to the radically different environment in which mammoths lived compared to modern elephants); and a sound archaeological base in mammoth assemblages. Mithen's particular non-linear dynamical approach was based upon the use of Leslie matrix models, a common method of

[1] Such as Bulliet 1992b, Knapp 1992c, Moreland 1992.

population modelling. He stated that his models faced the same set of problems as all archaeological simulations: limited and fuzzy data; the necessity to make assumptions; and the need to balance simplicity and generality against realism and complexity. He concluded that his model demonstrated that mammoth populations were highly sensitive to predation, suggesting that prehistoric hunters could have done little to avoid over-hunting and also that extinction would have been rapid if predation was severe enough to show an archaeological signal (1997: 176, 212). This seems a reasonable analysis, but there is perhaps an element here of a mathematical model being used to justify conclusions that could have been reached from a straightforward examination of the archaeological evidence: is it prudent to place so much weight on models which are based on such relatively broad assumptions?

Moving closer to the subject of this project, Zubrow used dynamic modelling and GIS to model the spread of major epidemic diseases into the unprotected populations of the New World. These were seen as spreading along trade routes from initial points of contact, resulting in a series of maps illustrating how diseases may have crossed the Atlantic and spread through Central America. Zubrow's work combined GIS, epidemic models and historical documents recording the logs and routes of voyages to the Americas between 1492-1650. He concluded that his results were preliminary and tentative, looking on them as creating detail and revealing relationships which might not otherwise have been recognised. The situation modelled was very complex and the full story had not yet been uncovered (1997: 216, 218, 249-250). The tentative nature of Zubrow's conclusions is a common theme of non-linear dynamics approaches, where models are seen not as replications of how things actually happened but more as tools to help archaeologists think about issues and problems. For example, Van der Leeuw and McGlade, in their non-linear dynamics inspired study of urban evolution, made no pretence of representing reality. They saw themselves as playing with hypotheses and trying to illustrate that there are other fun games out there which might provide new insight into archaeological problems (1997a: 331-332, 367).

Although not explicitly within the realm of non-linear dynamics, Going's work on pottery typologies shared the same inspiration from modern evolutionary theory. He noted how Welsh upland sites only have pottery on them during various short phases of the Roman period: this poses the question of whether the sites were abandoned or aceramic between these periods. Going concluded that, at times, people could not easily obtain pottery for their domestic needs over quite large areas of Roman Britain. Taking inspiration from the punctuated equilibrium model of evolution and assuming that development of pottery types showed changing rates of change, contexts with new forms of pottery can be read as equating to periods of rapid change and other contexts to periods of slower change. Further, pottery from contexts equating to slower rates of change ought to show more signs of repair, as lower availability would mean greater need for curation. However, the full pattern of this could only ever be visible on an ideal site where deposition was consistent in character and long-lived. Nevertheless, Going noted how his hypothesis was congruent with the cyclical models of economic historians and that it was well matched by patterns in the numismatic evidence. But what does this mean in practice? At times, ceramic dates must be out of alignment with their context and, as such, overly precise typological dates have distorted our picture of the past. Therefore, the curve of rate of change of ceramic production produced by Going could be used to calibrate typological dates: periods of higher rates of change would thus have much tighter typological dates than periods of slower change. For example, this would mean that contexts with late 4th century AD pottery (i.e. production) could actually date (i.e. deposition) to the 5th or even the 6th centuries AD (1992: 94, 96, 105, 107, 111-112). Going's work is very convincing and provides a strong route into solving some of the problems of archaeological dating discussed above.

McGlade saw the primary contribution of non-linear dynamics to archaeology in demonstrating the impossibility of meaningful long-term prediction (1997: 329), but I see far more potential in this idea of calibration (whether using economic data or something else). It may provide a powerful tool for approaching the integration of different archaeological dates and their multiple uncertainties (and is applicable to other dating methods beyond typology, such as numismatics). It also illustrates how theoretical conceptions of time can greatly aid everyday practice. However, little of the other work discussed above gives much sense of this: alternative conceptions of time still seem to be largely restricted to the realms of theoretical introspection.

The past in the past

> *Just as modern archaeologists are challenged by the survival of so many artefacts and monuments, people in antiquity could hardly have been unaware that they were living among the material remains of past generations, and those clues would have posed a challenge of its own.* (Bradley 2002: 7)

The study of past peoples' perceptions of their own past has been a growing trend within recent theoretical debate. Here we see the coming together in practice of several of the theories of time discussed in the previous chapter: Gell's A- and B-series (as noted by Bradley), elements of phenomenology, and Gosden's System of Reference. It is unlikely to be an area where great detail will be discovered, but past communities may have responded to earlier remains in similar ways to archaeologists (Bradley 2002: 8; Van Dyke and Alcock 2003: 5). This also relates to the theme of attempting to recover past peoples' perceptions of time through the study of possible methods of time indication, such as tally marks on Palaeolithic bone artefacts or the astronomical alignments of Neolithic monuments (Lucas 2005: 67-68, 71-73). Given the backwards-looking nature of archaeology, it is little surprise that the discipline should become increasingly focused on memory (Van Dyke and Alcock 2003: 3).

A key text of this archaeological movement is Bradley's work, *The Past in Prehistoric Societies* (2002). He considered in detail how past people negotiated with their own past within four broad themes (which I find useful to think of using Husserl's ideas of reproduction, retention and protention, and an idea that I will term *rediscovery*):

1. People in the past looking to their distant past - reproduction;
2. People in the past looking to their immediate past - retention;
3. People in the past looking towards the future - protention;
4. People in the past encountering the remains of ancient times - rediscovery.

Bradley considered the first theme (i.e. reproduction) through the example of the Linearbandkeramik (LBK) and its antecedents, noting that communities are often united by their belief in common origins. The LBK is often considered to epitomise the agricultural colonisation of Europe: it originated in central Europe and spread westward, where it maintained its distinctive identity for a considerable length of time before fragmenting into a series of regional traditions. LBK settlements implicitly recorded the past in several ways. In central Europe, the burning down of each generation of houses resulted in the build-up of tells. In the west, previous building plots were remembered, and the extension of houses was recorded by the presence of external borrow pits. Spondylus shells from the Aegean were buried with venerable members of the community. Finally, the doorways of LBK houses (and also burials) pointed back towards the area which had been previously been occupied. When the LBK ended, a range of mortuary monuments developed which echoed the form of the longhouses which were no longer built, although largely in regions just beyond the reach of the LBK itself. Bradley suggested that in this way the structural principles of the LBK were carried to new areas of Europe. He saw these as landscapes of memory, recalling an ideal existence from the remote past (2002: 17, 19-20, 24-25, 29-31, 33). Although alternative explanations could also be given for much of this evidence, the interpretations made by Bradley are attractive and, at the very least, illustrate how people in the past may have attempted to '*celebrate and codify*' (2002: 47) their origins.

For his second theme (i.e. retention), Bradley began with a more artefactual approach. He noted how hoards of metalwork could span huge timescales: for example, the Salisbury hoard contained objects spanning the early Bronze Age to the late Iron Age. Further, finer examples of early Bronze Age daggers from Wessex show signs of considerable wear and appear to have remained in use for much longer than more simple examples. Neolithic Grooved Ware pots show higher rates of repair than other contemporary ceramics, suggesting that the histories of individual vessels were important. Chronological typologies of Bronze Age beakers from Britain and Holland do not concord well with radiocarbon dates relating to their deposition, which could be explained by vessels entering the archaeological record a considerable period after their manufacture. From a monumental perspective, Bradley noted the alignment of LBK houses on older site features and gave a lengthy example of how field boundaries on Dartmoor respected (and sometimes incorporated) older cairns and stone rows. Bradley concluded that the lives of people in the past were clearly profoundly influenced by the tangible remains of their ancestors (2002: 54-55, 57, 65, 78-79, 80-81). This is the most straightforward of Bradley's themes and the most easily demonstrable archaeologically. This theme links back into the first, in that memories become increasingly corrupt over time until they can hardly be distinguished from myth (Bradley 2003: 221), passing (in my schema) from retention back to reproduction.

Moving on to his third theme (i.e. protention), Bradley suggested that constructing a monument helps build the fabric of society through the joining together of forces towards a single aim. However, the builders of a monument have no control over how their successors interpret it: meanings may be maintained or change (2002: 82, 84-85). Parker Pearson and Ramilisonina suggested that Neolithic monuments change their form in a predetermined sequence from timber, through stone to abandonment. Bradley considered that this sequence suggested that such monuments were always intended to have a finite history, following a pre-planned course of transformation: the remains may have remained for posterity, but they may not have been intended to remain static into perpetuity. For example, at Machrie Moor in Scotland, two settings of timber circles (or possibly buildings) were constructed. When they went out of use, they were ploughed over on at least two separate occasions, the ard marks avoiding the sites of the circles themselves. This cultivation **may** be seen as a deliberate eradication of the structures. After the cultivation, two stone circles were built in precisely the same position as the timber circles. This means either that the positions were remembered or that they were marked by small patches of unploughed land. The presence of at least seven fence lines running across site during the period between the timber and stone phases suggests that a considerable amount of time had elapsed. However, Bradley concluded that the sequence appeared pre-ordained and that the stone phase was always intended as the final stage of the monument (2002: 89-92). This is perhaps the hardest of Bradley's four themes to explore archaeologically. His explanations never venture far from description, although it is probably a forlorn hope to ever truly grasp the multiple motivations and plans of past people, especially those that never came to fruition.

For his fourth and final theme (i.e. rediscovery), Bradley considered the different ways in which we can see past peoples interacting with archaeological remains which they discovered. This demanded a response, whether ignoring the remains or attempting to interpret them - guided by tradition, literature and prejudice, much as we do as archaeologists (2003: 224-225). One of Bradley's examples of the rediscovery of past remains in the past was of the Irish royal site of Tara. The complex began with a Neolithic enclosure, henge, passage grave and cursus. This was followed by a Bronze Age cemetery of round barrows and the re-use of the passage grave. Many of these older remains were then incorporated in the following Iron Age royal centre. In the early medieval period, Tara (or Teamhair, pronounced T'yower) found its lasting fame in epic literature about this distant past (Bradley 2002: 142):

> *And the Tuatha de Danaan took possession of Teamhair.... And from that time it was above all other places, for its king was the High King over all Ireland. The king's rath lay to the north, and the Hill of the Hostages to the north-east of the High Seat, and the Green of Teamhair to the west of the Hill of the Hostages. And to the north-east, in the Hill of the Sidhe, was a well called Nemnach, and out of it there flowed a stream called Nith, and on that stream the first mill was built in Ireland.*
>
> *And to the north of the Hill of the Hostages was the stone, the Lia Fail, and it used to roar under the feet of every king that would take possession of Ireland. And the Wall of the Three Whispers was near the House of the Women that had seven doors to the east, and seven doors to the west; and it is in that house the feasts of Teamhair used to be held. And there was the Great House of a Thousand Soldiers, and near it, to the south, the little Hill of the Woman* Soldiers. (Gregory 1904: 31)

This passage is an early 20th century translation of a medieval description of an Irish prehistoric site: reinterpretation runs through the very veins of the history (and prehistory) of Tara. Almost every feature of the site seems to have been created in response to what was there previously, with structures reused and rebuilt. Bradley opined that Tara was of special interest because of the degree to which its prominence was rooted in the distant past, with the layout determined by visible Neolithic and Bronze Age earthworks. He suggested that the power of the kings of Tara was thus rooted in their links to the past (2002: 143-145).

Bradley's arguments on his fourth theme are convincing, but this clash between different pasts is likely only to be rarely visible archaeologically. Indeed, some of Bradley's examples are more historical than archaeological. Previously, Bradley had given a more archaeological example related to this fourth theme. He noted how Neolithic tombs in Cornwall tend to be similar in appearance to nearby natural rock outcroppings (the similarities are, indeed, very striking). It could be argued that the tombs were inspired by the granite tors, but they are also built in areas with no tors and the Cornish tombs do not appear to be the earliest examples. Thus, this similarity is perhaps better explained as people entering into the area (with their building tradition) and interpreting the natural landscape (in the light of that) as a decayed cultural landscape: newly constructed tombs then emphasised this perceived link. Bradley concluded that people who lived prior to the development of geology as a discipline would have had far greater difficulty in distinguishing natural and cultural features in that landscape, both for rock outcrops and tombs (1998: 18-21):

> *As well as ruined buildings, there were many ruined stones.*

Bradley later suggested something similar to Lucas's archaeology as recovery from amnesia (see Chapter 2) in respect of past peoples' negotiation with their past: a false memory syndrome featuring progressive loss of memory, change in memory, and recollection with new interpretations; a mixture of confusion and fabrication (2003: 226). This fractured past brings us back neatly to some of the themes discussed in the last chapter. Naturally, Bradley is not the only archaeologist to have approached this theme of the past in the past. In 1998, an issue of *World Archaeology* was devoted to the subject. Gosden and Lock wrote about the complex ways in which people negotiated their past. Their model of history differentiated between the genealogical and mythical past, whilst incorporating broader notions of action and ritualisation. Ritualised (and thus formalised) activities help people to create history, manifested archaeologically in repeated use (reinforcing meanings) and in later re-use (creating new meanings). People were not disconnected like archaeologists, but engaged with their landscape in this process (1998: 3-4). Unsurprisingly, this is reminiscent of Gosden's System of Reference (see Chapter 2). Mythological history (shaped by superhuman forces and limiting human agency) is contrasted from human history (open to the effects of collective and individual agency) in this model. Mythological history is temporally disconnected from the present (i.e. reproduction) whereas human history is continuous with it (i.e. retention) and primarily genealogical for prehistoric people (Gosden and Lock 1998: 4-5).

How can this be seen archaeologically? Gosden and Lock saw this broadly in the regular maintenance of the banks and ditches which carved up the prehistoric landscape and in the manner in which monuments would over time pass from remembered history into myth. More specifically, they gave the example of Rams Hill in Berkshire. Occupation there was discontinuous, but the Bronze Age enclosure was rebuilt twice on the same ground plan. They suggested that the people who rebuilt the enclosure would have remembered the previous builders in their genealogies and thus emphasised their links to the past through this act of rebuilding (a practical explanation is also possible: it is easier to re-cut an old ditch than to dig a new one). They interpreted this as illustrating that Bronze Age people maintained lengthy genealogical histories. In the Iron Age, these histories may have been much shorter and, thus, we see an alternative fascination with a mythological past in the re-use of Neolithic tombs and the aping of Neolithic pottery styles (1998: 6-11). A similar perspective was also taken by Mizoguchi in his paper on Neolithic / Bronze Age mortuary practices (1993). Gosden and Lock's ideas are intriguing, but are perhaps a little too abstract and lacking in detail. It may be that it is not possible to approach these questions in any more detail archaeologically. Certainly, the more detailed accounts given in a later paper by Gosden (with Marshall) are far more anthropological than archaeological (Gosden and Marshall 1999).

As another example, Williams wrote about the re-use of earlier monuments for Anglo-Saxon burial. *Beowulf* gives an impression of how the Anglo-Saxons conceptualised the remains of the past: the hero dies fighting in a barrow, described as the construction of ancient races and containing a hoard of treasure guarded by a dragon. As such, Williams argued that such monuments evoked memories of a distant, mythological past in ancestral homelands on the Continent. Predominantly, the Anglo-Saxons reused round barrows for burial, but they also reused many other types of monument (and even natural features that had the appearance of monuments). This was a widespread and frequent practice, not simply an elite phenomenon. The discovery of ancient remains when digging new graves may have encouraged these mythical interpretations. Williams suggests that, rather than veneration, this reuse was more emulation and appropriation. Some ancient monuments were elaborated or altered when reused. Further, the reuse of ancient monuments appears similar to the incorporation of mythical figures into their genealogies by Anglo-Saxon kings. Williams interpreted this evidence as a portrayal of legitimacy by immigrant Germanic groups as the rightful heirs of the ancient Britons, or perhaps as the evocation of memories of imagined homelands over the ocean (1998: 90-97, 101-104). Other archaeologists have considered how the reinterpretation of monuments continues today (e.g. Holtorf 1996; Dietler 1998).

A common theme running through much of this work on collective memory (whether real or constructed) is the connection to power: the use of the past to reinforce the legitimacy of present elites. Iconoclasm is the opposite process: the erasure of the remains of the past to remove any reminders of the illegitimacy of present elites (Lucas 2005: 88). This shows a strong influence from the thinkers explored so well by Gosden and encapsulated in his System of Reference: in the study of the past in the past we can see the practical application of the thinking of Husserl, Heidegger and Bourdieu to archaeological interpretation. Husserl is evident in Bradley's four different ways in which people encountered the past, which I summarised as reproduction, retention, protention and rediscovery. Bourdieu shows his influence in the strong connection to habit presupposed by the kinds of legitimising processes noted above. Finally, Heidegger provides his ever-present embodiment of past people in the world. Some of these studies err on the side of being a little too abstract or vague due to the difficulties presented by the unique nature of archaeological evidence, but there is still much to be admired and built upon. However, these ideas only really aid in attempting to explain past remains and do not help day-to-day archaeology in the field. As such (following on from my conclusions in the previous chapter), approaching these kind of questions computationally will not be possible until after a technological solution has been found to the difficulties of integrating archaeological data into a TGIS.

Gavin Lucas and biography

Geoff Bailey reviewed Lucas's *Time in Archaeology* by comparing it to a recent volume (Higham *et al.* 2004) on radiocarbon dating, the two books dealing respectively with time concepts and time measurement. He noted that the lack of overlap was part of his provocation for making the comparison. Bailey outlined some of the difficulties with radiocarbon dating discussed above, concluding that the vast numbers of new dates that had appeared since the invention of the technique had, if anything, created more uncertainty or at least made more explicit the inherent uncertainty of the radiocarbon method. He also noted that the collection of dates is meaningless without some reference to the problems that they are intended to solve or some awareness of the theoretical assumptions involved in their investigation. In considering Lucas's book, Bailey saw Lucas as placing archaeologists who use dating schemes (such as those provided by radiocarbon) within a position where they were bound to engage in explanations couched in terms of chronological time, thus subscribing to a unilinear, progressive, totalising view of history. Bailey felt that this was over-exaggerated, confusing chronology as frame-of-reference with chronology as explanation: he suggested that, in fact, scientific dating had restored a sense of their own history to peoples previously marginalised by western colonial visions of the past (2006: 717-719).

Most tellingly, Bailey noted how Lucas's concentration on a single jar for his main example (see previous chapter) allowed him to avoid the problematic and pervasive issue of contemporaneity. He believed that Lucas did not engage with the material dimension of palimpsests: the loss of evidence and resolution that is inevitable when moving beyond a single object. Although single events are visible in prehistory, one would not get very far in one's analysis if they were not compared to other events, which they might be separated from by thousands of years: changing scale of observation results in changing resolution, the central thesis of Bailey's time perspectivism. Further, he suggested that Lucas's best examples were historical and ethnographic ones (2006: 719), much like Gosden's work.

Nevertheless, there is some potential in the biographical approach (e.g. Gosden and Marshall 1999). Firstly, there is some common ground with the past in the past work discussed above. Examples of this include life histories (Holtorf 1998), and the interpretation of inscriptions not as narrative but as the creation of history (Barrett 1993: 236-238). Hodder also suggested that stylistic typologies could be used to infer large-scale narratives (1993: 269). Biographical approaches could apply equally to non-portable artefacts (Gillings and Pollard 1999: 180). However, I have not discovered any evidence of archaeologists attempting to take a biographical approach to a site or landscape in the whole. As such, if this methodology has potential, it is as yet untested.

Conclusions

In this chapter, I attempted to pull together the many varied temporal questions that archaeologist ask themselves. These start out from two basic archaeological puzzles: how old something is and what its temporal relationship with something else is. There are many techniques to answer the first of these conundrums, from relative methods (such as typology) through to absolute methods (such as TL dating). These different methods all carry different applications, problems and varied uncertainties. Further, all dating methods may only date very specific events and these may not relate closely to dates of deposition. These factors provide particular problems when we come to approach the second conundrum: how can we reconcile these varied types of date and combine them in a single chronology (or its multi-temporal equivalent)?

The predominant way in which archaeologists approach the second question is through stratigraphy. The Harris matrix is currently the most common method of recording this and has attractive elements in its (partial) multilinearity and the relative ease with which it can record complex stratigraphy. However, the matrix in some sense compounds the difficulties surrounding the integration of different dating evidence due to its overly simplistic take on context interfaces and formation processes: if the respective dates of two adjoining contexts overlap (or even are reversed), the unidirectional and absolute nature of the matrix is unable to cope. The incorporation of data on the formation of deposits and the degree to which they were sealed is an essential step if the Harris matrix is to carry any legitimacy into the future. Attempts to formulate post-processual field archaeologies have not solved this problem, as of yet. In time, finding a method of dealing with and integrating the various uncertainties of different dating techniques and their associated stratigraphies has to be a key aim of archaeological TGIS.

Building from this base, I then considered how archaeologists have attempted to apply the theories of time discussed in the last chapter to the practical questions that they wish to answer. *Annales* approaches were seen to perhaps carry some value in practice in the integration of archaeological and historical data, but had to date been rather too limited in scope and ambition. The theoretically similar non-linear dynamics approaches were seen to take a rather different path when approached practically. Although (as a whole) rather scientistic and rather too reliant on broad assumptions, positive practical assistance was found in one particular area: Going's use of economic cycles to calibrate typological dates. The extension of this idea of calibration may prove an especially useful tool in the key integrating process discussed just above.

I saw the combination of Heidegger with Husserl and Bourdieu in Gosden's System of Reference positively reflected in the study of past peoples' perceptions of the past. Bradley's work in this vein was considered at length and concluded to be important and intriguing. To me, the study of the past in the past provides the most interesting example of the influence of alternative conceptions of time in archaeology. It is an area with possibly huge potential, but with considerable difficulties arising from the nature of archaeological evidence: it is hard to reconstruct what people were actually thinking from the fragmentary remains of their material context. Perhaps, in the future archaeologists will be able to use TGIS to help approach these questions.

However, what is clear is that alternative conceptions of time do not appear to have had any impact upon day-to-day archaeological practice (in the field): all of the work discussed in the second half of this chapter rests on alternative interpretations not on alternative fieldwork. The raw materials (i.e. dates) and the techniques used to obtain them remain the same. This can be seen in Thomas's recent report on his excavations in Dumfries and Galloway (2007). It will be recalled that he is a prime advocate of taking a Heideggerian approach to archaeology (see Chapter 2). His final conclusions do contain some threads of interpretation that seem to draw in part upon that approach, specifically regarding life cycles, memory and place, and genealogies (although of people rather than institutions). Yet the chronological questions considered by Thomas are approached in an entirely conventional manner.

Bailey saw an uncharted gulf between developments in dating methods and the theory of archaeological time (2006: 719). This is perhaps not surprising due to the relative newness of the ideas discussed in Chapter 2. As they spread, perhaps these concepts will come to show more influence on archaeology at the coal face: dating specialists need to engage more with the theoretical implications of their work and theorists need to engage more seriously with the problems surrounding scientific dating (Bailey 2006: 719-720). Combined with the study of formation processes, ideas surrounding multiple temporalities ought to prove exceptionally useful in the improvement of the Harris matrix so that it **better** (i.e. not perfectly) reflects the true nature of archaeological assemblages. This is a process that is yet to take place. Further, at the basis of this structure and process of interpretation lies the archaeological dates themselves. As we have seen, these generally possess varying degrees of mathematical and interpretative uncertainty, and it is this fundamental fuzziness which is the key problematic as we try to improve our understanding of the evidence with which we work. Engaging more closely with this issue can help break out of totalising chronologies and help to produce archaeologies that are more representative of the evidence (and perhaps more reflexive). It is my contention that TGIS has the potential to aid in these arenas: to see how, we must move on to look at TGIS itself.

CHAPTER FOUR: TEMPORAL GIS

In this chapter, I will discuss the nature of temporal-GIS (TGIS) and try to illustrate how it can potentially aid archaeologists with the temporal questions discussed in the last chapter. Firstly, we will look at the use of IT in archaeology and the issues that arise from the growing influence of such approaches. Then we shall consider more specifically how GIS has been used in archaeology. Next, we shall discover the ways in which non-archaeologists have attempted to use GIS to answer temporal questions and explore how these approaches fall short for archaeology. Finally, we shall review earlier attempts to construct TGIS for archaeology.

Digital archaeology

Taking a positive approach to archaeological IT, Evans and Daly edited a volume on what they call *Digital Archaeology* (2006), whose central theme was the search for better ways to employ computer technologies in the discipline, based on an understanding on the strengths and limitations of the technology. Fairly sophisticated computer techniques are now available to non-specialist archaeologists, perhaps empowering them and bringing them into digital debate. Digital techniques might one day redefine the roles of everyone involved in archaeology, from the digger in the field to the general public (Daly and Evans 2006b: 3, 5), allowing real-time dissemination of results to the public and enabling them to share in the fieldwork process (Bradley 2006: 36). Further, digital archaeology should exist to aid us in the performance of archaeology as a whole: Daly and Evans argued that it should simply be seen as archaeology done well, not some arcane or distinct element, complementing but not replacing traditional approaches. They concluded that digital archaeology does not exist: what should instead exist are intelligent and practical ways to use computers in archaeology to better consider our theoretical and practical questions (2006b: 9; 2006a: 254).

Zubrow went into more detail in the same volume. He noted the gulf between narrative and digital archaeologies: the numerical world is seldom an agent of change but rather the group phenomena upon which processes act. This results in a tension between computerised models of the past and our narrative stories. Digital techniques may be viewed as being atheoretical or even anti-theoretical, but they can in fact provide a methodological foundation to theoretical perspectives. For example, the digital domain gives particular emphasis to both very large and very small things and thus allows the rediscovery of the individual agent as motivating force. Zubrow defined the following points of impact of digital archaeology on traditional methods:

1. It offers efficient ways to represent the world;
2. It allows one to count, do statistics, evaluate and analyse;
3. It allows efficient modelling of complex processes;
4. It allows the creation of virtual worlds;
5. It facilitates the transfer of results around the world at nigh-light speed.

This fifth element sees archaeologists rapidly becoming part of the global village, eliding the boundaries between professional, amateur and the general public (2006: 10-13), and aiding teaching and learning (see Lock 2006).

Zubrow believed that post-processual theory was incompatible with digital archaeology, with the interpretive, deconstructive and narrative nature of the former contrasting sharply with the analytic, reconstructive and measured nature of the latter. Unlike digital methods, favourite post-processual subjects such as art, insight or belief do not follow algorithms. Indeed, Zubrow noted that processes that follow rules and could be described algorithmically, such as causation or common sense knowledge, are rejected by post-processualists. By contrast, these factors are present in artificial intelligence, because humans and computers both begin with information of a common sense character. Further, parallel computing makes available non-sequential, multi-linear models and methods (2006: 17-19). This recalls Zubrow's interest in non-linear dynamics (see Chapters 2 and 3). His resultant position on the incompatibility of post-processualism and digital archaeology is not, however, ultimately sustainable. As discussed in the last chapter, Hodder's attempt to create a post-processual field archaeology at Çatalhöyük placed digital technologies at the very core of his collaborative approach, making powerful use of Zubrow's fifth element above. In fact, Zubrow has not discovered a dichotomy between digital archaeology and post-processualism, but rather one between post-processualism and numerical modelling (his third element). This is just one aspect of digital archaeology and it does not preclude post-processualists from using digital techniques in the broader sense. By contrast, Zubrow saw the emphasis on public participation as de-emphasising digital archaeology and thus ignoring the value of digital techniques for modelling and understanding past human thought (2006: 27).

Zubrow identified three problems with digital archaeology, which he believed to either be challenges or to be rapidly fading away:

1. Data size. There is never enough data to satisfy the rapacious data appetite of digital analysis.

2. Complexity. Digital archaeology relies on the flawless integration of multiple components: increasing complexity results in increasing difficulties over implementation and operation, and over tracing cause and effect. It becomes hard to tease out which effects are caused by the model and which by past processes.
3. The toys issue. There is a tendency to use digital techniques simply because they are available, many archaeologists finding emotional comfort in digital technology.

These problems are particularly prevalent in the non-linear dynamics inspired areas in which Zubrow works, demonstrating that digital archaeology is anything but atheoretical, in that it enables and impacts particular theoretical positions (2006: 25-27). The first and second problems can be mitigated by restricting the amount of numerical modelling built into our digital techniques.

However, the toys issue is seen by digerati[1] such as Zubrow as a key area of concern for digital archaeology: they see a danger that digital archaeologists create problems that require digital solutions (2006: 15). Only appropriate digital techniques should be used and not without thought or just because they are there: a computer will produce results even if the technique used is inappropriate (Evans 2006: 87). Also, caution should be taken in using new-fangled gadgets on important sites where there are no clear-cut benefits (Bradley 2006: 44). Backhouse concluded that digital archaeology should be about using technology well; in other words, with careful forethought. He suggested that failing to take such precautions would involve the sinking of good data in the morass of bad (2006: 57). This argument is common sense and is unlikely to find much disagreement amongst archaeologists more sceptical about archaeological IT.

As noted above, Evans, Daly and Zubrow are all firmly on the positive side of discourse on the benefits and perils of digital archaeology. Thomas took a more sceptical view. In particular, he questioned the claims made that GIS (see below) and virtual reality are able to take us beyond traditional 2D representation, and examine perception and subjectivity in the past. He suggested that such systems are embedded in the modernist perspective and, thus, anachronistic when applied to the past. Thomas saw these approaches as overwhelmingly visual, with their approach to perception involving subjective humans observing an object world. In other words, the GIS representation of the world is considered perfected, with our task simply being to comprehend its past perception. As a result, he argued that this formulation left intact the modernist oppositions between subject and object, and mind and matter, resulting in a circularity of argument: models of things past and human behaviour become the unquestioned grounds for digital reconstruction. If these assumptions are discounted, then the virtual past begins to crumble (2004: 198-200).

Thomas saw a contextual approach as the way out of this problem, embedding objects in structures of intelligibility; in other words, a background of meanings and practices. He argued that digital archaeology removes context and views objects as geometrical forms, creating simulacra and reducing the past to a pattern of pixels, '*...viewed on the screen of modern rationalism*' (2004: 201). The influence of his previous work on Heidegger is very clear here (see Chapter 2), although that work also carried the same separation of perception and object world seen by him in digital archaeology. Thomas also opined that exclusively visual media could not achieve the metaphor, allusion and ambivalency that natural language allows. While I would agree that such media do currently struggle to express such concepts, there is no reason to assume that they are **unable** to do so. Thomas concluded by questioning how far digital archaeology could take us in the development of a sensuous and experiential archaeology of place and landscape, due to the shortcomings that he had elucidated (2004: 200-201). However, a contextual approach can be at the core of digital archaeology: GIS analysis is **about** putting people and things into their context - in the sense of integrating multiple sources of data (Lock 2003: 166) - and, as such, it is more a question of improving that context and focusing on the particular areas of it which aid us in answering our specific questions (see below). Increasing our ability to handle - and, in some sense, create - uncertainties in our data and taking a more fluid approach to our framings, such as time, will provide the tools for us to do so.

A more considered view has been offered by Huggett (2000). He began by noting how archaeological IT can be viewed in a myriad of contradictory ways: computers are tools like trowels, yet trowels are not generally the subject of conferences; computers can be instruments of revolution or of conservatism; computers can be neutral or value-laden; computers can facilitate communication or can increase isolation. Huggett also suggested that the reporting of digital archaeology is very often so positive as to verge on the evangelical: reticence to admit failure can hide hours of frustration and confusion. Furthermore, such writing tends to focus on **how** applications should be undertaken, ignoring whether computers should be used at all. Huggett suggested that this application-driven approach results in the lack of more sceptical examinations of digital archaeology. He argued that the ubiquity of computers makes the asking of questions about whether to use them important: this is not something that can be ignored by archaeologists. More widely, attitudes to IT range between optimism and pessimism, with optimism increasing as level of analysis focuses from the global to the particular. Huggett suggested that breaking out of the techno-evangelical mould is very difficult, even when one tries to adopt a more neutral approach to IT: it is all too easy to intuitively take this type of technological deterministic view, with nigh-powerless users pulled along by

[1] i.e. those who are keen advocates or even evangelists for the use of digital technologies.

technocratic computerisation movements and manufacturers (2000: 6-7, 9-10). As we shall see, this is a particular problem when attempting to apply conventional TGIS to archaeological data, as we are forced to tailor our questions to their particular functionality.

Huggett thus asked to what extent digital archaeology applications are determined by the tools available and archaeological questions determined by the tools applied (e.g. archaeological GIS researchers tend to focus on landscapes over intra-site data). He suggested that computers are seen as malleable tools which adapt to applications rather than changing applications, but this is counter-intuitive. Archaeologists are content to accept that our data are theory-laden, but loathe to accept that the computers used to analyse that data are themselves value-laden (e.g. computers might be seen to skew interpretations towards quantitative methods). Huggett judged that tools may be content-free and widely applicable, but they are not value-free. Descriptions of change driven by computing are often characterised by technological utopianism, an enthusiastic and polemical style of writing which inevitably triggers anti-utopian responses. Although an element of this creeps into writing about archaeological computing through the need to justify positions and grants, most archaeological writing tends more toward social realism, in which emphasis is placed on empirical data about actual applications. The growth of digital archaeology is a rational trend, but one that Huggett argued is based upon essentially utopian or deterministic views of technology. Thus, justification for this growth focuses on perceived benefits and, as a result, most assessments of archaeological computer usage are essentially flawed (2000: 11-12).

Huggett opined that most arguments for computerisation are based upon increased productivity, but he questioned whether this was justified. Most data used to answer this question are rarely more than anecdotal. Outside of archaeology, this issue has raised much concern. IT methods are not necessarily more efficient and, even when they are, this may be offset by social processes such as sending long emails where once a quick telephone conversation would have sufficed. Because of these side-effects, Huggett concluded that computers more often than not do not increase productivity (2000: 13-15). Related to this are concerns over people not creating enough metadata. Metadata are data **about** data, explaining what it is and what it means. Without metadata, digital data will eventually lose its meaning (Backhouse 2006: 55):

> *As a result, it is easy to end up with a mountain of millions of bits of meaningless nothing. This nothing is then sent off to be stored*

Data standards are needed to maintain consistency between and within databases. This may be aided by thesauri (which restrict rather than expand the terminology available), as advocated by Baines and Brophy. They suspected that theoretically informed archaeologists distrust the idea of data standards, due to their restrictive nature, but they argued that data standards are essential to effective digital archaeology. They suggested that data standards can in fact create a freer language for archaeology, liberating expression and justification of our ideas and opinions. However, a data standard thesauri should be taken only as a tool with which to attempt to order and work on archaeological material and should not be taken implicitly to be representations of an ordered and structured past. Using non-specific natural language terms, rather than convoluted typological jargon, is perhaps the best way forward as they retain the characteristic fuzziness of archaeological discourse (2006: 236-240, 246). However, whilst creating metadata does help preserve the original data, it itself results in more data. This could lead to exponential growth and informational entropy (Huggett 2000: 15). The argument that data standards encourage expression is also unconvincing. Bodies like the AHDS[1] require metadata for their archiving and it is when archiving data that these issues attain particular relevancy.

Humans only have a limited capacity to handle information and IT can push that limit, reducing our ability to process that which we do know: relying on digital solutions to answer this growth in data simply emphasises the disempowerment of users. Huggett has concluded that archaeologists fail to apply the critical analysis to their methods that they apply to the past. They may become beguiled by technology, so that the need to be seen to be using IT becomes all-important. This beguilement can become deterministic when technological solutions are given primacy over archaeological questions, or utopian where overly optimistic or unrealistic claims are made about digital archaeology. He suggested that, so long as this is the case, archaeologists will be in thrall to technology, which will be blamed for problems that may in fact be archaeological. Huggett reassured us that there is still time to rectify this situation. What is needed is for archaeologists to go beyond description of digital techniques and to discover an awareness of the problems and issues surrounding the impact of IT on the discipline (2000: 15-19).

Lock also usefully considered digital archaeology through two main themes: the potential of computers to act as active agents for thought rather than just as passive tools; and the symbiotic relationship between the development of IT and archaeological theory. He began by querying what we mean by archaeological data and what their relationship is to computer data. He suggested that data are theoretical appropriations of the archaeological record and, to render this digital, there must be a structural link between this data as theoretical

[1] The Arts and Humanities Data Service, see: http://ahds.ac.uk/metadata/index.htm

objects and data in the computer. Lock suggested that most archaeologists choose a pragmatic middle ground between various competing theoretical positions, but that data collection and interpretation are still their bread and butter. When it comes to building models, those built up from data tend to be more mathematical than those constructed from theory. Computer models add another level of abstraction to this, possessing their own reflexive relationships with data and theoretical models. As a result, Lock argued that computers cannot be simple passive tools, but are integral with the process of interpretation (2003: 1, 3-7).

This never-perfected and never-closing hermeneutic spiral is where mediation takes place between the unknowable past - whose vestiges form the archaeological record - and interpretation in the present. Lock asserted that digital models force the presupposed conditions of data models onto the surface, the structure of these presupposed conditions forming the basis of theoretical models. The mediation of the digital model is reliant upon the digital representation of data and theory models, which is itself dependent upon the richness of the software tools, data structures, and reflexive questioning capabilities. This richness can be paralleled with the contextual approach (2003: 8-9) advocated by Thomas and others. Indeed, Lock raised the question (2003: 13):

> *Are we just doing the same things as before but more easily and quickly or are the digerati correct in their ultra-positivist view of the future where the only limit to computing is our own imagination?*

As Huggett queried, whether we are even doing the same things more easily and quickly is not a proven case (2000: 13).

To return to these very digerati, Daly and Evans suggested that there is a myth of digital infallibility (although I doubt someone struggling to upgrade their Windows installation would agree). They suggest that using digital technology without understanding or considering the theoretical implications can be disempowering: unthinking reliance on technology leading rapidly to disappointment. Yet digital archaeology used well can be empowering, highlighting the work of excavators, accelerating and expanding the dissemination of results, and altering the traditional boundaries between archaeologists and others. Daly and Evans opined that digital archaeology is resonating throughout our discipline and on the verge of launching the next great leap forwards: where that leap leads is up to us (2006a: 254-255). Although this is somewhat hyperbolic, I agree that digital methods will continue to aid archaeologists, but they should only be applied where appropriate. This appropriateness encompasses whether IT can help answer questions both more easily and more effectively, and whether the data dealt with is suitable for incorporation into digital databases.

In the context of this project, the questions I wanted to help answer are those discussed in the last chapter that form the foundation of archaeological interpretation: essentially, the assessment of contemporaneity, the integration of different dates and dating methods and their varied uncertainties, and breaking down the monolithic formation of chronologies. Digital archaeology is an appropriate route into dealing with these issues: the use of computers can allow archaeologists to conceive of spatial relations and multiple temporalities in a nuanced way that would be very difficult to perform with pen, paper and brain alone. To be clear, however, I did not wish to build software that answers these questions **for** archaeologists, but software that **helps** archaeologists to more ably answer these questions themselves.

GIS in archaeology

GIS is generally taken to stand for Geographic Information Systems, although increasingly the S is taken to refer to Science.[1] GIS are computerised systems for collecting, storing, analysing and disseminating information about geographic space: they are toolboxes designed to be effective and efficient for certain purposes (Chrisman 1997: 5; Wheatley and Gillings 2002: 9), with a huge number of uses (Laurini and Thompson 1992: 10); they are commercial products that change rapidly - with no little hyperbole - to fill profitable market niches (Lock 2003: 165). GIS have made spatial analysis much easier than with paper maps (Jones 1997: 4), and are unique due to their ability to perform such analysis on combined spatial and non-spatial data from different data sets (Kraak and Ormeling 1996: 8). Most archaeological data has an important spatial component and archaeologists have thus been long aware of the importance of space to their studies. Routinely, archaeologists deal with enormous amounts of spatial data, from continental distributions, through the landscape and down to within an archaeological site. Thus, GIS has spread rapidly to become widely available to archaeologists and has revolutionised the way in which we implement our spatial methodologies (Wheatley and Gillings 2002: 1, 3). Today, GIS has become a fundamental tool for any archaeological project that handles spatial data (Lock 2003: 182).

Following and extending Marble (1990), Wheatley and Gillings defined five main subsystems of GIS:

1. Data entry;
2. Spatial database;
3. Manipulation and analysis;
4. Visualisation and reporting;
5. User interface.

[1] As in the *International Journal of Geographical Information Science.*

The third is what separates GIS from other tools and gives it its unique identity, although the boundaries of this identity are being elided (2002: 10-11, 13). When compared to conventional paper maps, GIS facilitates spatial analysis (Jones 1997: 4) and provides an environment for thinking and exploring ideas. Whilst most GIS methods have paper-based precursors, their easy availability in GIS has revolutionised these and opened up new avenues for research. In essence, GIS spatial analysis powerfully extends our observational equipment so long as it is not seen as producing complete solutions (Wheatley and Gillings 2002: 18, 20, 125, 147-148).

GIS allows archaeologists to handle vast quantities of data in ways once unimaginable, but it is all too easy to become overly enamoured of the aesthetics of GIS output, allowing this factor to dominate interpretation. For example, GIS enables the plotting of archaeological distributions against the background topographic, environmental and palaeo-environmental situation, but this can lead to merely stating the obvious or at worst to the promotion of an uncritical environmental determinism (Gaffney *et al.* 1995: 211). Taking a more contextual approach to analysis, involving carefully thought out use of available data, may help to escape from this incipient and discredited model (Gaffney and Van Leusen 1995: 378; Gaffney *et al.* 1995: 227). As with all digital archaeology, GIS applications should be based around specific archaeological questions and upon theoretical considerations (Wheatley and Gillings 2002: 237), and the challenge facing archaeological GIS is whether it represents a change in the way archaeologists approach problems or is simply a new way of doing old things (Lock and Harris 2000: 2).

Laurini and Thompson (non-archaeologists) listed archaeology as only a user of GIS, not an innovator or theoriser (1992: 21), but archaeologists have always been **innovative users** of IT (Lock 2003: 10) as the following examples illustrate. Claxton suggested the use of neural nets, virtual reality and artificial intelligence as possible sources of answers to the difficulties encountered in applying GIS to archaeology (1995). Further, Daly developed a GIS using industry standard software with little modification that was able to simultaneously analyse horizontal surface data and vertical sub-surface data at multiple scales of resolution and quality (2003: 5, 249). There are also possibilities in bringing much of this GIS work forward onto excavations, enabling spatial analysis during the excavation process (Backhouse 2006: 56-57). This site-based computing and the contextual approach suggested above has much in common with Hodder's work at Çatalhöyük (see Chapter 3) and is one way in which archaeological GIS is forging itself a unique identity. This is not to say, however, that there is not scope for archaeologists to be innovative in GIS development. Archaeology will likely only ever be a niche interest to the large GIS software developers - due to the restricted opportunities for financial profit - and, as such, we must strive ourselves to forge GIS that better deal with our unique data and questions. One obvious way in which contemporary commercial GIS fall down for archaeology is in their integration of time. As a result, innovative and inexpensive solutions are required from archaeologists to handle this aspect better.

Temporal GIS

Time and space are irretrievably linked through the realisation that all movement in space is also movement in time (Gosden 1994: 79-80). This intrinsic temporality separates geography from geometry (Langran 1992: 11) and, thus, GIS that are ignorant of time will prove ultimately inadequate for spatial analysis (Lock and Harris 2000: 10). Archaeologically, this linkage between space and time is perhaps even more apparent, specifically through the relationship between past time and present stratigraphy. Raper argued that taking a multidimensional approach to GIS (i.e. extending GIS into the third and fourth dimensions) allows us to escape from the limitations of flat, static maps and '...*engage the user in a reflexive cycle of conceptualisation, representation, reflection and evaluation*' (2000: 1). Conversely, he argued that dealing only with timeless space limits conceptualisation and thus representation: richer spatial and temporal structures are needed (2000: 124). As a result, contemporary conventional GIS will fade into obsolescence as newer systems are developed that give fuller account both to three-dimensional space, and to time. This is true both of geography and archaeology, but ought to be especially the case for the latter due to our discipline's fundamental temporality.

To counter this, temporal-GIS (TGIS) have been and are being designed to deal with time alongside space and, thus, liberate the potential to provide deeper models of the world - and, if used archaeologically, the past. TGIS is both a field of study and the systems that result from that study. TGIS have potential application in many fields besides archaeology, including environmental change, transportation, demography, health, governance, administration, and defence (Worboys and Duckham 2004: 359). However, most work on TGIS has been undertaken by geographers and their allies. Therefore, we shall examine the geographical corpus on TGIS first.

The geographers' view

The first geographer to really attempt to get fully to grips with TGIS was Langran in her seminal book *Time in Geographic Information Systems* (1992). The fast pace of technological change means that this work is now somewhat obsolete,[1] but it forms the conceptual basis of much that has followed. Langran noted that a reasonable goal for GIS would be the ability to trace and analyse spatial change, but that this was not possible for atemporal GIS. She went on to outline six fundamental functions of a TGIS (1992: 4-7):

[1] For example, Langran imagines a time when GIS are widespread (Langran 1992: 1) - that time is now here.

1. Inventory - the ability to store a complete description of a study area and to account for changes in the physical world and in computer storage. A TGIS should be able to supply a complete lineage of a feature, the evolution of an area, and the state of a feature or area at any given time.
2. Analysis - the ability to explain, exploit, or forecast components of and processes at work in a study area. TGIS should extend the analytical capabilities of GIS by including specific reference to change.
3. Updates - the ability to supercede outdated information with current information.
4. Quality control - the ability to evaluate logical consistency between new and previous data.
5. Scheduling - the ability to identify or anticipate threshold database states, triggering pre-defined responses.
6. Display - the ability to generate a static or dynamic map or table of temporal processes at work in a study area. Langran saw the ability of TGIS to illustrate real time occurrences or to respond to queries with animated maps as one of its more rewarding abilities.

These functions highlight the defining contrast that separates geographical from archaeological TGIS (as we shall come to see): geographers have often found themselves largely concerned with numbers 3-5 (see Chrisman 1997: 86), such as the problems and solutions surrounding dealing with dynamically updating contemporaneous spatio-temporal data (such as traffic flow, public utilities, etc.). For archaeologists, these concerns are less pressing, as our data does not generally carry this need to be **constantly** up-to-date:[1] of this specific list, we should be more concerned with numbers 1-2 and 6.

Langran went on to define five technical requirements for TGIS (1992: 8):

1. A conceptual model of spatial change.
2. Treatment of aspatial attributes.
3. Data-processing logistics.
4. A spatio-temporal data access model.
5. Efficient algorithms to operate on the spatio-temporal data.

As we shall see, these have been elaborated and worked upon by others since. Langran then noted four temporal constructs that require mapping (1992: 22):

1. Qualitative change - what changed?
2. Quantitative change - by how much?
3. Composite change - mapping more than one time slice (e.g. paths, time sequences, diffusion, cycles).
4. Space time ratios - describing space-time interactions (e.g. travel times, cognitive spaces, rates of spatial change).

Temporal maps effectively add a dimension to the geographic information mapped and, as such, require multidimensional display formats. To do this, Langran defined four classes of mapping technique (1992: 23):

1. Time sequences (e.g. multiple editions or time series).
2. Change data (e.g. textual, graphic or digital amendments to a base representation).
3. Static maps with thematic symbols of a temporal theme (e.g. symbols depicting dates, rates, paths, or order of occurrence).
4. Animations, where both space and time are scaled to depict the metamorphosis of a region.

Again, below we shall see how these ideas have been elaborated upon since.

Langran acknowledged that temporal maps are cartographic statements of theories about geographic development processes. However, she then proceeded to state that as they are solely concerned with representation rather than duplication of reality, cartographers could concentrate on representing the effects of time rather than get involved in debates about its nature: she suggested that cartographic time should be a pragmatic and generic distillation of real time. In Langran's model, time and space are separate and time is divided along temporal boundaries in a similar manner to spatial objects. This simplifies time. Topologically, her cartographic time is a 1D line - representing duration of a state, punctuated by 0D points - representing instants of change (1992: 24, 28-30, 32-33, Chapter 3). She defined three main ways of achieving this practically, alongside various other methods of representing change to database state which are irrelevant to this project (1992: 37-41):

1. Space-time cube - representing one time and two spatial dimensions, but unable to handle a third spatial dimension.
2. Sequent snapshots - time slices representing states, but not representing change events and with wasted storage issues in recording the entire state for each slice.
3. Base state with amendments - similar to time slices, but uses less storage and records change events.

[1] Whilst updating of archaeological databases is undoubtedly important, it does not carry the same time pressures involved in dealing with the dynamic updating of real-time data.

This third method was preferred by Langran, although the snapshot approach does reflect well some methods of data collection, such as remote sensing (Peuquet 2001: 14). In ignoring events, however, snapshots cannot be used to understand underlying processes (Hornsby and Egenhofer 2000: 210).

Peuquet was the next geographer to make a serious attempt to come to grips with TGIS, taking a more considered approach in contrast to Langran's pragmatism. She suggested that previous work on TGIS just served to reveal that the representation of space and time was much more complex than the representation of space alone. Previous models, such as Langran's, were extensions of traditional GIS representations which were based in turn upon the primitive elements of static map display. Peuquet attempted to take inspiration from other fields, such as artificial intelligence and perceptual psychology, to deliver a more encompassing perspective by recognising the much needed distinction between representation in an abstract conceptual sense and visual presentation. She proposed a set of alternative interrelated and largely non-visual mappings of reality that organised information in different ways, facilitating the seeking of answers to alternative questions whilst retaining the possibility of integration with more traditional cartography (1994: 442).

Peuquet noted that despite their continuous nature, space and time are broken down into discrete units for the purpose of measurement. For time, this is conventionally achieved through the use of intervals and events. Ideally, the resolution of these units should be related to the phenomena being measured, as distinct patterns at one resolution might dissolve into chaos at another. Thus, temporal analysis is scale dependent. Returning briefly to archaeology, there is very little choice available to archaeologists when it comes to temporal resolution: the resolution of our data are determined by the accuracy of our dating techniques and the stratigraphy of a site. Further, Peuquet noted how time is conventionally modelled as a fourth dimension, a position supported by the expression of time using spatial metaphors in natural language. However, Peuquet suggested that this was not sufficient due to the important differences between space and time in their properties and in their referential bases for potential queries. She also noted how the base state and amendments model of Langran was problematic in practice, due to data storage becoming unwieldy with increasing complexity and with objects that appear, disappear, then reappear over time. The method by which time is represented is inextricably linked to the methods by which it is analysed and used. Peuquet proposed a model that incorporated distinct location, time and object based components (1994: 445-446). This triad framework allows the user to pose three basic questions (Peuquet 1994: 448):

1. when + where → what
2. when + what → where
3. where + what → when

Raper argued, however, that the separation of time and space on which the triad framework was based could only be successfully applied to entities with discrete spatial and temporal identities (2000: 125). Despite this, Peuquet's model is a more detailed and potentially useful one than Langran's. Her critic, Raper, himself concentrated mostly on the third spatial dimension in his work on multidimensionality and gave time little detailed examination.

Peuquet went on to note five intrinsic characteristics of temporal distributions (1994: 451):

1. Temporal cohesiveness - a tendency towards smooth evolutionary-type change.
2. Temporal similarity - different objects influenced by the same process will tend to exhibit similar rates of change.
3. Temporal continuity - distributions of events across time influenced by a specific process will tend to exhibit an organised temporal pattern resulting from a chain of cause and effect.
4. Hierarchical organisation - distributions of events across time are often generated by a number of different processes operating at different temporal scales.
5. Incompleteness - our knowledge of the changing state of objects and locations is always partial.

Points 2, 4 and 5 are all logical and ought to be familiar to archaeologists. However, points 1 and 3 are clearly contradicted by some of the alternative models of time that inhabit archaeology. Peuquet acknowledged this, noting that capturing the true complexity of spatiotemporal relationships would require a powerful and very flexible method of description (1994: 454). Specifically, the smooth temporality implied by temporal cohesiveness does not match the punctuated temporality of many archaeological distributions. Similarly, the straightforward chains of cause and effect implied by temporal continuity are not easily identifiable in most archaeological data. These are the types of temporality that non-linear dynamics inspired archaeologists try to model or that those inspired by *Annales* try to simplify (see Chapter 2). Of the above list, points 4 and 5 more closely reflect these archaeological models of time and form stronger base assumptions for any archaeological TGIS.

Peuquet advocated describing spatio-temporal patterns as combinations of relationships between her triad: locations, times, and / or objects. These relationships could be expressed through (a) metric and topology; (b) Boolean operators; and (c) generalisation (1994: 454). Temporal topology can comprise (Jones 1997: 26):

x before y

x immediately before y [1]

x equivalent to y

x overlaps y

x immediately after y

x after y

Boolean operations combine different types of temporal distributions or different values within the same type. Generalisation is dependent upon the metric used - such as seasons, months, days, years, etc. - and scale. For example, generalisation may be necessary to relate a temporal dataset at one resolution with another at a different resolution. Peuquet noted that one difficulty with temporal reasoning is the frame problem, that is knowing which attributes of an object will change due to an event; for example, the location of a migrant worker changes when he/she migrates but his/her gender does not. Generalisation allows this type of information to be stored as rules rather than wastefully recorded explicitly for each instance (1994: 455-456). An archaeological example of temporal generalisation might be for the comparison of palaeoenvironmental data recorded at a yearly (or seasonal) resolution with economic data recorded from coinage / pottery production with a resolution of several decades.

In 2001, Peuquet stated that evolving research into space-time representation had focussed on five areas (2001: 15):

1. The ontology of space and time.
2. The development of robust and efficient space-time database models and languages.
3. Inexactness and scaling issues.
4. Graphical user interface (GUI) and query optimisation.
5. Indexing techniques for space-time databases.

The first concerns the need for a common conceptual framework, such as Peuquet's triad framework and or Langran's base state with amendments. On the second, Peuquet noted that extending the relational database model to include linear time worked well, but became complex when space was introduced. It ultimately results in inflexible and inefficient representations for space-time data, but using these familiar paradigms does allow rapid testing of new ideas. Moving beyond conventional database structures to object oriented[2] approaches, Peuquet believed that these had not yet met their potential, but were conceptually strong due to their ability to separate out space, time and attributes and consolidate them in a modular fashion as distinct elements. On inexactness and scaling issues, Peuquet noted that this was inherent in all temporal and spatio-temporal databases, but that they had been ignored in early work into TGIS. These issues could be divided into fuzziness and uncertainty: fuzziness concerns inherent imprecision in data and fuzzy categories that might be largely based on subjective choices; uncertainty concerns lack of information, in other words things that are not exactly known (2001: 15-16, 19-20, 23). Archaeologically, uncertainty in this sense would concern things like mysterious bone rings whose purpose can only be guessed at. Fuzziness is of more relevance to temporality, concerning issues such as the statistical uncertainty of scientific dates or unclear context relationships. To avoid confusion for the purposes of this project both the terms uncertainty and fuzziness can be taken herein as referring to this type of fuzziness.

Peuquet also saw a fundamental framework developing for TGIS that separated implementations between discrete and continuous views. The discrete view is entity based, in which distinct entities have spatial and temporal attributes. This view has the advantage of being able to handle non-spatial temporal events, such as election results. The continuous view is field based, in which individual objects are denoted as attributes attached to a given location in space-time. The two views are duals of each other and the same phenomena may be modelled using either paradigm (2001: 24-25). They parallel the vector and raster spatial models used in GIS. In conclusion, Peuquet saw four areas of TGIS research that still needed to be tackled (Peuquet 2001: 27-28):

1. The development of a theory of space-time representation, including the relationships between linear / cyclical time, continuous / discrete time, multi-linearity, etc.
2. The development of strategies for handling inexact and complex objects and relationships. Peuquet rightly accused current TGIS of imposing a false level of certainty and precision on data.
3. The development of a means to deal with missing data, filling in the gaps in historical data using secondary sources of evidence, etc.
4. The development of a means to deal with multiple times and alternative histories.

Although written in 2001, geographers have, as of yet, failed to deal with these issues. It is thus fair to say that geographers' TGIS remain insufficient for archaeology: archaeological data exhibits all of the above and for TGIS to be applied successfully and widely to archaeology it needs to be able to draw on developments in these four areas. For example, multiple temporalities are present in non-linear dynamics and *Annales* theoretical approaches (see Chapter 2) and, to some extent, in the Harris matrix practical approach; almost all archaeological dates carry some form of uncertainty (see Chapter 3); furthermore,

[1] i.e. minimum of y = maximum of x.

[2] In object oriented databases, features exist as entities not just as attributes and acquire their general character from defined classes and subclasses; sometimes also combining methods in an integration of data and operations (Laurini and Thompson 1992: 624-626, 631-632).

archaeological data are never complete and interpretations are never perfected. All of this makes progress on at least some of the above agenda essential for archaeological TGIS.

So, where is geographical TGIS at the time of writing? An example of a commercially available TGIS is Discovery Software's STEMgis, which is capable of producing animated maps and the querying of a whole range of dynamic spatio-temporal phenomena: their website gives examples of mapping changing traffic volumes, sea levels, and tidal currents, amongst others.[1] However, the basic temporal model appears to be based upon snapshots and clock time. Raper drew a distinction between what he called spatio-temporal GIS with two spatial dimensions and one temporal, and four dimensional GIS with three spatial dimensions and one temporal (STEMgis falls between these two categories, as it is able to deal with depth data). He argued that only the 4D approach can express the full multidimensional character of the world, yet early TGIS approaches such as Langran's were based solely on extending existing 2D GIS (Raper 2000: 160, 163). Whether full 4D TGIS would be necessary to help answer most questions that might be asked of such software is not, however, a proven case as of yet. Nevertheless, true TGIS remains primarily a research project, albeit with commercial GIS products being extended to offer some practical temporal functionality. Current TGIS, such as STEMgis, are only able to answer questions based on what / where / when. Worboys and Duckham suggested that a wholly new TGIS would be needed to extend these questions to explore instead what happened / where / when / how and why, providing the functionality to analyse causality. They noted that representation had moved from static maps to time slices to object lifelines[2] to events, actions and processes, and that change could be spatial or attribute based. In their opinion, the final stage of TGIS evolution would be a '*...full-bodied treatment of change, in terms of complexes of events, actions, and processes.*' (2004: 359-361, 363-364).

To Worboys and Duckham, one of the key limitations in achieving this concerns the distinction between transaction time (i.e. the time when data entered a database), and valid time (i.e. the time when an event took place in the real world) (2004: 367). To any archaeologist, it is valid time that is the most critical, however, geographers often appear more fascinated by the former (e.g. see Egenhofer and Golledge 1998). When modelling time, Worboys and Duckham noted that it could be treated as a timeline based upon either continuous (real), dense (irrational) or discrete (integer) numbers, with discrete being most commonly used. A chronon or tick is the basic unit of temporal measurement, with duration built out of a whole number of chronons, but with no fixed place on the timeline. They also noted that timelines could be linear, branching or cyclical, but that linear time was the most common and most useful (2004: 372).

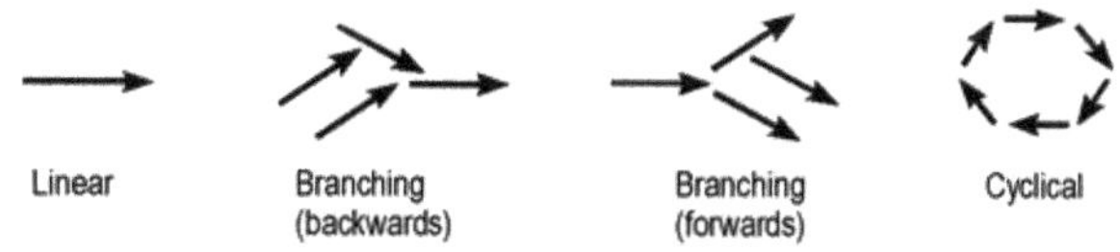

FIG 4.1 - LINEAR, BRANCHING AND CYCLICAL TIME (AFTER WORBOYS AND DUCKHAM 2004: 373).

All three of these models of time resonate with various ideas discussed in the previous chapters to a greater or lesser extent. Specifically, branching and cyclical models of time have much in common with the multiple temporalities of non-linear dynamics. Also, linear and cyclical time are familiar - if not always useful - archaeological chronotypes (see Chapter 2). The Harris matrix could be conceived of as a kind of branching time, though operating with no particular backwards or forwards direction (see Chapter 3). As such, a TGIS for archaeology would **ideally** be able to handle all three models if it is to attempt to achieve general application.

A temporal database also needs extended query languages in order to be queried effectively. Worboys and Duckham suggest that Temporal SQL2 is an emerging standard for such (2004: 380). In reviewing the state of temporal query languages (i.e. query languages for databases containing temporal information), Chomicki declared that it was impossible to describe in one paper the vastness of temporal data models and query languages that researchers have proposed.[3] Instead, he attempted to formulate a unified framework for temporal databases, defining the following issues:

1. Choice of temporal domain - points or intervals; linear or branching; dense or discrete; bounded or unbounded time.
2. Definitions of abstract and concrete temporal databases.
3. Query languages - formal semantics, expressiveness, implementation.

On the first point, Chomicki stated that a truly general temporal database ought to support many different temporal domains. The temporal domain contains a temporal ontology, such as points or intervals (points currently dominate in databases), and some kind of mathematical structure imposed on the time instants (most commonly order). To handle multiple time granularities, multiple interrelated temporal domains are required (1994: 506-508).

[1] http://www.discoverysoftware.co.uk/STEMgis.htm

[2] Object lifelines represent changes in the state of an object and relationships between objects (Worboys and Duckham 2004: 361).

[3] Non-spatial temporal databases having been in existence since the 1980s (Peuquet 2001: 13).

Temporal query languages have the following properties:

1. Declarative semantics - semantics which assign meaning to a query without referring to the way in which the query is evaluated.
2. Representation-independence - the answer to a query should be the same for every two concrete databases representing the same abstract database.
3. Expressiveness - a measure of how many different queries a language can handle.
4. Data complexity of query evaluation - the computational complexity of the set of finite databases for which a fixed query evaluates to true.
5. Efficient implementation.

Unfortunately, most temporal data models are incompatible (Chomicki 1994: 514, 524). As such, a temporal query language that is well known, such as TSQL2, may not prove to be applicable to archaeological data and a simple bespoke query language may have to be constructed, following Chomicki's guidelines. For example, the study of change - a key question to be assessed using archaeological TGIS - requires more complex queries than traditional temporal query languages can address (Hornsby and Egenhofer 2000: 211). Finally, Chomicki noted that incomplete temporal information may be dealt with by abandoning linearity and developing a temporal logic based upon events. However, he also suggested that this was much easier for an abstract query language than a concrete one (1994: 524-526).

Animation is usually considered a key factor in developing TGIS. Displaying spatio-temporal change is difficult with a static map and animation may provide a better alternative. However, animation should be interactive and rendered in real time, or it does not add much to a series of static maps on a topic. The user should be able to influence the flow of the animation and to play with the time line: forwards / backwards, slow / fast, pausing. More control is required for data exploration than for data presentation. When animating temporal change, there has to be a direct relationship between time in the database and display time for the user. DiBiase defined dynamic variables for computer cartography as duration, order, and rate of change. MacEachren added frequency, display time, and synchronisation. Further, animations are narrative in character (Kraak and Ormeling 1996: 192, 195, 197). Essentially, animation should be treated with the same care as static cartography. Selection of appropriate colour, symbology and dynamic variables should be undertaken with care. However, moving animated display into a fifth dimension (i.e. three spatial, one temporal, and an attribute such as probability) is considerably more complex (Stead 1998: 218).

In his GIS textbook, Chrisman (a geographer) wrote that time is simpler than space due to its linear, ordered nature (1997: 17). As well as disagreeing strongly with the ideas discussed in the previous chapters, this illustrates once more the major difficulty with archaeological appropriation of geographers' TGIS. As Lucas commented, GIS has not grasped its potential to handle more fluid temporalities, generally contenting itself with phased temporal models (2001: 128). Some geographers, however, have begun to consider alternative models of time in this context, although I have seen little evidence of them building functioning TGIS that incorporated such.

For example, Frank wrote about different types[1] of time in GIS. He argued that different types of time might have different applicable operations or the same operations might produce different outcomes. By distinguishing applicable operations and expected outcomes, Frank believed that you could define a type of time, allowing links to be built between models of time and people's experiences. This could be defined formally through category theory and algebra. He wrote that, whilst feasible, treating time as calendar time ignores most of its important properties. He suggested that building support for temporal reasoning into GIS requires an understanding of the conceptual structure of time and formal models of temporal representation. Essentially, using calendar time simplifies the problem, providing access to powerful mathematical models and enabling integration with current GIS, but excludes all temporal information that is not in that form, especially data that only has temporal order and no temporal fix (1998: 40-42).

Frank suggested that different types of time could not be categorised hierarchically, as they can be combined into more complex meaningful types. He divided the taxonomy of time in several ways:

1. Events / intervals.
2. Linear / cyclic.
3. Ordinal / continuous timescale.
4. All events completely ordered / sequences of events that are not completely related / variants of planned future actions (branching) / multiple non-congruent sets of knowledge about the world.

Current GIS cannot cope with time that is not defined in real numbers or integers and is thus unable to deal with ordinal information that carries no exact dates. Interestingly, Frank noted that branching time could carry relevance to archaeology, with sequences displaying alternatives which indicate which hypothetical initial situation might have led to the current one. He also noted

[1] A type being a set of objects with uniform behaviour (Frank 1998: 44).

that models of time can never be totally synchronised with reality. In conclusion, Frank stated that a large class of models of time could be built from a small number of variants, but that the difficulty was in combining these alternatives consistently (1998: 42, 44-46, 54, 58). Noting the lack of a systematic treatment of change in TGIS research, Hornsby and Egenhofer suggested that a model based upon object identity might prove useful in studying processes that change over time, by tracking transitions to the identity of objects. Identity separates objects from one another and changes in this can be continuous (e.g. the flow of water), or discrete (e.g. change of ownership) (2000: 207-210).

To conclude, geographers have formulated a useful agenda for the development of TGIS and have developed software that is of use in many areas. However, they have failed to get to grips practically with any conception of time beyond modern clock time (Wheatley and Gillings 2002: 242). They have acknowledged this as a limitation and suggested ways in which it might be approached, but done little about it practically. Presumably, clock time is sufficient for the majority of their purposes. It is not sufficient for archaeology. As we have seen over the previous chapters, archaeologists have been wrestling with alternative conceptions of temporality and a TGIS that is to be useful to archaeologists ought to embrace this, especially as these alternative conceptions may serve to provide a route into integrating the varied uncertainties of archaeological data. In attempting to do so, however, we separate ourselves from the TGIS models of geographers: snapshots / base state with amendments are still the normal basis of geographical TGIS, but these are frameworks that will not sustain more complex models of time. However, in some areas geographers have provided considerable aid, in particular in the language of animated cartography and in providing agendas and ideals for creating TGIS and temporal query languages. Ultimately, as Wheatley and Gillings suggested, the fruits of mainstream research into TGIS have had little impact on archaeology (2002: 242), but this is not to say that the seeds they contain may not be profitably cultivated.

Experiments in temporal GIS in archaeology

> *The issue of integrating time into GIS is clearly one where archaeologists can make a significant contribution to the discipline of GIS as a whole.* (Wheatley and Gillings 2002: 242)

In 1992, Castleford became the first researcher to consider in detail the potential application of TGIS to archaeology. He began by considering the nature of time and its relationship to space. Castleford suggested that it might be pragmatic to ignore the true physical natures of time and space, as we require representation, not reality. He noted that most TGIS research was centred on the western concept of linear time, but that it might prove useful to pursue more cross-cultural studies into notions of time and space. He also took some inspiration from the time geography of Hägerstrand regarding time budgeting and human decision making. Moving on to TGIS itself, Castleford saw problems centring around three areas: temporal modelling, temporal databases, and GIS architecture. After considering some of the same issues as geographers (points against intervals; temporal topology), he suggested that introducing TGIS to archaeology required more than just waiting for technical specialists to produce what we need: if we want to develop alternative models that facilitate the analysis of spatio-temporal phenomena, we need to co-operate with technical specialists in creating the technology to build these models. Archaeology's temporal requirements are idiosyncratic, and as such many features built into generic TGIS are not relevant to archaeology (1992: 95-101). However, as discussed above, the problems of applying geographical TGIS to archaeology go beyond the simple redundancy of specific features: there is a fundamental incompatibility due to the complexity and fuzziness of archaeological time.

Castleford pleaded for archaeologists to provide case studies to TGIS researchers and engage in the construction of new TGIS, swimming against the trend of archaeologists merely adopting external methods uncritically. After considering some of the same database issues as geographers (query languages, object oriented databases), Castleford suggested that base state and amendments models such as Langran's might be used for the study of some archaeological phenomena, such as the spread of farming or diffusion of material culture. He opined that ingenuity, imagination, and many case studies would be needed to produce an operational TGIS. Castleford concluded with a request that archaeologists clarify how we want time represented, so that we might attract the attentions of TGIS and database designers (1992: 101-104). However, archaeology can only ever hope to be little more than an afterthought for commercial GIS developers: the money to be made by selling your product to archaeologists is paltry when compared to the weighty purses of the military, government and public utilities. It is notable in this regard that the central critiques of Castleford still strike home nearly two decades on.

In 1999, Daly and Lock wrote a paper explicitly conceived as a follow-up to Castleford's foundation. They suggested that whilst much work had been done on temporal databases and live updating of GIS, little attention had been given by researchers to the actual modelling of time. They noted that the snapshot and base state with amendments models could provide little insight into processes causing change and that the model of Peuquet was designed for instant-specific change and thus difficult to apply to archaeology, due to its being structured as a specific ordering of events based upon an absolute temporal scale. Daly and Lock suggested that too much research into TGIS was conceptual and not enough was practical, especially within archaeology (1999: 287-288).

They also noted the restrictive result of attempting to apply commercial TGIS to archaeology designed as they are to deal with modern issues with their focus on modern linear clock time. Archaeology very rarely has access to data of a similar quantity and temporal quality as conventional TGIS presupposes its users to possess. This restricts archaeologists to using tools designed for alternative purposes, removing archaeology's agency in the building of tools and restricting usage to paradigms defined outside the discipline. Daly and Lock suggested that the subtle difference between experienced and measured time had not translated well into computer science (1999: 288-289). Furthermore and by way of example, when I attempted to load my main Willington data layer (see Chapter 6) into STEMgis[1] (see above), it refused to accept the data (expressed as negative years BC) as it was not a valid date format.

In the opinion of Daly and Lock, the basic problem of archaeological GIS is the inability to incorporate temporal attributes without compromising the integrity of other aspects of our data or models. They speculated that similar GIS approaches to work on human experience[2] could be adopted for uncovering past persons' temporal structures, especially where correlations are discoverable between peoples' spatial and temporal characteristics. Daly and Lock also made the important realisation that access is only available to archaeologists to a few GIS packages and that, as such, research into archaeological TGIS ought to focus on how best to use these particular packages to model time.[3] They saw the way forward for archaeological TGIS in either object oriented databases / GIS, animation, or in use of the z-axis. Object oriented databases have the strongest potential, but are not widely available and require high expertise to make use of. Animation is a powerful visualisation tool, but provides little analytical capacity and is hard to incorporate in a final printed volume - although this is only a problem if a paper end product is desired. Use of the z-axis also carries potential, but using it to represent time restricts the representation to two spatial dimensions. Daly and Lock concluded that the growing influence of GIS on the definition of archaeological categories means that we need to get a strong handle on the temporal nature of these categories or mistakes will inevitably be made (1999: 289-292).

In the same volume, Lock and Daly made a pragmatic attempt to undertake spatio-temporal modelling within the requirements of a single project, using readily available, low cost GIS and well-known remote sensing methodologies. Through quantified comparison of snapshot images, they produced a useful methodology for identifying change and continuity over time for surface survey data. The data used were pottery distributions for different periods, which would normally be examined visually. Using GIS, a more quantified approach is possible which facilitates identification of smaller areas of gradual change. The pottery distributions were divided along spatial and temporal lines: temporal categories were defined from typologies and were thus highly variable; spatial categories were defined via thresholding, a computerised method for identifying spatial boundaries. Image differencing then produced a map of change and continuity between two time periods: by subtracting the earlier (T1) distribution from the later (T2), a positive value shows positive change, a negative value negative change, and zero shows no change (1999: 259-260).

However, Lock and Daly noted that this method was somewhat crude: pottery that was more important in interpretive terms might be much less common than other types and would thus be subsumed within the overall broad brush results produced. Dividing the later distribution by the earlier - image ratios - normalises the distribution and gives greater weighting to lower sherd counts. An alternative would be to reclassify the quantitative data into qualitative categories - low, medium and high sherd densities - and to then compare the distributions using contingency table or cross-tabulation analysis. Comparing the change between the categories at the two time periods produces four results: "present at T1 and T2", "present at T1 not T2", "present at T2 not T1", and "not present at T1 or T2". This complements the quantitative analysis: four categories are more sensitive than three and highlight aceramic areas. Further, the results may all be tested for statistical significance (1999: 261-262). Lock and Daly's work illustrates the potential of a very simple TGIS operation. However, their particular methodology is only applicable to the specific case of datable artefact distributions from field survey. I can see little application of these methods to intra-site analysis for example. Furthermore, although their results do illustrate change on the one level, they give no information about change events or underlying processes: was change sudden or gradual (Lock and Harris 2000: 6)?

Harris and Lock suggested that current GIS fossilise geography, thus restricting the study of time as a continuum. They advocated the use of topology to explore spatio-temporal relationships: earlier, later, or contemporary (1995: 356-357). In 2000, Lock and Harris expanded upon their brief earlier thoughts. They opined that archaeologists were comfortable with categorical time in GIS, because archaeologists' temporal conceptions were largely limited by similar categorical forms (i.e. typologies and ordered periods). However, as we have seen in previous chapters, categorical time is an oversimplification. When visualising data in a GIS, this shortcoming becomes opaque: how do you colour code a site that spans more than one period? Lock and Harris suggested that our basic problem was that time is continuous and must thus be analysed as such. They noted how spatial depth and temporality are closely related in archaeology, depth of deposition having a (complex) relationship with temporal sequence (2000: 4-5), as per Harris's Law of Superposition (see Chapter 3).

[1] Downloaded from http://www.discoverysoftware.co.uk/STEMgis.htm
[2] Such as viewsheds or movement through landscape (Daly and Lock 1999: 290).
[3] I believe ArcGIS, MapInfo and GRASS to be the most commonly used of these, at the time of writing.

Having previously modelled time in GIS using the third dimension (Daly and Lock 1999: 288), Lock and Harris came to realise that this was insufficient as a fully 3D space was needed. They suggested that an archaeological TGIS could model 3D space using voxels,[1] with temporal relationships represented topologically. Using such a system has significant potential for intra-site analysis: for example, Harris matrices could be generated automatically from the topology. At a regional level, the landscape could be deconstructed in terms of inter-site temporal relations. Using a true 3D model with a continuous temporal z-axis allows accurate temporal spans of sites to be recorded: it is then possible to query the GIS to produce, say, all sites within 50 years of AD 500, in the same way as current GIS may be queried to produce all sites within 50 miles of a place (2000: 5-6). Lock and Harris then went on to discuss the fundamental need to integrate uncertainty into archaeological TGIS (see Chapter 5).

Ceccarelli and Niccolucci used 3D modelling and GIS to explore the changing coastline of an Italian lake throughout history. They worked from various reconstructions of the landscape at key moments in its history, based upon historical, topographical, archaeological and geological sources. They attempted to create a 3D reconstruction of the dynamic process of landscape change over time, with animation being used to convey time. The changing relationships between human settlement and the geography of the lake illustrated how geography is not a static stage upon which humanity acts. Ceccarelli and Niccolucci's methodology was to reconstruct the ancient landscape, develop a methodology to incorporate time into GIS, then to distribute the results over the internet (2003: 133-134). Herein lies a problem: constructing your data before attempting to integrate time into your GIS will inevitably mean that the TGIS that results is specific to your dataset with restricted application to other datasets. If an aim of TGIS research in archaeology is to produce something widely applicable, then consideration of TGIS construction must come **before** data are introduced. This is not, of course, to imply that the TGIS cannot be altered once given practical application.

Ceccarelli and Niccolucci encountered problems in both constructing their model and representing time. Their solution lay in combining a temporal database with GIS spatial modelling, using a large amount of software, including AutoCAD, Access, ArcGIS, GRASS, PostgreSQL, 3D Studio Max, and Quicktime (2003: 134-135). The use of Quicktime to view their final animation demonstrates that it was not interactive, and the number of packages with which fluency was required to produce that animation suggests that their solution would not be particularly straightforward for later users to engage with. Further, their temporal model was confined to modern western clock time. Time appears in their final output to be a continuous variable, but this is in fact misleading. Time in their database was represented categorically and the continuous time of the animation was interpolated from these base time slices. Ceccarelli and Niccolucci concluded that, although simplified, their system was a suitable solution, historically and technically. They suggested that their work was an important preliminary step forward for archaeological TGIS (2003: 135-137). However, much like Lock and Daly's work, Ceccarelli and Niccolucci only succeeded in introducing TGIS functionality that is limited to a very specific data type. Interpolation from time slices may be justified for evolutionary geographical change, but cannot capture the more random nature of cultural or historical change.

Temporality in TGIS does not have to limit itself to consideration of dated material, but can also be extended to the relative temporality of seasonal practices and the study of formation processes. Palmer and Daly surveyed Bedouin campsites and then used GIS analysis of artefact distributions to demonstrate differences in material culture discard due to seasonality, and how post-discard processes modify cultural assemblages over time (2006: 107). Frachetti used digital analysis to model the spatio-temporal patterns of movement of prehistoric pastoralists, noting that survey maps presented as synchronic time-slices may actually represent generations of occupation. He stated that digital tools are incredibly powerful for the simulation of past conditions and facilitate the exploration and testing of strategies possibly employed in the past, across multiple scales and reflecting change in spatial and temporal boundaries. For example, trends in mobility patterns may be linked to seasonal ecology, local patterns of settlement, burial and ritual, and changes to the environment. Furthermore, repetition of patterns results in the creation of institutionalised activity, such as Kazakh horse games (2006: 128-133). Frachetti's work is important in highlighting the possibilities for TGIS in the study of seasonal and other short term patterns of change.

Arroyo-Bishop and Lantada Zarzosa, in their attempt to create a European archaeological information system, stated that fundamental to this aim was structuring a spatio-temporal component to enable analysis of data over time, as objects, space and time are all needed to reach valid archaeological conclusions. The conventional temporal organisation of archaeology into phases and sequences is handled well by the human brain, which is able to bridge gaps in information structure. Arroyo-Bishop and Lantada Zarzosa believed that this system was quite satisfactory and simply required formalising into part of an analytical system that manages the temporal dimension. They suggested that such structured temporal entities could handle any level of chronology and its associated data, and that archaeologists should prepare their data so that it could one day be integrated into a TGIS built around this model, believing that the arrival of an object-space-time GIS was inevitable (1995: 44, 49-50, 52). There does not appear to be any suggestion in their paper that archaeologists ought to involve themselves in the process of TGIS creation, but simply that we should be ready for when geographers

[1] A voxel is a 3D pixel.

provide us with the finished model. As noted above, this is short-sighted.

Finally, the most extensive work into TGIS in archaeology is that of Johnson and his TimeMap project, which it has been suggested is taking the first steps towards the management and manipulation of spatio-temporal data (Wheatley and Gillings 2002: 243). At the Computer Applications in Archaeology Conference in 1997, Johnson outlined his aims for TimeMap. He suggested that it was necessary to completely rethink GIS rather than modify existing software and that this may have scared many previous researchers off. Johnson noted that animated maps had been produced by archaeologists for museum displays and the like, but that these were all one-off productions or case studies with no pretensions towards a generalised methodology. By contrast, the TimeMap project's main aim was to allow the interactive control of a GIS-based temporal database and the interactive generation of animations from that database. Johnson suggested that there was a submerged iceberg of graduate students pursuing TGIS (1999: 1.2.2, 1.2.5), but I can see little evidence of these appearing as yet.

Johnson believed that archaeology required a TGIS solution that did not require one-off programming efforts. Thus, the ad hoc or special purpose applications that had appeared prior to TimeMap were not especially useful (1999: 1.3, 1.4). He outlined the elements needed for handling spatio-temporal data, identified as the core of TimeMap - alongside data availability concerns (1999: 2.1):

1. A temporally-explicit methodology for recording information about changing features and / or changes in the landscape.
2. A means of interpolating intermediate conditions between known points in time.
3. A means of recording uncertainty and diffuseness in the spatial and temporal extent of features.
4. A means of displaying information which has a fourth (temporal) component in addition to the normal 3D spatial components on a 2D screen.

I agree with the first, third and fourth, but am unconvinced by the second. Temporal interpolation of archaeology is complex and very uncertain. It is perhaps an act best left to the imagination, as computerised interpolation can give false authority to very speculative models. Also, as we shall see, TimeMap has never fully come to grips with these four elements, having become largely sidetracked by the data availability issue noted above.

More specifically, Johnson stated that the TimeMap project aimed to develop a software system, backed by an explicit spatio-temporal recording methodology incorporating spatio-temporal fuzziness, that allows (1999: 2.2):

1. The superimposition of data layers in a time-enabled GIS interface.
2. The querying and display of data layers with explicit support for temporal information and spatio-temporal fuzziness.
3. The derivation of data layers by query from remote datasets across the internet without reformatting or case-by-case programming.
4. The generation of data-based in-line and off-line animations and virtual reality worlds.
5. The navigation from features on the screen to web-based information resources.

Items 3 and 5 are not truly TGIS issues but in fact issues relating to wider GIS, yet they have dominated much of Johnson's TimeMap research. The data model used by TimeMap is a snapshot-transition model, where snapshots are recorded of features at known points in time with transitions between snapshots also recorded. Johnson thus avoided recording spatial or temporal topology for the sake of reducing data complexity. He suggested that this data model paralleled the state of our knowledge of the past, with fairly detailed understanding of features at particular moments in time but little understanding of what lies outside their boundaries. Interpolation is then needed to bridge the gaps between these snapshots (1999: 2.3-2.4). I disagree with Johnson's characterisation of our knowledge of the past: archaeology only appears similar to snapshots when dealing with agglomerated period data. Archaeological temporal data in their raw state are instead comprised of fuzzily positioned points (i.e. dates) and topology (i.e. stratigraphy) and a snapshot model cannot represent this without considerable oversimplification.

Johnson noted that animation provides a more effective, though not necessarily more informative, manner of depicting time than static maps such as time slices, symbolism or difference maps. Rather bizarrely, however, he suggested that animation suffered a stigma due to its association with Disney and computer games. He also suggested that animation ought to be generated from data on-the-fly according to a user's needs, rather than being pre-rendered. Johnson acknowledged that there was a need to be able to display temporal depth without animation, but he had not yet considered this issue in detail. When animating long-term sequences of time, problems arise due to the punctuated nature of historical events. Johnson believed that smoothing out such sequences was misleading, but accurate portrayal resulted in a largely static picture. To solve this problem, he suggested varying the speed of playback around critical historical events or, for a more comprehensive approach, by considering historical events through a multi-scalar portrayal. The multi-scalar approach is difficult to combine into a single animation, but Johnson proposed overlaying detail at different scales: for example, the urban development of a city could be depicted as an animation at 1 year : 1 second, with animation textures overlaid to show the building of particular landmarks or the daily commuter cycle (1999: 2.6.1 - 2.6.4). This multi-scalar approach conveys a flavour of some of the theories of multi-temporality considered in the previous chapters.

Laudably, TimeMap is aimed at a broad audience from researchers to the general public, with potential applications envisaged in museum information terminals and in the dissemination of data over the internet. As such, Johnson distinguished TimeMap from other archaeological experiments in TGIS due to its having no specific target application in mind. He believed that this flexibility was facilitated by its database / GIS base - as opposed to being display oriented - which allows the application of GIS tools beyond features explicitly programmed into the user interface. As such, although TimeMap might appear less slick than bespoke software, Johnson suggested that it was far easier to incorporate new data into TimeMap. He went on to suggest the following possible applications of TimeMap: coastline evolution; climate change; expansion and decline of empires; settlement patterning on a landscape; settlement layout and urban growth; trade routes and exchange; military campaigns and battles (1999: 3.1). It is notable that several of these applications are more historical than archaeological or based upon interpolative models. I have never seen any evidence of Johnson attempting to use TimeMap to deal with the intra-site data that forms the basis of much archaeology: TimeMap may aim for generality, but it is a generality that captures applications around the edges of archaeology and avoids embracing the fundamental core. As a result, although Johnson has acknowledged the difficulties with representing uncertainty in archaeological TGIS (e.g. 2004a; see Chapter 5), he has never touched upon the problems surrounding integrating different sources of dating evidence. I consider the solving of these problems to be fundamental to the creation of an archaeological TGIS.

In 2002, Johnson expanded further upon the use of interactive mapping. Johnson suggested that an interactive map viewer (i.e. not a full TGIS) would provide a subset of the following capabilities (2002: 2):

1. Opportunities to explore data and encourage experiential learning.
2. Delivery of large amounts of data that may be teased apart through zooming, layers and symbology.
3. Linking between symbols and areas and richer textual or graphic information.
4. Overlay of different sets of data with potential for analytical functions.
5. Filtering by time or animation to illustrate events, change and processes.
6. Little cost in using colour and in publishing different views - there is no need to compromise on illustration to reduce costs.
7. Relevant additional data may be downloaded from the internet.
8. Data layers may be exported for statistical or GIS analysis.

Of these capabilities, only 5 is a purely TGIS concern and many - particularly 1, 3, 6, 7 - carry only passing relevance to TGIS. Johnson saw the strengths of interactive mapping in enhancing the ability of users to move beyond canned interpretations through free interrogation and visualisation of data, and in providing the ability to combine time and all three spatial dimensions in a visualisation - nigh impossible for paper maps - through computerised calculation of perspective and the rapid generation of new versions. The latter ability provides the opportunity for the former exploration (2002: 2, 4). The majority of the rest of this paper concerns itself with the provision of interactive mapping over the internet (2002: 5-7).

Johnson's next TimeMap paper was again largely concerned with the web delivery of interactive mapping and with historical maps (Johnson and Wilson 2002). Johnson and Wilson covered similar ground again in 2003. They noted that initial work on TimeMap had been on the methodology, user interface and generation of time-based maps, but that this remit had expanded after 2001 to encompass issues surrounding access to distributed data - including cultural data in general, not just archaeology - alongside the Electronic Cultural Atlas Initiative[1] (ECAI) (2003: 125). This suggests to me that perhaps Johnson took a pragmatic decision to concentrate on historical over archaeological data due to its easier integration into TGIS (e.g. less uncertainty, accurate dates, more complete datasets). Johnson and Wilson believed that interpolation was still necessary for the animation of historical datasets, but noted that the results, though smooth, would not necessarily match past reality. As such, they worked on improving their shape tweening methods (2003: 128). As noted above, however, I do not believe that interpolation is the right way to approach temporal gaps in archaeological data: presenting a false completeness regarding our knowledge of the past has too great a potential to mislead.

Other papers by Johnson (2005) and Wilson (2001) also concentrated on presentation of historical data to the public and carry little relevance to archaeological TGIS. However, the apparently overwhelmingly positive public reaction to this work does suggest that interactive mapping can work well in museum environments for public enlightenment (2001: 48). In 2004, Johnson boasted of the widespread adoption of TimeMap by non-archaeologists: mobile telephone networks, municipal councils, cultural atlases, computer games, environmental

[1] http://www.ecai.org

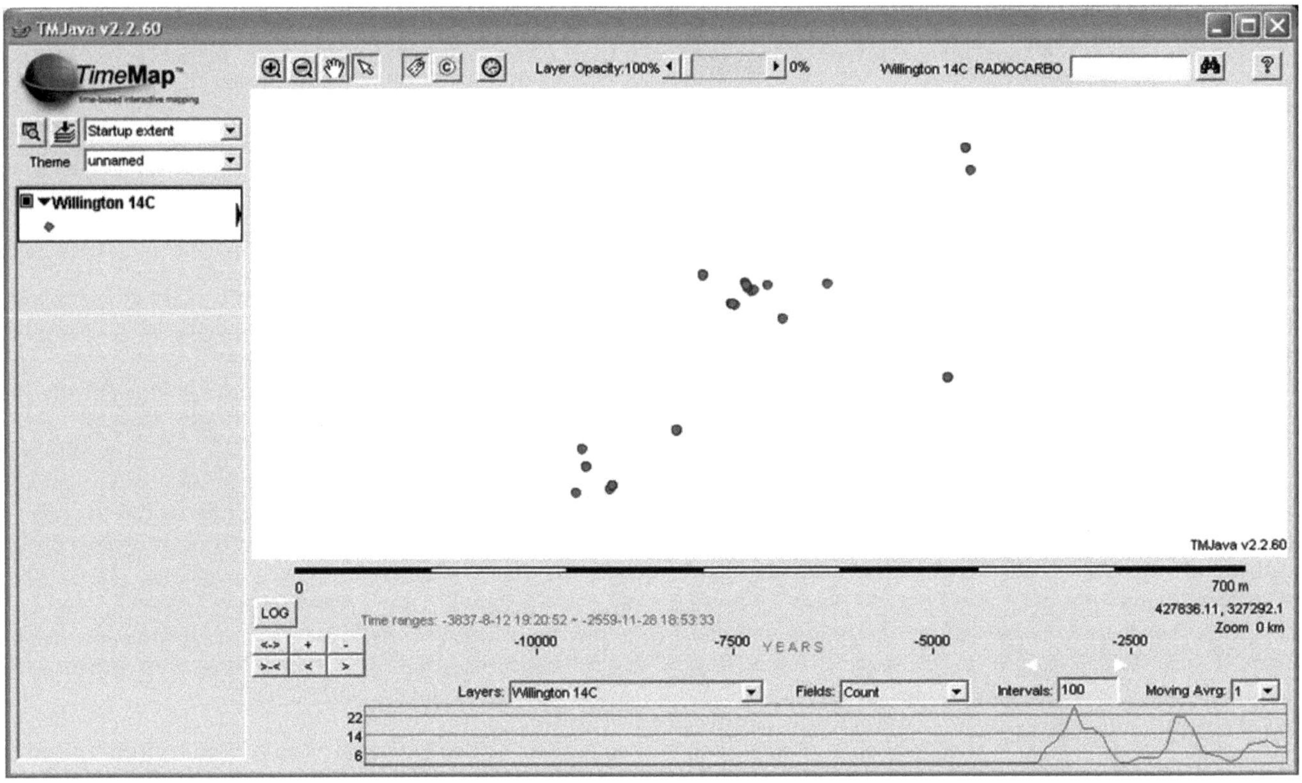

FIG 4.2 - TIMEMAP OUTPUT OF WILLINGTON RADIOCARBON DATA, SHOWING DATES PRESENT 3837-2559 BC AS POINTS. NOTE THE USEFUL TIMELINE OF SITE ACTIVITY ALONG THE BASE.

mapping, epidemiology, remote sensing, etc. (2004b: 29). It is notable that he is unable to make a similar boast about its adoption by archaeologists themselves, the original intended users. TimeMap's greatest successes have been in areas outside of core TGIS issues for archaeologists and whilst using non-archaeological data.

As I had tried and failed to do with the conventional TGIS software STEMgis (see above), I tried to load my Willington data (see Chapter 6) into TimeMap,[1] this time with more success. TimeMap proved able to cope with dates expressed as a negative year BC and also coped with objects possessing start and end dates. I was able to examine simple temporal change across the site and was also able to output a simple interactive animation showing change over time in century slices (Fig 4.2). However, TimeMap is not able to consider dates as different types or to provide any indication of the probabilities of dates appearing during a selected period. Objects are either present during a particular time-frame or absent, but this is not how archaeological dates usually work (see Chapter 3). When dealing with radiocarbon dates, as with the Willington dataset, the actual event being dated falls somewhere **within** the probability span. Through its inability to acknowledge this fact graphically, TimeMap presents a misleading account of the site's temporality. Nevertheless, for an initial examination of temporal data or if animation or web dissemination is your primary concern, it is undoubtedly a useful tool.

Finally, it should be noted that it is possible to study chronology using GIS without relying on TGIS function. In his spatial analysis of a late Eneolithic / early Bronze Age cemetery at Holešov in the Czech Republic, Ladislav Šmejda tracked broad chronological change through change in the artefacts deposited, through material type (stone to copper to bronze) then content (increasing percentage of tin in bronze). With a large enough dataset, mapping the density of these factors can show spatial trends. This is, of course, facilitated by GIS, but does not require any temporal function as it involves no actual dates, just a chain of typological change. This type of study can only show the average pattern and exceptions should be expected. Furthermore, it is possible that the spatial density of finds might more reflect the density of burial over chronological factors. To avoid this, artefact density can be normalised by burial density, but this can lead to an excessive reduction in quantitative variability. It is possible to work around this issue by comparison of the actual distributions against filtered mean maps for each type. Using this methodology, Šmejda was able to demonstrate that stone and copper alloy artefacts clustered on different parts of the site, thus demonstrating that the spatial analysis of artefact distributions can express something of the chronology of a site. However, he also noted that these distributions might also reflect other factors, such as social structure (2004: 59, 60-63).

To conclude then, there has been significant - if to date niche - interest in TGIS from archaeologists. Archaeologists such as Castleford, Daly, Lock and Harris have considered the problems and potentials of archaeological TGIS in detail, but little of this work

[1] Downloaded from http://www.timemap.net/

moves beyond theoretical consideration (Wheatley and Gillings 2002: 242). Other archaeologists who have attempted to approach TGIS practically have limited themselves to case specific applications with little wider use outside of their limited milieu. Johnson's TimeMap project acknowledged the shortcomings of such an approach and attempted to produce a generalised methodology. However, if anything TimeMap is too generalised in its approach: by failing to concentrate on archaeological issues and usages and through an over-emphasis on issues only tangentially relevant to TGIS, TimeMap has ended up having limited application to day-to-day archaeology. The key issues regarding applying TGIS to archaeology are the need to integrate different sources of dating evidence and the need to handle the uncertainty of that dating evidence. There is no TGIS in existence today that is able to meet those needs and, as such, there is no TGIS in existence today that archaeologists may easily use for the analysis of their raw data.

Conclusions

In this chapter, I have considered discourse over what we have called digital archaeology, the use of GIS in archaeology, and finally previous work on TGIS by geographers and archaeologists. Drawing from the ongoing lively debate on the nature of digital archaeology, I concluded that we need to apply forethought to and understanding of the technology that we use. It is essential to question **whether** the use of IT methods in a given context is appropriate, not simply consider **how** to apply technology to our problems. In the long run, digital archaeology **may** provide a springboard for all kinds of new archaeological theories and methods, but equally it may not. The argument that IT increases productivity is not really the key one, and is rather debatable. In fact, the key question to ask is how digital methods can aid us in the answering of questions in a more fluid and nuanced fashion: not faster but **better**. Nevertheless, due to the fundamentally spatial nature of archaeological data, GIS has revolutionised many areas of archaeological research. However, care must be taken not to get seduced by the attractive output of GIS, and we must promote an awareness of the false authority that the precision of computers can lend to data that would otherwise be understood to be of more dubious quality. Archaeological data are also fundamentally temporal and, as such, conventional GIS can only take us so far in our spatial analyses.

TGIS will provide us with answers to this problem. Geographers have undertaken much work on TGIS and made important theoretical realisations about the nature of temporal topology, query languages, user interfaces and animation. However, in practice geographers' TGIS have stuck resolutely to a model of time as modern clock time that is not adequate to deal with archaeological data. Furthermore, much of their work has followed avenues which carry little relevance to archaeological TGIS, such as dynamic updating of real-time data. Finally, the data models used by geographers to construct their TGIS, primarily the snapshot model and its variants, do not reflect well the temporal structure of archaeological material. As we have seen, the temporality of archaeological data can be very complex, multi-linear and fuzzy (see Chapters 2 and 3). A TGIS designed to cope with mapping traffic patterns in a modern city or to trace supply patterns and bottlenecks in public utilities will always struggle to handle archaeological data: in essence, our time is too different a time from that of geographers and others.

Archaeologists have also experimented with TGIS, but mostly on a theoretical basis. These archaeologists have identified the shortcomings of geographers' TGIS and suggested ways in which archaeologists might take TGIS forward along their own lines. However, in the rare instances where practical work has been undertaken on archaeological TGIS it has largely consisted of one-off case studies with little wider applicability. The most important work with pretensions towards generality undertaken on TGIS in archaeology has been Johnson's TimeMap project. However, Johnson has largely concentrated of late on issues that are somewhat peripheral to core archaeological TGIS concerns. TimeMap is also restricted by its snapshot-transition data model and its weakness in dealing with uncertainty / temporal fuzziness (and thus intra-site archaeological data).

I identified two key issues that need to be addressed to expand the applicability of archaeological TGIS and bring it closer to the goal of a general tool that will one day aid archaeologists in their day to day work. These arise out of the ideas discussed in Chapter 3. Firstly, a need to deal simultaneously in a TGIS environment with disparate sources of dating evidence, with varying precisions and accuracies. Nobody has yet approached this problem using TGIS and sources for solutions may be found in the ideas discussed in Chapter 2. The second issue is that of dealing with the uncertainties of these varied dating techniques. Much work has been undertaken on dealing with uncertainty in recent years (see Chapter 5). In a TGIS context, however, much of this has been concerned with identification of errors in real-time dynamic databases: this is not the type of uncertainty we need to grapple with. The uncertainty of archaeological dates is fuzziness. The concept of fuzzy sets captures something of the nature of combining different dates and has great potential if used in the construction of typologies. As we shall see in the next chapter, there are possible routes into solving our fuzzy problems using statistical methods, but I shall show that these run the risk of telling us very little or of providing a spurious authority to data that might not merit it.

To conclude, an archaeological TGIS needs to be able to deal with fuzzy dates and to be able to integrate or compare said fuzzy dates. If geographers' TGIS fall down on these fronts for archaeological purposes due to their restrictive conception of time and their inappropriate data models, we need to build our own TGIS. This should aspire towards general application across archaeology and, as such, ought to be based around GIS already widely available to archaeologists. To be of greatest use in considering the questions considered in the previous chapter, it perhaps ought to contain tools to aid in the assessment of contemporaneity and site phasing, although only if this is undertaken with forethought and explicit acknowledgement of its constituent uncertainties and subjectivities. To reiterate, my aim is not to build software that answers questions **for** archaeologists - click a button and up pops your site phasing - but to construct software that **aids** archaeologists in answering their temporal questions in a more fluid and nuanced fashion than ought to be possible with brain, paper and ink alone. Raising these issues once more whilst creating working solutions might then encourage commercial TGIS developers to build features into their new software that would make them more applicable to archaeological data.

CHAPTER FIVE: THE CREATION OF THE TGIS

In this chapter, I will begin to explain how I attempted to come to terms in practice with the issues discussed in the previous chapters. Based on my analysis of the nature of TGIS and the nature of archaeological temporalities, I wrote a TGIS extension for ESRI's ArcGIS that caters to specifically archaeological needs, addressing the key datasets and concerns associated with the foundation of interpretation (Chapter 3) and temporal theorising (Chapter 2). First, I shall outline at length how other archaeologists have attempted to consider temporal uncertainty, identified as the key issue for archaeological TGIS in the previous chapter. After that, I shall explain the reasoning behind the software environment choices made. Then, I shall describe the TGIS temporal model and implementation of uncertainty, and discuss how it is able to take forward the analysis of temporal uncertainty. Finally, I shall briefly consider the cartographic choices made within the TGIS.

Understanding uncertainty

As argued in Chapter 4, fuzziness is one of the key problems in applying TGIS to archaeology (Ceccarelli and Niccolucci 2003: 135). Unlike archaeological data, computers are inherently precise:[1] as such, a poor quality paper map can acquire false authority when digitised (Miller 1995: 320-321). Whereas geographers have often approached uncertainty in the sense of discovering errors in databases (e.g. Heuvelink and Burrough 2002), our archaeological TGIS needs to take account of inherent uncertainty in the original data - fuzziness (Peuquet 2001: 20). As such, it needs explicit representation, and careful use of fuzzy space and time has potential in this vein (Harris and Lock 1995: 357).

Fuzzy logic

> *Far from simply offering apologia or attempting to minimize the ambiguity, uncertainty is embraced and incorporated into the analytical environment.* (Gillings 1998: 129)

It is easy to deal with fuzziness in natural language, but more difficult in computer science where the principle that things may only be true or false is deeply embedded (Burrough and McDonnell 1998: 265-267). Reality is very often not so crisp and the term fuzziness is used to denote the vagueness concerning description of the semantic meaning of events, phenomena or statements (Zimmermann 2001: 3). Fuzziness is a measure of possibility of membership, not probability - defined intuitively or from mathematical uncertainty. Essentially, it is a way of defining the imprecision of objects with no sharply defined boundaries. When grouped, the objects form fuzzy sets, a form of classification that is appropriately used where we have to deal with ambiguity, vagueness, or ambivalence in our empirical models. With fuzzy sets, individual objects may be partial members of different, overlapping sets. Standard database query languages have been extended to allow the logical query of fuzzy data (Burrough and McDonnell 1998: 265, 268, 273) and this querying process may be undertaken on mixed datasets of conventional and fuzzy objects (Jones 1997: 223).

Fuzziness is defined as a membership degree (Laurini and Thompson 1992: 646), but is a multi-faceted phenomenon that requires considerable preparation to model. Indeed, membership degree may itself be fuzzy (Zimmermann 2001: 4, 24). Fuzzy set theory is useful in complex situations with ill-defined variables (Gillings 1998: 130). However, this concept of fuzzy sets does not really capture the nature of the uncertainty of individual archaeological dates, although it does have its uses in environmental modelling (e.g. Gillings 1998). The uncertainty of an archaeological date is fuzziness, insofar as it is an expression of the possibility of that date relating to a particular real date, but there is no specific set for it to be a member of. Fuzzy sets become relevant to archaeological dating when we come to consider the contemporaneity of different dates: a phase of an archaeological site is analogous to a fuzzy set, with individual dates and contexts possessing various possibilities of belonging to one or more phases. Indeed, this characterisation of phase as fuzzy set may offer a method of moving beyond the totalising chronologies criticised in Chapter 2.

Castleford sounded a note of caution regarding fuzzy TGIS, when he suggested that if spatial fuzziness is a problem of conventional GIS then expanding GIS to include fuzzy time would magnify the effects of the problem (1992: 100). However, this is only the case if we wish to deal with both fuzzy space and time together. For the majority of archaeological data, fuzzy time is probably more important than fuzzy space and, as such, for the sake of simplicity we can perhaps afford to deal with fuzzy time alone for the time being at least. An exception to this would occur, however, where spatial fuzziness and temporal fuzziness are related, such as where an object's spatial location is uncertain and, as a result, its position in the site's temporal sequence is uncertain. Lock and Harris noted a problem in that digital capture of fuzzy archaeology was difficult, yet they felt it necessary to the effective use of GIS. They also noted that TGIS and probability models would create possibilities for testing new theoretical models such as continuous time, fuzzy boundaries, and temporal topology (2000: 6-7, 10), despite earlier suggestions that models generally do not cope well with uncertain data (Daly and Lock 1999: 290).

[1] Precision being a measure of the quality of data measurement / storage - for example, an instrument may measure a value to the nearest 10 millimetres. Precision is importantly distinguished from accuracy, which refers to how close the measured value of data are to their true value - such as a TL date expressed at ±10% (Wheatley and Gillings 2002: 83).

Several archaeologists have attempted to deal with uncertainty practically. For landscape archaeology, Lock and Harris advocated the use of a continuous time axis with a calculation of the probability of use - p(use) varying between 0.0 and 1.0 - for each site. Such a model can handle both gradual and sudden change. The p(use) could be calculated by combining different sources of evidence for each site, assigning a different value and weighting to each dating element - with the closest dating evidence to the present day given greatest weighting. The actual values would be assigned at the discretion of the archaeologist on the basis of intuitive knowledge. A timeline would result which formed a series of values that would fluctuate considerably according to available evidence. Lock and Harris suggested that calculating averages from this timeline would not be useful in practice, as archaeological sites can be in use in more than one period and being in use at a particular period does not necessarily imply that it was in use in another period. They did advocate the use of such a model for simulation exercises, however (2000: 7-9). I am not convinced that this p(use) model is of much benefit: by basing it upon the archaeologist's intuitive assessment of the evidence, is it not doomed to simply replicate the intuitive model that already exists in his/her head?[1] Put simply, why go to the trouble of calculating probabilities.

Hermon and Niccolucci applied a fuzzy logic approach to typology. The underlying argument is that using fuzzy sets can return to a computerised database the subjectivity that existed prior to its digitisation: losing this subjectivity can lead to dangerous approximations and fallacious conclusions, in that borderline human decisions become fossilized as orthodoxy. Construction of typologies is a highly subjective process: by using fuzzy sets, an object can be assigned to several groups if we are unsure of its designation. As a result, fuzzy typologies have great potential for improving the reliability of typological research (2003: 307-308, 310) and will hopefully become more widespread in the future. In another study, Bagg and Ryan produced a prototype TGIS for modelling historical change in Corsica. Uncertainties in their dates were recorded as ranges of dates, providing a basis for the modelling of imprecise historical change. They recorded precision using its granularity and accuracy using an optional uncertainty figure based on a normal distribution. Comparison between two objects using Boolean operators then produced a probability that the comparison was true. However, this project was not an attempt to produce a fully functional TGIS, but instead an experiment to test the extensibility of databases to handle spatio-temporal applications. They concluded that such applications were only possible using object-oriented type databases (1997: 45-46, 53-54), putting it out of reach of most archaeologists. As the results of this work hopefully demonstrate, this need not necessarily be the case.

[1] Astute readers will of course be aware that all archaeology is interpretative, but this does not invalidate the point.

Aoristic analysis and spatial autocorrelation

Returning to TimeMap, Johnson's original intention was to record temporal fuzziness as a text attribute attached to snapshots and transitions (1999: 2.5). In 2004, he attempted to go beyond this to assess the relative contemporaneity of archaeological data. Johnson stated that archaeological interpretation relies upon knowledge of contemporaneity if it is to avoid convoluted arguments or conjecture: the challenge being to move from a loosely dated assemblage of archaeological objects to estimation of past occupation patterns and, if possible, change in those patterns (2004a: 448). Johnson saw a route into this in the form of a methodology used for the analysis of crime (Ratcliffe 2000) - aoristic analysis. This methodology works by evenly distributing the probability of events across the time span during which they may have occurred: these probabilities may then be summed to produce the probability of all events, with more closely dated events providing a stronger weighting over their range than less closely dated events (Johnson 2004a: 449).

As such, the sum of probabilities of one possible event might be higher than the individual probability of another possible event. For archaeology, weightings are applied to artefacts or features according to dating evidence: higher weightings are applied to more tightly defined time ranges. The method allows artefact distributions to overlap across different time intervals and escapes the situation where data are swamped by the mass of undiagnostic material. Aoristic analysis of archaeological material maximises the use of information available by sacrificing temporal precision for the sake of temporal pattern (Johnson 2004a: 449-450) and avoids the difficulty of having to assign an instant event to a date range (Ratcliffe 2000: 670).

Johnson attempted to apply this analysis to field survey material from Kythera, with a specific interest in testing the hypothesis of periodic abandonment of marginal environments. Such data are often subjective due to the lack of tools for quantifying and analysing temporal spread, especially when combined with spatial distribution. Johnson's methodology was as follows:

1. Divide the overall time frame of recorded material into equal intervals (e.g. 100 years).
2. Allocate a weight to each artefact (aoristic weight per interval = interval size ÷ time span for artefact type) for each interval within which its possible date falls.

Johnson recorded a problem with this methodology in that it naturally gave a much higher weighting to more closely datable modern artefacts. Possible solutions to this were seen in dividing the weightings by the median date of the period - although this over-amplifies the actual probabilities - or preferably by increasing the size of intervals in proportion to age. For the next stage, finds, chronotypes and periods were joined to generate time ranges and aoristic weights for each find, and the weights then summed for each survey unit. The map produced was shaded by survey unit according to aoristic weight for the time period in question. Despite the use of aoristic analysis, Johnson's results for his Kythera material were, however, still swamped by the mass of undiagnostic material that formed a large percentage of the dataset. Furthermore, he noted that the system could only cope with one source of dating evidence - in this case typology (2004a: 448, 450-451). This issue of including less closely dated material in our analyses is discussed more fully in Chapter 7.

This example of aoristic analysis is representative of the problems with Johnson's approach. His aim was to use it to provide a more statistical analysis of field survey and settlement pattern, as opposed to selective interpretation of observations (2004a: 452). Attempting to remove the subjectivity from archaeological interpretation is highly reminiscent of the positivist aspirations of the early New Archaeology: however, it is not possible - nor necessarily desirable - to remove subjectivity wholly from interpretation, and perhaps dangerous to assume that you have. The typological dating schemes and dating of objects on which the analysis was based are themselves already highly subjective and, furthermore, the statistical output seems to have produced little useful interpretation: in my opinion it would require a more nuanced human understanding of the dates, intervals and patterns to produce useful output. The paradox is this: we must be careful not to simply reverse the situation; if we allow archaeologists to iteratively tune such models then they may simply shape them to their own pre-existing interpretation of the data, the result hidden behind a computer model that lends a false objectivity to their results. Yet human input appears necessary to produce results that can take interpretations forward.

Another example of taking a statistical approach to chronological uncertainty is provided by local spatial autocorrelation, a tool for the analysis of spatial patterning in variate values. Premo (2004) applied this technique in an attempt to provide a quantitative method for the study of the Classic Maya collapse. He recorded the terminal long-count date for a series of Maya sites, that is the date of the most recent known stone monument constructed on a site (i.e. an indicator of when elites stopped raising monuments), and quantitatively characterised the landscape signatures using local spatial autocorrelation. The assumption behind the technique is that closer locations ought to be more similar than distant locations, and the statistics test the degree of similarity between adjacent locales. If the assumption is proved true, the data are classified as having positive spatial autocorrelation, and vice versa. Where no pattern is seen, there is no autocorrelation. However, the other examples of this technique given by Premo involve artefactual, rather than explicitly temporal distributions. Furthermore, there is little sense of the intrinsic uncertainties within each date (as they are historically dated from inscription, rather than more normal, fuzzier archaeological dates). As such, this is perhaps more of a technique to apply to data derived **from** a TGIS, rather than an alternative to TGIS. For example, this methodology could be applied to the output of my TGIS, which would then give a stronger emphasis to the complex uncertainties of the original data than in Premo's example, resulting in more representative final output in which one could take greater confidence. What is clear is that these statistical approaches to uncertainty in archaeological data should only be undertaken with care and whilst maintaining clarity over the level of subjective interpretation included.

Radiocarbon dating and Bayesian logic

As might be expected, I am not the first person to have attempted to study radiocarbon dates using GIS. As part of a wider collaborative project, Russell attempted to use GIS to study the spread of farming across Europe. She used a spatial database constructed from over 2,500 later Mesolithic / early Neolithic radiocarbon dates from across the continent, collected after 1980 and of a standard deviation of less than 150 years. Where more than one date existed on a site, the earliest was ideally used (2004: 3, 6; Gkiasta *et al.* 2003: 47-48). Her initial analysis was effectively a reworking of older analyses with a new, larger dataset, mapping the modal value of each date. Weighted regression was used to minimise the influence of dates with a large error band. The results showed a pattern of agricultural diffusion out of the east, broadly similar to that suggested by earlier analyses (Russell and Steele 2002: 511; Gkiasta *et al.* 2003: 50).

Russell was aware that using the modal value undersold the complex probability of a calibrated radiocarbon date, but believed that doing so was necessary in order to produce isochron and chronological maps to visualise the broad patterns in the data. In an attempt to resolve this issue, she calibrated her dates and calculated the probability of each falling within a succession of 100-year time-slices. These probabilities were displayed using the symbol size for the dates and an animation produced. The pattern previously seen in the modal distributions became more fluid and much less distinct, and apparent co-occurrence could be seen between Mesolithic and Neolithic activity (2004: 118; Russell and Steele 2002: 512; Gkiasta *et al.* 2003: 50, 54). Russell also attempted to move her analysis beyond visualisation by lowering her time-slice intervals to twenty years, squaring the probabilities to give greater weight to higher values and to diminish lower values, then conducting a Bayesian analysis (see below) of which resultant date was most likely to be the earliest on a site. She then used these models to work out an origin point for the Neolithic and to calculate its rate of spread (2004: 120-124).

Russell's final conclusion was that the quality of the radiocarbon database was not sufficient to produce any firm results without further analysis of each site's stratigraphic sequence. She suggested that the large size of the database meant that broad patterns overwhelmed the noise caused by these quality issues, but also noted that the error bands intrinsic to radiocarbon dates would in turn overwhelm any rapid spread which might have taken place (2004: 191-193). There are elements to be admired in this work, insofar as it acknowledges some of the nuances in working with radiocarbon data, but there are also significant difficulties. Firstly, working with modal values may perhaps be necessary in order to produce isochron / chronological maps, but the fact that working with calibrated date spans elides the patterns visible in the modal results demonstrates clearly how any patterns visible in the modal data may be false. Furthermore, as noted, the nature of a radiocarbon date is probabilistic: the actual date may fall at **any one** point within the probability band. Working with the calibrated data does reintroduce some of this probabilistic nature, but any apparent co-occurrence between periods may simply be a factor of this and not representative of actual co-occurrence in the past. Furthermore, reducing the probabilities to fixed 100-year time-slices still short-changes the variability in probability visible at finer temporal resolutions.

Russell's attempt to engage with Bayesian analysis is an example of a growing trend in archaeology since the late 1980s. Bayesian thinking was devised by the Rev. Thomas Bayes in the 18th century and is a methodology used to support making good decisions about probabilistic data (Aldenderfer 2005: 530-532). Recent work on re-dating the Neolithic of southern Britain has illustrated the power of Bayesian analysis and has also highlighted the complexities of working with radiocarbon data (Bayliss *et al.* 2007; Whittle *et al.* 2007). Bayliss *et al.* wrote that chronology is fundamental to archaeology, enabling the separation of time-dependent variations from other variations, and the comparison of contemporary sites to track changes in the material record through time. Chronology remains fundamental to archaeological questions, but is itself sterile without those questions (2007: 1-2), whatever theoretical viewpoint you take. However, that is not to say that conventional, totalising chronology does not need improvement: we must build our chronologies in a fashion that is more representative of the nature of our data and without allowing undue influence to evolutionary models (see Chapter 2). Methodologies that give greater account to the fuzziness of archaeological dates, such as Bayesian analysis, are one particularly fruitful way in which chronologies can be improved in this manner.

In order to capture a sense of past experience of time and the tempo of cultural change, a framework of calendar dating is necessary, '*...the order **and** spacing of the beads on the string*' (Bayliss *et al.* 2007: 2 - original emphasis). This allows the assessment of factors such as duration, which can in turn be carried through into interpretation, for example with longer lived rites perhaps having greater significance in past social memory. Bayliss *et al.* suggested that archaeologists have tended to universalise past human experience and deny the particularity of individual contexts due to the temporal remoteness and imprecision of people in the past. They stated that archaeologists advocate the importance of reliable chronology, but do not translate this into serious, scholarly attention. We still work with timescales in which a millennium can be considered in three or four distinct blocks: we can do better than this, so why are we not doing so (2007: 1-3)?

Radiocarbon calibration is now widespread, but we still struggle with imprecise chronologies. Bayliss *et al.* suggested that this is partly due to a lack of awareness, skill, and funds for Bayesian radiocarbon chronologies, and partly due to the method falling between the stools of processual (with its suspicion of models with more than one answer) and postprocessual (with its distaste for modelling) archaeology. The Bayesian algorithms themselves are theory-free for archaeologists, but theory-dependent choices are made at every step. Furthermore, objectives need to be explicitly defined at the start of the process and models built to query them: as such, the models are fundamentally dependent upon the research framework determining the selection of objectives (2007: 3-4). In other words, Bayesian modelling of radiocarbon dates requires archaeological interpretation throughout, including questions beyond the individual sites themselves (Whittle *et al.* 2007: 123). Using it enables the construction of chronologies that operate along timescales that are relevant to past human experience (Bayliss *et al.* 2007: 4). Returning to the discussion in Chapter 2, these new chronologies are what our TGIS must be built to deal with, rather than the conventional chronologies of our hypothetical C-GIS.

Using Bayes' Theorem in an archaeological context, new data collected about a problem (the standardised likelihood) is analysed in the context of our existing knowledge of the problem (prior beliefs). This produces a new understanding of the problem incorporating both the new and existing data (posterior beliefs). Today's posterior belief becomes tomorrow's prior belief, informing the collection of new data and its interpretation. Thus, Bayesian logic enables explicit and quantitative implementation of the hermeneutic spiral by which human decisions are made. Chronology is the most commonly used arena for Bayesian analysis in archaeology, through the integration of radiocarbon data (as the standardised likelihood) with relative chronological information such as stratigraphy (as the prior beliefs), based upon the premise that radiocarbon dates accurately estimate the age of samples but that it is the dates of the events associated with them that are of importance. In other words, dating an object is a means to an end: dating its context. The posterior beliefs output by the analysis are expressed as probability distributions (posterior density estimates). They should be taken not as absolutes, but as interpretative estimates which can mutate as further data come to light (Bayliss *et al.* 2007: 4-5):

> *This production of alternative models, none of them definitive, is simply a means of creating multiple pasts and is entirely congruent with the post-modern perspective.* (Bayliss *et al.* 2007: 5)

Bayliss *et al.* used the Bayesian modelling system implemented in the calibration software, OxCal[1] (v.3.10). This model utilises the radiocarbon results, the explicitly specified known relative ages of the samples, and a necessary assumption about the mathematical distribution of the archaeological events to counteract the statistical scatter caused by the dates' error bands. Leaving out this assumption causes archaeological activity to appear earlier and disappear later than was actually the case. OxCal assumes that the events are distributed uniformly. The software then attempts to reconcile the probability distributions of the radiocarbon dates against the prior information, by repeatedly randomly sampling each distribution to build up a set of solutions consistent with the model structure. It is also able to produce estimated dates for undated events that fall between two radiocarbon dated events in the sequence. Comparing these probability distributions provides quantitative estimates of the durations of phases of activity and hiatuses between phases. OxCal provides estimates of the reliability and stability of the model, in order that the user might assess the quality of their results. All models are wrong, but we must assess what is importantly wrong about them to understand if they are useful. Varying the input elements allows the determination of which factors are critical to the date estimates produced: elements which are wrong but not critical to output do not cause the model to fail as importantly wrong (2007: 5-7).

Probably the most popular method for interpreting radiocarbon dates has been visual inspection of their calibration curves. This is not problematic so long as it is understood that the curve represents the time during which activity took place at some point, as opposed to the erroneous view that the curve (for a single date) represents the temporal extent of activity. The latter view confuses the scatter and uncertainties of radiocarbon evidence with the longevity of past activity. Unfortunately, this problem is very real. Awareness of this problem led to the adoption of *floruit*, in which calibrated date probabilities are added and then normalised to an integral of 1.0. This summed distribution is then integrated until a certain proportion of the curve has been excluded, the remaining distribution forming the date range. However, simulations of this methodology show that it still does not take enough account of outliers and statistical scatter. Further, it implicitly assumes that the uncertainties are independent, which is manifestly untrue for a coherent archaeological phase. As such, although popular, using *floruit* is wrong. According to Bayliss *et al.*, the only correct archaeological usage for summed probability distributions is to provide an estimate for the frequency distribution of dated events in a phase (and only then in combination with Bayesian analysis), but this is not usually of particular archaeological interest (2007: 8-11).

Returning to Bayesian analysis, Bayliss *et al.* discovered that the answers produced were reassuringly correct more often than expected, and for any portion of the calibration curve. However, its precision is limited by the number of radiocarbon dates and the number of stratigraphic relationships between them. They then tested how wrong input had to be to produce inaccurate results. They concluded that accurate stratigraphy was essential, that the distribution model had to be included but could vary in its form, that variation in the calibration curve was not unduly problematic, and that the radiocarbon dates and their laboratory errors have to be accurate. Bayliss *et al.* concluded that Bayesian modelling was still a developing area, but that it ought not to prove ultimately wrong in any important way. In comparison, less formal methods proved frequently to be importantly wrong, giving misleading impressions of contemporaneity and duration (2007: 11-24).

Turning to the archaeological results of the same project, Bayliss *et al.* were able to consider Neolithic sites within the context of human lifespans, or even generations, and the order and duration of events in their local and regional context (2007: 4). Previously, the radiocarbon dating of Neolithic monuments such as long mounds had been very imprecise, with many post-processual interpretations based upon agency and experience, happy to float free of chronological context (Whittle *et al.* 2007: 124-125). The drastic improvements in chronology produced through the Bayesian modelling enabled them to begin writing much more detailed histories for the Neolithic in southern Britain. Timescales for usage of monuments was reduced to four to six generations at most, allowing this usage to be considered on a much more human scale within the personal memories of a few generations and involving particular people at particular times (as opposed to timeless agents). For example, Neolithic long mounds have been interpreted as expressions of longevity or domination. However, reducing the use-life for these monuments weakens interpretations based upon exclusivity of burial and reduces their density at any period with the result that they appear less convincing as tools for social control. A gap of seven centuries has also appeared between the end of the European LBK and the beginning of long barrow building in Britain: either the new chronologies reduce the plausibility of long-term social memory in the Neolithic, or they reinforce it: myth may persist for a long time, but detailed comparisons between long barrows and the LBK become unconvincing (Whittle *et al.* 2007: 131-140; also see Chapter 3 above).

The TGIS implementation

The next section of this chapter will discuss the implementation of the uncertainty calculations and analytical tools in the TGIS. Consideration will also be given to the temporal model incepted and the cartographic decisions made.

[1] http://c14.arch.ox.ac.uk/oxcal.html

Why ArcGIS?

Fundamental to this project was the desire to produce a software solution that was widely accessible to archaeologists. If we accept that archaeological GIS without any temporal element is insufficient, TGIS ought in the future to move from being the exception to the rule. As such, software solutions need to be accessible to archaeologists working with GIS and should be adoptable without too much difficulty. This is one reason why previous bespoke solutions have never met their full potential. As a result, software simplicity was enshrined as a central tenet of this project. I decided that I would start with what seemed the simplest solution and only increase the complexity where it seemed necessary or useful.

To maintain this aim, it was important to work under the aegis of an existing GIS package. This helps keep the project within its democratic aims and also maintains the presence of the wider suite of GIS tools routinely included within commercial GIS software. There were two obvious choices of software package. GRASS GIS[1] is a free, open source GIS originally developed by the US Army, but now very widely used across many different industries.[2] The advantages of GRASS lie in its lack of cost and in its highly customisable nature. However, my experience at the outset of this project showed it not to be the most user-friendly of software packages, though efforts are continually under way to enhance and develop the software - a reflection of its open-source ethos. By contrast, ESRI's ArcGIS[3] environment is a high cost commercial product but is again very widely used. By its very nature, ArcGIS is a more polished and user-friendly product than GRASS, but not without its own foibles. Partly due to my own greater familiarity with the software, I decided to use ArcGIS as the basis for my TGIS. All other things considered, it was deemed a good place to start in moving on from the bespoke, but limited C14View (see Chapter 1), and the ArcScripts[4] provision of the ESRI website routinises distribution. It would appear to be widely available to archaeologists working at universities and at archaeological units, although is less accessible to smaller organisations. Fundamentally, however, it is simply easier to work with than GRASS and rather more comprehensible for the average Windows-entwined user.

Working within the bounds of ArcGIS itself, another choice arose: that of programming language. The simplest solution was to write the scripts in the Visual Basic for Applications (VBA) editor included within ArcGIS.[5] This solution means that the scripts are all saved within a map template and the user simply has to create a map using that template to get access to the TGIS functionality. It also means that users have all the code within easy reach and are able to customise it further to their own requirements or to examine it in detail and learn more about the underlying mechanics of the TGIS. The difficulty of working with VBA is that it is limited in its functionality: I was prepared to move on to a higher programming language if the design requirements of the TGIS reached the integral limits of VBA, but I hoped to avoid introducing any unnecessary complication where possible. In the event, VBA and ArcGIS proved quite capable of handling everything I was seeking to implement.

The temporal model

The series of dates that the user wishes to examine should be entered into the GIS as a shapefile. Various other types of vector layer format (e.g. data added directly to the map from a table, or saved as a layer) do not function correctly, presumably because of issues with regard to writing to their data tables. This is not a major issue, as any layer can be easily exported to a shapefile. Amongst whatever other data are included in the layer, its data table has to contain two fields: one to give the minimum possible date of each item and one to give the maximum, with the minimum / maximum recorded as an integer year. Negative integers refer to BCE / BC / BP and positive integers to CE / AD. All of the different dating layers used in the TGIS should be expressed in the same calendar system (BCE / CE or BP). This temporal model creates a minor problem in the existence of an erroneous year zero, but this is too small an error to require any complex workarounds.[6] Dates that apply only to a single year should be given that year for their minimum and maximum. As previously discussed (Chapter 3), dates of an accuracy of better than one year are rare in archaeology and the sub-year portion of the date will usually feed into different interpretative threads than the year portion. Thus, they should be taken as single year dates of equal minimum and maximum. The vast majority of other archaeological dates should easily fit into the integer time period temporal model enshrined in the TGIS. As such, the temporal model implemented in the TGIS scripts is actually rather simple in execution and should not require any major conceptual restructuring on the part of the average user: they are treated as attributes of the dating layers in the GIS. The question that then arises is how to model the internal uncertainty of each time period.

Implementing archaeological uncertainties

The initial probability calculation method implemented was a straightforward percentage, which we shall call the standard probability. The user selects the time period in which they are interested, then the software compares this to each item's date range and calculates the percentage by

[1] http://grass.itc.it/intro/general.php
[2] http://en.wikipedia.org/wiki/GRASS_GIS
[3] http://www.esri.com/software/arcgis/index.html
[4] http://arcscripts.esri.com/
[5] Unfortunately, this editor has been removed from ArcGIS 10, so the TGIS will only function in ArcGIS 9.

[6] I was also concerned that including any workarounds would have compromised the mathematical calculation undertaken by the software and produced unpredictable errors.

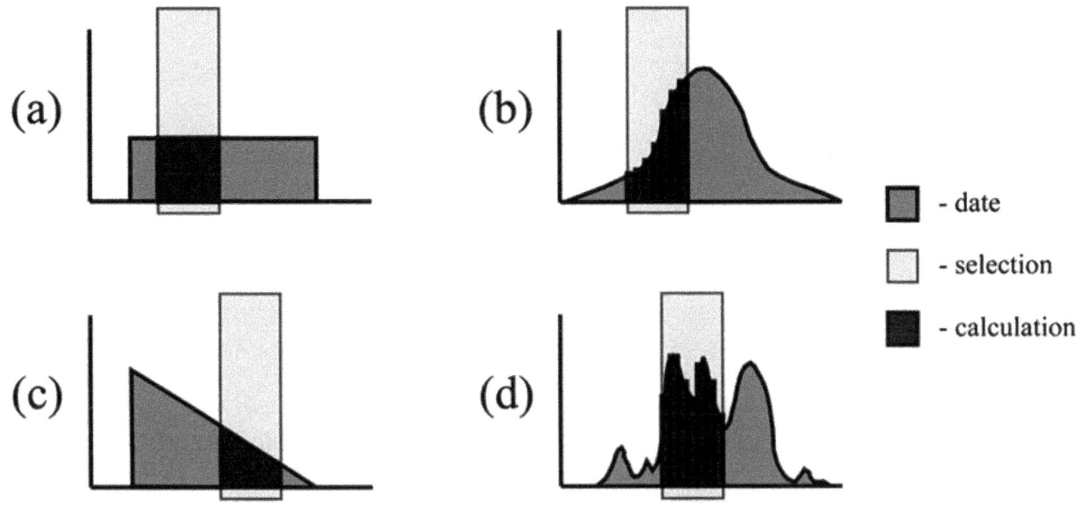

FIG 5.1 - GRAPHS OF THE DIFFERENT PROBABILITY MODELS IMPLEMENTED IN THE TGIS, SHOWING THE NATURE AND VARYING COMPLEXITY OF THEIR DIFFERENT PROBABILITY CURVES AND THE VARIOUS CALCULATION METHODS: (A) STANDARD; (B) NORMAL; (C) TERMINUS POST QUEM; (D) OXCAL. NOTE THAT THE NORMAL AND OXCAL METHODS ARE SIMPLIFIED FOR CLARITY. THE HORIZONTAL AXES REPRESENT TIME, THE VERTICAL AXES PROBABILITY, AND THE AREA UNDER EACH GRAPH WOULD TOTAL 1.0 (100%) PROBABILITY.

which they overlap (if at all). These percentages are then written to the layer's table for each item, expressed as a probability of between 0.0 (0% probability) and 1.0 (100%). This is also true for the other probability calculation methods. Some users might be unused to fractional weightings (Ratcliffe 2000: 678), but archaeologists accustomed to working with radiocarbon dates ought to grasp these intuitively. The probabilities recorded in the layer table in the TGIS are rounded to two decimal places: any probability variations of less than 1% would be too precise to be of any great archaeological relevance. Any probabilities given in the examples in the rest of this book will be expressed to the same degree of precision.

This standard probability calculates faster than the other probability models introduced later and, as such, it is ideal for making initial calculations and assessments. Primarily, however, the standard probability is intended for forms of date with no known internal probability model, or models not integral to the TGIS. This would include, in particular, ceramic and other typological dates, and calibrated radiocarbon dates where the full internal probabilities are unknown. If the user wishes to examine different types of date as part of the same layer, then this also would be the recommended probability model for that scenario. However, a more elegant way to handle multiple date types is also offered by the TGIS, as we shall see below.

The standard model is similar to the aoristic methods discussed above. However, there is one key difference. An aoristic probability model is constrained by the temporal resolution chosen to make the snapshot divisions. For example, an aoristic evaluation at 100 year resolution of the date range AD 340-520 would be divided into thirds according to a binary decision of presence or not in each snapshot, and given the following aoristic weights: AD 301-400 - 0.33 (present in one of three snapshots); AD 401-500 - 0.33 (as previous); AD 501-600 - 0.33 (as previous). In truth, however, the probabilities for each snapshot should in fact be: AD 301-400 - 0.33 (61 years [AD 340 to 400] out of the 180 year span for the date); AD 401-500 - 0.56 (100 years out of 180); AD 501-600 - 0.11 (19 years out of 180). This is the result that the standard probability calculation in the TGIS would produce. As will be apparent, the aoristic method has greatly reduced the true probability of the period AD 401-500 and greatly exaggerated the probability of the period AD 501-600. Naturally, an aoristic model produced at a resolution of 1 year would produce the same result as the standard model, but the probabilities output in each snapshot would be too small to be of any realistic use (0.006 for each year in the example above). Later work by Ratcliffe (2002) did correct the mathematical inaccuracy of his original method (but this improved methodology was not applied by Johnson).

The next probability calculation method included was designed to capture an internal probability that obeys a normal (or Gaussian) distribution, which we shall call the normal probability. The normal probability is intended to cater for scientific dates expressed as a date with an error margin, such as thermoluminescence dates and uncalibrated radiocarbon dates. The TGIS assumes that the minimum and maximum given for each date in the layer's data table are expressed at ±2 standard deviations - containing 95.44% of all possible solutions (Orton 1980: 90-91), as this is conventionally the case when quoting such a date; a future adaptation of the TGIS might be to include the ability to handle dates expressed as a mean and error band. When the user of the TGIS selects a period of interest, the TGIS works out the mean and standard deviation, then calculates the probabilities for a series of slices through the date's normal curve[1] where the normal curve overlaps the selected period. These slices are of 1/20 of a standard deviation, resulting in an output precision of better than 1%. Calculating this proportion directly using one formula would be far too complex to embed in usable software. When using the normal method, the TGIS works out the probabilities between ±3 standard deviations, to incorporate 99.74% of all possible answers. Archaeologists should always bear in mind, however, that there will always be a tiny percentage chance of the true date falling even outside of this bracket - or indeed a bracket of any size (Orton 1980: 91-92).

The next probability calculation method implemented was designed to capture the internal probability of a *terminus post quem* date, which we shall call the T.P.Q. probability. The T.P.Q. probability is intended to cater for dates which may be very accurate in themselves, but where the event of archaeological interest falls after that date. It is particularly intended for numismatic, dendrochronological, and other similar dates. Orton modelled the probability of *terminus post quem* dates as a curve that dropped off rapidly then levelled out to infinity (Orton 1980: 99). This could be the case in some instances, but I do not believe it would be the most common circumstance. The Orton model implies that most items would have been deposited immediately after manufacture, but I believe that deposition is likely to peak some time later (the length of time being dependent on the type of item in question, perhaps a few years for a pot or some decades for a coin). In my opinion, this probability ought to be expressed as the right-hand half (or perhaps two-thirds) of a normal curve, with a period of slowly declining probability, that then drops off quickly before again levelling off to infinity. This is, however, little more than a hunch and without any quantitative evidence to support my instinct, the T.P.Q. probability is modelled as a triangle of constantly decreasing probability - as is the current convention amongst chronologists (Derek Hamilton, *pers. comm.*).

Thus, when defining dates of this nature, the user must make an assessment of the likely period during which the item could have been deposited. This date should be given as the maximum and the date of the object itself as the minimum. Naturally, if the user is confident that the object date equates to the deposition date then the same year can be used for minimum and maximum. In that event, the TGIS will always output a probability of 1.0 if the date falls within the selected period, no matter which probability model is used. In the T.P.Q. method, the probability of the date falling within the selected period is calculated using some elementary trigonometry. We know that the total area under the triangle is 1.0 (100% probability) and so three simple formulae can be used to calculate the probability, depending on whether the selection overlaps the minimum, the maximum, or falls within the probability triangle. Naturally, if the selection subsumes the entire probability triangle, the probability is simply 1.0.

The normal and T.P.Q. methods have been included in an attempt to maximise the usability of the TGIS software. They provide more realistic methodologies for assessing the probabilities of a whole raft of dating methods. In a neat reflection of the comparison between aoristic probabilities and the standard method, comparing the normal and T.P.Q. methods to the standard method shows that they better reflect the internal probabilities of appropriate date types. It should be noted that the normal method takes longer to calculate than the standard method, but not prohibitively so; the main obstacle to fast calculation would appear to be time spent writing to the layer table. Both of these latter models have been carefully tested mathematically through comparative manual calculation and function as intended. However, appropriate dates for these models do not feature in my two main datasets and, as such, they will not feature strongly in the discussion that comprises the remainder of this book.

The final probability calculation method introduced was designed to capture the internal probability of calibrated radiocarbon dates and Bayesian modelled dates, which we shall call the OxCal probability, after the calibration software mentioned above. The nature of radiocarbon dating and calibration was discussed in Chapter 3. The key element to recall is that mapping the normal distribution of an uncalibrated date against the vacillations of the tree-ring curve results in a calibrated probability that can oscillate up and down, like the decks, guns and smokestacks of a battleship. This is by far the most complex probability model implemented in the TGIS. Fortunately, OxCal is able to output these internal probabilities in a series of slices, by default five years thick (the TGIS is able to discover this thickness on calculation). To utilise this output in the TGIS, these probabilities should be copied into a table for the entire series of radiocarbon dates in the layer, along with an identifying code for each sample's series of probabilities. This identifying code, conventionally the laboratory code, should also be in the data table for the shapefile in the GIS. The user then has to add this probability table to their map and link the date layer to it using a one-to-many relate based upon the identifier common to both tables.

When the TGIS compares the layer to the currently selected period, it no longer performs any calculations based on the layer's attributes. Instead, for OxCal probabilities it simply has to query the probability table linked to the layer table, and sum the probabilities for each item of any time-slices that fall within the currently selected period. Some small recalculations are also undertaken to proportionally split the probability for areas where the selected period maximum or minimum falls

[1] The formula for a normal distribution may be found in Lipschutz and Schiller 1998: 182-183.

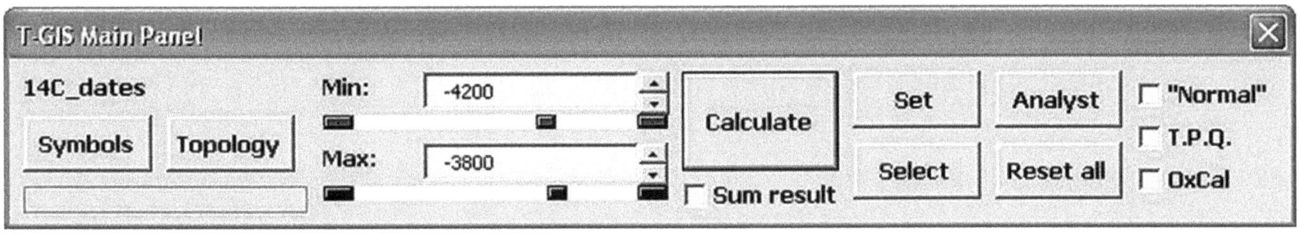

FIG 5.2 - THE MAIN TGIS USER INTERFACE. THE MIN AND MAX AREAS ARE USED TO SET THE PERIOD OF INTEREST AND THE CALCULATE BUTTON LAUNCHES THE CALCULATION PROCESS. THE CHECKBOXES ON THE RIGHT ARE USED TO SELECT THE PROBABILITY MODEL FOR THE CURRENT LAYER. THE CURRENT LAYER NAME IS GIVEN AT THE TOP LEFT.

within the span of a time-slice. The OxCal probability is significantly slower to calculate than the other methods, due to the extensive data that the TGIS has to work through to find the appropriate time-slices. As such, if the user is working with a very large dataset, it is recommended that they make initial calculations using the standard probability and only use the OxCal probability when they require greater precision in their uncertainties. Another restriction of the OxCal probability is that it does not function with the analytical tools described below: they are based upon various methods of summing the probabilities, which is simply not appropriate with radiocarbon data (see above).

To summarise the basic function of the TGIS, when a user wishes to examine their spatio-temporal data, they have to add that data as a shapefile layer to an ArcMap document based upon the TGIS template. They then select the layer in question and click on the TGIS button. After selecting the appropriate minimum and maximum date fields (see above), the main TGIS user interface will appear (Fig 5.1). They may then select whichever probability calculation method they believe is most appropriate for their data (or make no selection for the standard method), selecting the appropriate fields in the sub-table for OxCal probability. They should then select their period of initial interest and ask the TGIS to make the calculations. The TGIS will then work through the layer's data table, make the appropriate probability calculations and write the results to said data table as a decimal between 0.0 and 1.0 for each item. The time taken to make these calculations and, in particular, write the results to the data table means that any animation of the results is not possible. The same applies to calculation on the fly, hence the Calculate button. The four probability calculation methods integrated into the TGIS provide the functionality to handle the vast majority of conventional archaeological dates, as discussed in Chapter 3. Furthermore, the standard probability stands in as a useful approximating catch-all for any dating methods that behave differently. Although conceptually straightforward, this system for probability calculation represents an important step forward beyond the limits of pre-existing TGIS methodologies designed to handle temporal uncertainty, insofar as it gives far greater account to the internal probabilities of dates and escapes from the limitations of snapshot timeline methodologies.

Temporal topology and multilinearity

It is important to note that as well as calculating probabilities, the TGIS also writes a second piece of useful information to the layer's data table when calculating the probabilities: that is, the temporal topological relationship between each date and the selected period. This topological system is an adaptation of that given in Chapter 4 (page 47). The results are expressed with regard to the date's relationship to the selected time period. As such, they may be:

- Date falls **before** the time period (i.e. 0.0 probability);
- Date **overlaps the maximum** of the time period;
- Date **overlaps all** of the time period;
- Date falls **within** the time period (i.e. 1.0 probability);
- Date **overlaps the minimum** of the time period;
- Date falls **after** the time period (i.e. 0.0 probability).

These six states capture all of the six important temporal topological relationships identified by Jones (1997) that may exist between any date and the time period of interest. The recording of this topology enables the user to be aware of how any particular date relates to the selected time period and also gives a strong clue as to temporal change across the dataset. The immediately before and immediately after states have been excluded on the grounds that they do not add much beyond a very subtle nuance and would be very rare occurrences when dealing with data of relatively large temporal scale, as in most archaeological cases. However, an expression of the degree to which an object fell before or after the selected period would be a useful piece of interpretative information (when considering temporal continuity / discontinuity) that could be included in later versions of the TGIS.

The other important element of the probability calculation within the TGIS is the ability to examine dates across multiple layers. Thus, if a user has an intra-site dataset combining pottery, radiocarbon and perhaps coinage evidence, they can separate each type of date out onto a separate layer. Each layer can then be loaded into the TGIS and switched between at will. As such, different

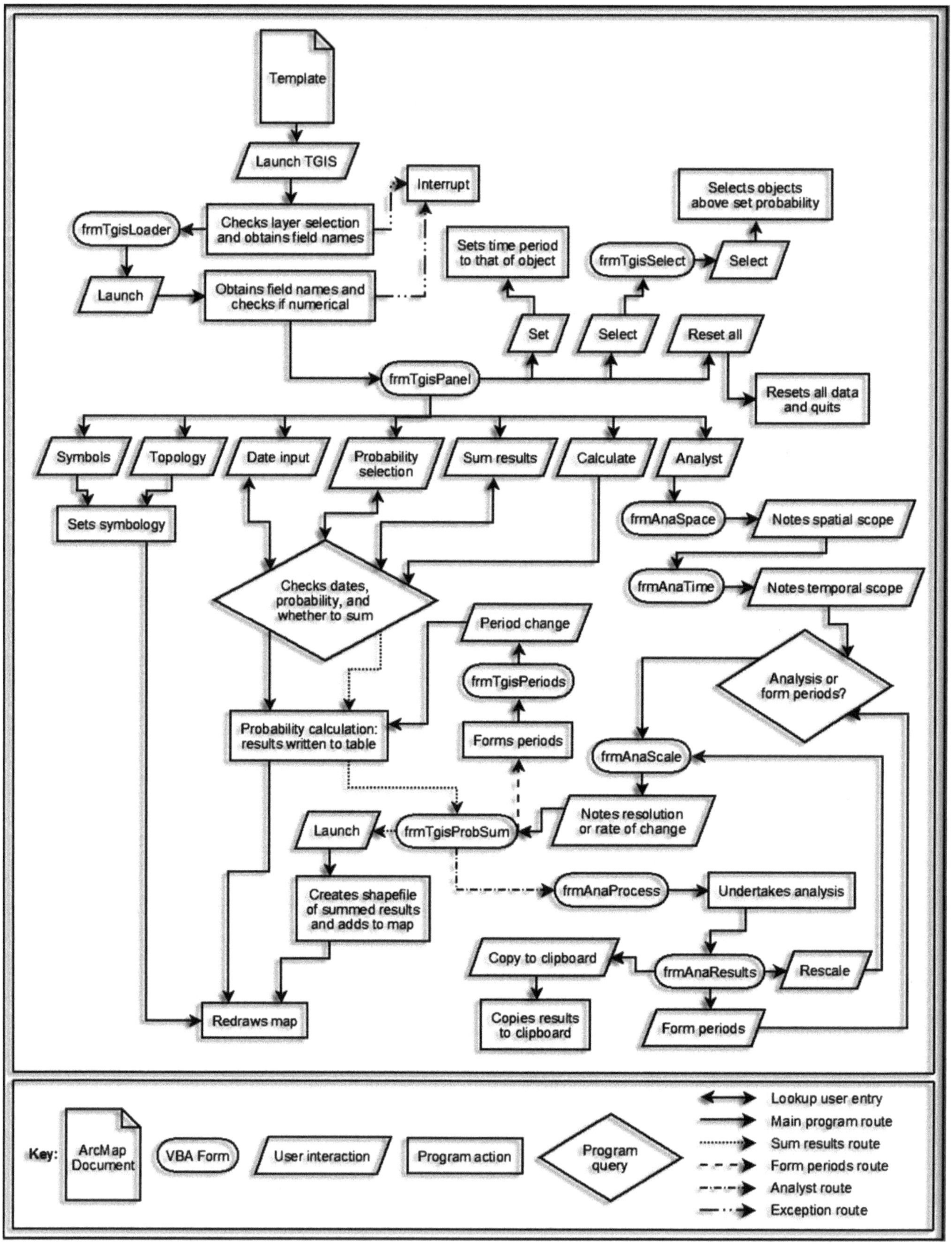

FIG 5.3 - FLOWCHART SHOWING DECISIONS MADE AND PROCESSES UNDERTAKEN IN THE FUNCTIONING OF THE TGIS.

probability calculation methods can then be applied on each layer. The TGIS will remember the settings for each layer already loaded into the system until the Reset All button is clicked (which clears all saved TGIS data). Therefore, it is also possible to examine different time periods on different layers within the same work environment: this provides an element of multilinearity to the TGIS. For example, this could be used to examine any intrinsic bias across an entire dating type, by comparing temporally disjointed but apparently related patterns between different map layers. An improvement to this system might be in the implementation of the ability to examine several layers concurrently, perhaps with offsets to counter discovered bias. An omission from this multi-layered system is the recollection of settings when switching between OxCal layers, as this was deemed too complex to implement compared to the benefit: it is trivial enough to quickly reselect the fields when selecting a radiocarbon layer. The multi-layered approach also allows the separation of dates on a thematic basis: the power of this approach will be demonstrated in Chapter 6.

Analysing archaeological uncertainties

Several methods for analysing archaeological uncertainties are built into the TGIS. Looking back to the discussion in Chapter 3, a common question that might arise when analysing the chronology of an archaeological site is "how does this date (or group of dates) relate to the other dates in the site?". It is possible to provide rapid quantitative evidence to help answer this question using the TGIS. Firstly, the user must create a selection of a date (or dates) on the map, using whatever method they wish (e.g. spatial selection using the appropriate tool, selection from the data table, via a query, or using the Select button in the TGIS to select dates of above a certain probability). Then, if the user clicks on the Set button in the main TGIS user interface, the minimum and maximum of the current period will be set to the minimum and maximum of the selected date or dates. Adjustments may be made manually to increase (or indeed decrease) these buffers. Calculation of the probabilities will then provide quantitative evidence as to the probability of the other dates on the site being contemporary with the full probability period of the selected date or dates. A 100% probability would not mean that the dates were necessarily contemporary, but it would mean that they both fell somewhere within the selected period. This provides a simple, yet powerful form of temporal buffering as a methodological route towards answering such questions.

Also, summing probabilities can be a very useful tool when analysing non-radiocarbon dates (as mentioned above, summing radiocarbon dates is not necessarily a worthwhile endeavour). As such, two methodologies for summing probabilities are built into the TGIS. Firstly, by clicking on the Sum Result checkbox, the results of any calculations made will be summed where they occur in the same spatial location and the results written to a new shapefile (and added to the current map). During the processing, these results can be multiplied according to another field in the layer's data table. This methodology is intended to be used with large datasets where many items fall upon the same location. The weighting (called normalisation in the TGIS) by another field is intended to provide a way of giving greater weight to certain item probabilities where appropriate.[1] The obvious example where such a methodology would be of use is with large field-walked pottery datasets: many different fabric types might fall upon the same locations but with differing sherd counts / weights / densities. The importance of this approach will be demonstrated in Chapter 7.

The second methodology for summing probabilities is provided by the analytical tool built into the TGIS. By clicking on the Analyst button and following the instructions, the user can sum all of the probabilities across their site in a series of time-slices. This analysis can be undertaken using a subset of a layer (selected by the user), a whole layer, or even using every layer loaded into the TGIS, and using either the currently selected time period or the maximum temporal extent of the layer / layers. It will also make the calculations using the probability model last used for each layer, excluding the OxCal model (the standard would be used instead) as this tool is not primarily intended for use with radiocarbon data. The width of the time-slices is selected by the user.

The output table then gives the summed probability for each time-slice and also expresses it as a percentage of the maximum summed probability in the table. This percentage is important for comparing the overall probability pattern of large and small datasets, as larger datasets will naturally have a larger summed probability but with a temporal pattern that a user might wish to compare to that of smaller datasets. It is then possible to copy the output of this tool to a separate spreadsheet package for further analysis. The TGIS is also able to form periods from this data by aggregating the time-slices according to the percentage rate of change between each time-slice when compared against the maximum summed probability. Again, the potential of this tool will be demonstrated in Chapter 7.

Moving beyond the tools specifically implemented in the TGIS, probably the most important way to analyse uncertainties using the TGIS is to make use of the probabilities written to the layer's data table when calculations are undertaken. These figures can be used as a normalisation field for any spatial analysis undertaken in the GIS: an analogy for this would be the interpolation of a terrain model from spot height data, albeit with spot heights that only vary between 0.0 and 1.0. Writing probabilities to the layer table slows down the calculations and makes animation impossible, but is

[1] Conceptually speaking, we are in fact weighting the variable using the probability: in other words, a sherd count of 10 is weighted via multiplication by the decimal probability 0.66, giving the weighted sherd count of 6.6.

essential if users are to be able to work with the probability data in any meaningful fashion. In any event, the way in which ArcMap updates its map display currently precludes animation: it re-draws maps on-screen (and somewhat slowly at that), whereas smooth animation conventionally requires the writing of frames to an off-screen buffer which are then copied to the visible display. I believe that this writing process is also the reason why the TGIS will only work with shapefile data, as it would appear that some restrictions apply in ArcGIS as to what may be written to the data tables of other layer types. Nevertheless, this recording of probabilities is the main *raison d'être* of the TGIS and is the manner in which I have approached the problem of extending spatial analysis into the temporal dimension. The power of this analytical approach will be demonstrated in Chapters 6 and 7.

The cartography of uncertainty

Clear graphical representation has to accompany the database-oriented aspects of a TGIS (Ratcliffe 2000: 674), but how do we display temporal uncertainty graphically? Fuzziness is essentially a fifth display dimension and designing a 5D display module is a major challenge. Stead suggested that with the three spatial dimensions represented isometrically and time represented by a clock, probability could be represented by rotating a map element into the fifth dimension to show a battleship probability curve similar to those used for calibrated radiocarbon dates (Stead 1998: 218). There are also prospects for animation of snapshot data, although that can make low probability but long-lived objects appear more important than they might be, or for the generation of surfaces depicting temporal weighting (Ratcliffe 2000: 675-677).

However, these systems seem overly complex, as in practice we may extend our display into further dimensions using simpler graphical devices. Bertin has defined six graphic variables (Kraak and Ormeling 1996: 124):

- Difference in size;
- Difference in lightness (grey) value;
- Difference in grain or texture;
- Difference in colour hue;
- Difference in orientation;
- Difference in shape.

Other suggested variables include differences in colour saturation, in regularity / irregularity of arrangement, in focus (Kraak and Ormeling 1996: 125), and in transparency (Zubrow 2006: 124). These variables are naturally perceived in different ways by users: for example, difference in colour hue can be used to convey qualitative difference, but only where degree of lightness is kept fixed. Careful use of these variables enables cartographers to communicate with their audience in a manner which makes intuitive sense to that audience (Kraak and Ormeling 1996: 124-125).

Sound can be used in a similar fashion to complement visualisation. People are generally able to spot changes in audio patterns without concentrating and may even process multiple audio dimensions simultaneously if properly presented. For sonification to be effective, sonic symbols must draw upon users' preconceptions so that their metaphoric representation forms a natural cognitive link between the data's characteristics and the user's conceptions (Weber 1998: 74-75, 78). The use of sound thus has great potential for conveying temporal fuzziness to TGIS users, perhaps with harmony / disharmony being used as a natural analogy for possibility. However, noisy software has great potential to irritate users, and sound is also difficult to localise onto a map. The use of colour is the most straightforward method of representing temporal fuzziness and ought to prove intuitively understandable for users. Pop-ups could be used to provide a more detailed picture of uncertainty curves, or they might be represented outside the TGIS at the database level.

As the TGIS created for this project treats the uncertainties of its dates as attributes, the simplest way to express this cartographically is through colour. It is always open to the user to use ArcGIS's integral tools to adjust how uncertainties are displayed, and users may edit the TGIS code to force the native cartography to comply with their own wishes. The cartographic system inherent to the TGIS displays uncertainty as a colour ramp: no colour (i.e. transparent) conveys 0 probability, then the colour ramps up through from yellow through to red as probability increases to 1.0.[1] This provides a display method that paints the dates according to an intuitive scale of probability hotness. By contrast, the TGIS natively displays temporal topology using clashing colours to reflect the categorical rather than scalar nature of those variables. This should also be intuitively understandable to users.

Conclusions

In this chapter, I have discussed the different ways in which archaeologists have previously attempted to analyse the uncertainties of our unique data, and have described the implementation of my new TGIS designed for similar purposes. This new TGIS was implemented in ESRI's ArcGIS environment, as that system is widely available to archaeologists working at universities or for commercial units. Thus, the software produced ought to be intuitively understandable and accessible to most GIS-aware archaeologists without them having to gaze too far beyond their current software horizons. Fundamental to this project is the desire to communicate the advantages and need for TGIS within archaeology and the ArcGIS implementation helps to engender this scheme.

[1] The maps reproduced in this book use an alternative, monochrome colour scheme.

In Chapters 3 and 4, I argued that the key to providing a firm foundation for new chronological interpretation was to incorporate the inherent uncertainties of the dating techniques that provide our base data. Previous to this project, archaeologists have approached temporal uncertainties using several different methodologies. The use of fuzzy logic has potential in dealing with site phasing and typology building, but is less applicable to more basic dating questions. Aoristic analysis has provided a first step in understanding temporal uncertainty, but is restricted by the lack of complexity in its probability models and its reliance on a snapshot model of time. Study of the intrinsic probabilities of radiocarbon dating has naturally been an area in which greater progress has been made, although some work has fallen short of a full understanding of the nature of that technique. Probably the strongest candidate for the successful implementation of archaeological temporal uncertainty has come from the area of Bayesian analysis. That work has achieved an excellent comprehension of the intricacies of working with radiocarbon evidence and probabilities, but has lacked any spatial aspect beyond stratigraphic structure.

The current project has attempted to forge a new agenda for working with archaeological temporal uncertainty, building upon an integration between the spatial methodologies of previous TGIS work on uncertainty and the stronger grasp on uncertainty of non-spatial methodologies. The TGIS is fully flexible as to consideration of time periods of interest (as opposed to restrictive snapshot approaches) and has excellent functionality in regard to the internal probabilities of archaeological dates. It is able to make calculations based upon four probability models: a standard percentage, a normal distribution, a model of *terminus post quem* probability, and finally calibrated radiocarbon probabilities exported from OxCal. The four models provide the ability to make internal probability calculations for the vast majority of archaeological dates, with the standard percentage standing in as a catch-all for any date types omitted. It also records the important temporal topological relationship between each date and the time period of interest. With these two pieces of data, users are able to visualise and analyse the probabilities of items falling within a particular period of interest and the broad patterns of temporal change across their sites.

Beyond this, the TGIS also provides tools for the further analysis of temporal uncertainty. The ability to perform separate calculations across multiple layers provides flexibility and a welcome sense of multilinearity. Further, simple temporal buffering capabilities are also provided. Also, two tools enable the examination of summed probabilities for non-radiocarbon data, summed either spatially or temporally. Finally, the writing of probabilities to the data table of the layer studied in the GIS enables the extension of spatial analysis to take account of time, through the use of these probabilities as a normalisation factor in analyses undertaken. The power of the TGIS and these analytical approaches will be demonstrated and illustrated by the case studies that follow in the next two chapters. The first provides an example of TGIS use with an intra-site, radiocarbon-oriented dataset (Chapter 6). The second takes a wider spatial perspective and provides an example of the TGIS in use to examine a regional ceramic dataset (Chapter 7). Chapter 8 will analyse the successes of these two experiments and consider ways in which improvements might be made to the TGIS to produce even stronger results.

CHAPTER SIX: RADIOCARBON DATING AT WILLINGTON, DERBYSHIRE

In this chapter, I shall demonstrate how the TGIS can be used in the analysis of an intra-site archaeological dataset. That analysis consisted of an examination of a suite of absolute dates taken during excavation in their spatial context, paying particular attention to their internal probabilities and with the aim of discovering any spatio-temporal patterning. Returning to the two fundamental temporal questions discussed in Chapter 3, the main questions consist of a consideration of when activity took place and an assessment of the possible spatial relationships between activity on different parts of the site. I shall demonstrate how the use of the TGIS can greatly aid with those two enquiries.

Background

The dataset in question contains the results of excavations undertaken by University of Leicester Archaeological Services (ULAS) during 1998-1999 prior to the extension of a sand and gravel quarry at Willington in southern Derbyshire. The site report was published in the 2009 volume of *Derbyshire Archaeology* and was edited by Matthew Beamish of ULAS, with the scientific dating section undertaken by Marshall *et al.* The large number of radiocarbon dates obtained (more on which below) make the dataset ideal for examination in the TGIS.

The excavations revealed remarkable evidence for prehistoric activity on the floodplain of the River Trent. Activity on the site began in the Upper Palaeolithic, and continued into a Mesolithic that was well represented in the flint assemblage, but which possessed no identified cut features. During the fourth millennium cal BC (Early to Middle Neolithic), activity focused on areas of low, wooded, gravel islands surrounded by active streams - fourteen palaeochannels have been recorded to date. The people predominantly used Peterborough Ware and their non-animal diet was dominated by wild foodstuffs, alongside some evidence for dairying. The subsequent Late Neolithic phase was aceramic on this site (ceramics were present a kilometre to the east), and consisted of fire clearance of the woodland - begun during the latter half of the third millennium cal BC. A well-formed burnt mound was also created on a low island, perhaps related to feasting and alongside possibly associated pits and post-holes. Then, as clearance continued, the landscape underwent alluviation and became largely abandoned. However, a grave pit was dug and a ceremonial ring ditch constructed during the Early Bronze Age. Several centuries later, further burnt mound activity took place alongside a stream in the east, in a now fully open landscape (Beamish 2009: 17, 20, 137).

The site was arranged into spatial groups associating features, finds and environmental evidence, in the hope that increasing the size of the groups would increase the chance of identifying similarity or dissimilarity between assemblages. This appears to have been done based upon an intuitive assessment of spatial proximity. However, the radiocarbon dating painted the site as prehistoric palimpsest, meaning that the broad groups contained material of many different ages and that deposits thus had to be understood on an individual basis (Beamish 2009: 20). Analysis of the radiocarbon data through the TGIS reintroduced the possibility of extracting this spatial patterning hidden within the palimpsest.

The radiocarbon evidence

A total of 65 radiocarbon determinations were undertaken on samples of charcoal, waterlogged macrofossils, wood, and charred residues on pottery sherd interiors from across the site. Four different laboratories[1] were used and all but five dates were obtained using the AMS methodology (see Chapter 3) - those five being obtained using the gas proportional methodology (designated with the GrN prefix). Quality assurance processes for all of the laboratories indicated no laboratory offsets to the results. Calibration was undertaken in OxCal v.3.10 using the Reimer *et al.* 2004 curve. The dating program had six objectives:

1. To provide a chronological framework for interpretation of the environment sequence in the palaeochannel deposits;
2. To date and ascertain the significance of human activity in the vicinity of fallen trees;
3. To date the fire-clearance of the floodplain;
4. To provide overall estimates of the start, end, and duration of the use of the burnt mounds;
5. To date the alluviation;
6. To provide precise dates for the Peterborough Ware (and its sub-styles) ceramic assemblage.

In an attempt to discover the chronology of use for the burnt mounds and the start of use of the pottery types in the ceramic sequence, Bayesian analysis (see Chapter 5) of the dates was undertaken using the recorded stratigraphy and ceramic typologies (Marshall *et al.* 2009a: 62-63; 2009b: 1-2).

The first stage in the dating process was to identify short-lived, demonstrably non-residual material. The difficulty with this process is that determination of the manner in which a sample came to arrive in its context of discovery is an interpretative decision, rather than a matter of certainty. The main material judged to meet this criteria was charcoal from short-lived species found in apparently freshly-deposited contexts. Charcoal from post-holes, charcoal adhering to ceramics, and bulked waterlogged plant remains from the palaeochannels were also used

[1] Scottish Universities Environmental Research Centre (16 samples - designated SUERC); Oxford Radiocarbon Accelerator Unit (27 samples - designated OxA); Centre for Isotope Research at the University of Groningen (17 samples - designated GrA/GrN); Poznań (5 samples - designated Poz).

(Marshall *et al.* 2009b: 2). I shall now outline each phase of the site identified by the excavators and discuss the dating evidence obtained and the interpretative conclusions reached.

Fallen trees and associated features:

The majority of Early to Middle Neolithic activity took place in uncleared woodland. Under the specific conditions present on the site, every tree could be expected to have left some form of soil deformation at the end of its life: in such an environment, tree-throws form the major contributor to the formation of subsurface deposits. The excavators encountered some difficulty in recognising and fully understanding these deposits, causing problems with their differentiation, insofar as it is critical that excavators understand the mechanics of tree-throw formation in order to answer the specific questions that could be asked of this evidence. The possible derivations of tree-throw contents are complex and need to be fully understood to allow an understanding of chronological sequence. The size and form of the deposit depends upon the nature of the fall event and the soil conditions, specifically the amount of material pulled out of the ground and the amount that remains *in situ*. An ancient giant of the forest will produce a noticeable pit, dragging soil horizons up to the surface which then weather back into the void. Such a tree will also displace soil where its trunk falls, creating the appearance of a cut feature on excavation, perhaps surrounded by an up-cast ridge of earth. The leading edge of the root ball will be dragged down into the ground, carrying and rotating its associated soil horizons, and forming a deposit often undercutting the up-cast ridge. As such, the down-drag deposits may contain artefacts that pre-date or are contemporary to the tree, whereas the root void may collect artefacts as it naturally fills after the death of the tree (Beamish 2009: 20, 138-140).

Radiocarbon samples were taken from eleven contexts associated with root-void silts see (Fig 6.1a-t). Group 802 contained four dated samples, taken from a burnt deposit over a possible root-void (OxA-15116) and from charcoal fragments from the fill of a small pit (OxA-15127, OxA-15128, SUERC-7607). The dates were not statistically consistent and the Group was thus interpreted as containing material of different ages. OxA-15127 and OxA-15128 were consistent and were thus combined in the Bayesian model (labelled [291]<25>A). Group 803 contained two dated samples (OxA-15899, GrA-31799), taken from the fill of a small pit-type feature to the west of a probable root-void. The dates were statistically consistent and thus could be of the same age. Group 809 contained one dated sample (OxA-15047), taken from residue adhering to a sherd of Peterborough Ware from a probable root-void (Marshall *et al.* 2009b: 2-3).

Group 2503 contained ten dated samples: GrA-31770 was taken from a fragment of hazel charcoal from a lens of charcoal-rich clay within the fill of a pit to the north-east of a probable root void. GrA-31786 came from a fragment of blackthorn from the upper fill of a plausible post pit. GrA-31785, OxA-15110, and SUERC-7597 came from the fill of a pit classified as a root-void silting - the first two dates came from the same sample and were statistically consistent, but the third was not and thus the context was deemed to contain material of different ages. OxA-15045, OxA-15109, SUERC-7596, GrA-31803, and OxA-15900 were taken from a lens of charcoal within a pit fill. OxA-15045 and OxA-15109 were consistent and were thus combined in the Bayesian model (labelled [1328]<64>). However, the five measurements as a whole were not statistically consistent (Marshall *et al.* 2009b: 3-4).

Group 2508 contained two samples (OxA-15084, SUERC-8156), taken from the charcoal-rich fill of a pit sealed by deposits containing Peterborough Ware and lithics. The dates are statistically consistent. Group 2509 contained one sample (OxA-14484), taken from residue adhering to a Peterborough Ware sherd from a shallow pit or post-hole cutting. Group 2504 contained one sample (GrA-31801), taken from a fragment of charcoal from a substantial spread with concentrations of charcoal and Peterborough Ware, sealing a small pit of fire-cracked stones. Finally, Group 2541 contained one sample (OxA-14483), taken from residue adhering to Peterborough Ware from the fill of a wide gully or pit. Interpretation of these dates suggested that this evidence amounted to a clear phase of occupation in the vicinity of fallen trees lasting from c.3500-3000 cal BC, and that any material of later date was probably intrusive from clearance activity (Marshall *et al.* 2009b: 3-4).

Based on the dating evidence outlined above, the excavator concluded that Neolithic occupation became more substantial around 3500-3000 cal BC, and was associated with various forms of ceramic tradition (Early Neolithic Bowl, Mildenhall, Ebsfleet, Peterborough Ware) and a small assemblage of lithics (pointing to task-specific occupation, such as perhaps butchery, plant harvesting, or hide scraping). Activity took place within a probably uncleared woodland landscape with areas of grassy vegetation: or, at the very least, any woodland cleared during the Neolithic had regenerated prior to the later fire-clearance of the site. Life within this arboreal landscape would have resulted in a spatial relationship between trees and settlement: however, the continuing life of the forest resulted in problems for the archaeologists in respect of residuality and intrusive artefacts. The finds associated with fallen trees were in much lower quantities than on other similar sites, suggesting that deposition may have been deliberate. Beyond this, structural remains of this period were scant and difficult to interpret with confidence, other than some short arcs of postholes; features of very different date being juxtaposed horizontally and vertically. As such, although evidence of Neolithic occupation is substantial, there is no compelling evidence for permanent roofed buildings. The riverine woodland habitat would have provided a rich source of wild resources, and the demonstrable diet consisted of gathered plant resources alongside dairy foods, and ruminant and pig meat (Beamish 2009: 140-144).

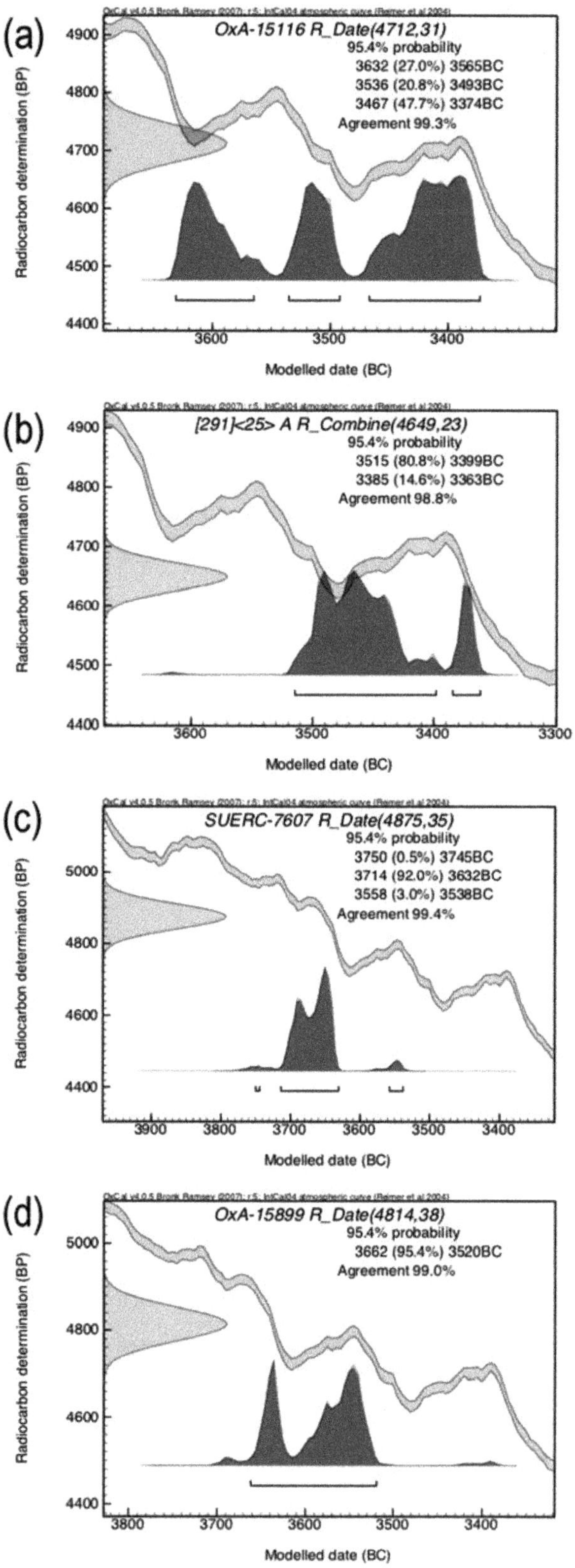

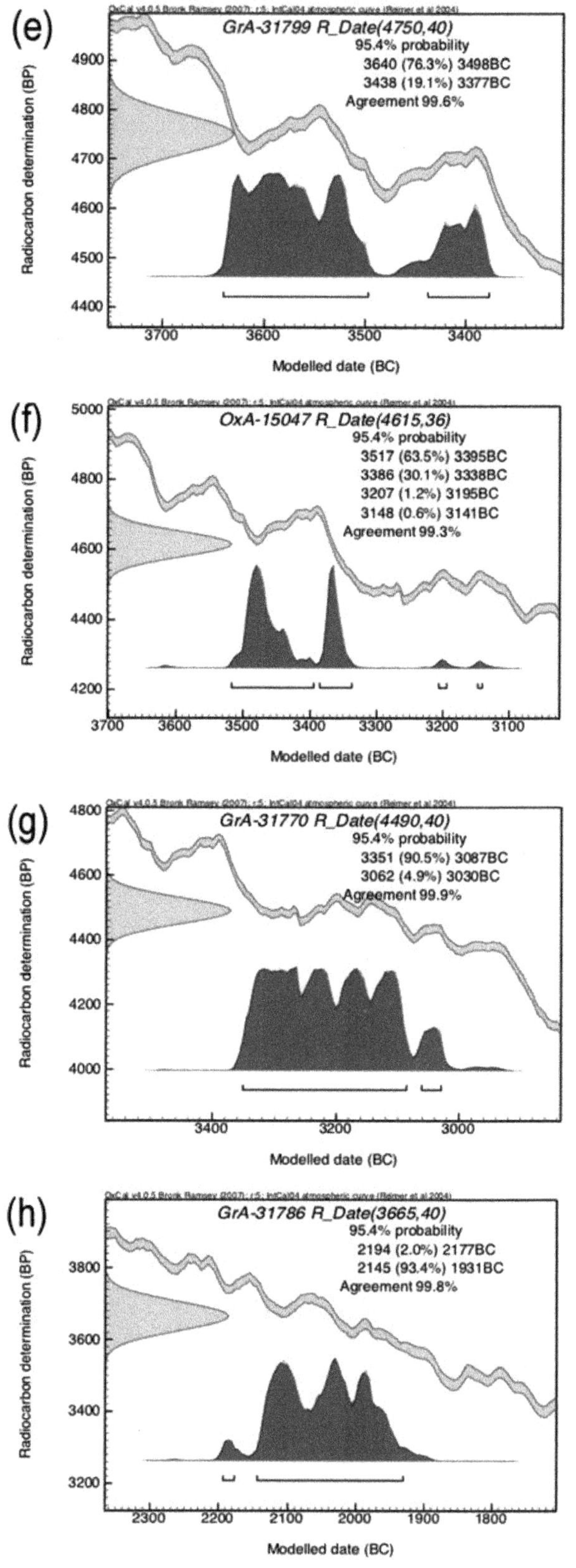

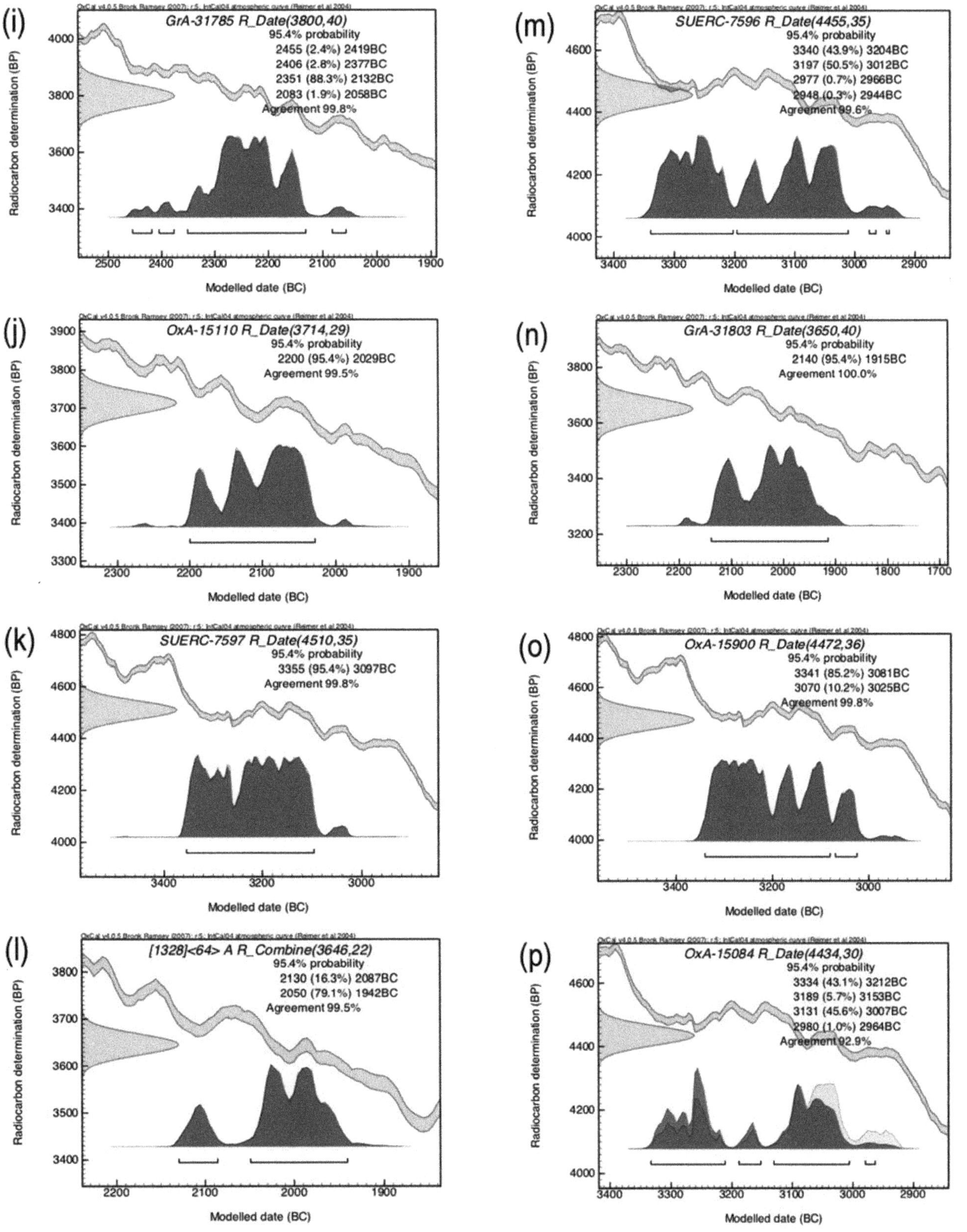
(i)
GrA-31785 R_Date(3800,40)
95.4% probability
2455 (2.4%) 2419BC
2406 (2.8%) 2377BC
2351 (88.3%) 2132BC
2083 (1.9%) 2058BC
Agreement 99.8%
(m)
SUERC-7596 R_Date(4455,35)
95.4% probability
3340 (43.9%) 3204BC
3197 (50.5%) 3012BC
2977 (0.7%) 2966BC
2948 (0.3%) 2944BC
Agreement 99.6%
(j)
OxA-15110 R_Date(3714,29)
95.4% probability
2200 (95.4%) 2029BC
Agreement 99.5%
(n)
GrA-31803 R_Date(3650,40)
95.4% probability
2140 (95.4%) 1915BC
Agreement 100.0%
(k)
SUERC-7597 R_Date(4510,35)
95.4% probability
3355 (95.4%) 3097BC
Agreement 99.8%
(o)
OxA-15900 R_Date(4472,36)
95.4% probability
3341 (85.2%) 3081BC
3070 (10.2%) 3025BC
Agreement 99.8%
(l)
[1328]<64> A R_Combine(3646,22)
95.4% probability
2130 (16.3%) 2087BC
2050 (79.1%) 1942BC
Agreement 99.5%
(p)
OxA-15084 R_Date(4434,30)
95.4% probability
3334 (43.1%) 3212BC
3189 (5.7%) 3153BC
3131 (45.6%) 3007BC
2980 (1.0%) 2964BC
Agreement 92.9%
Radiocarbon determination (BP)
Modelled date (BC)

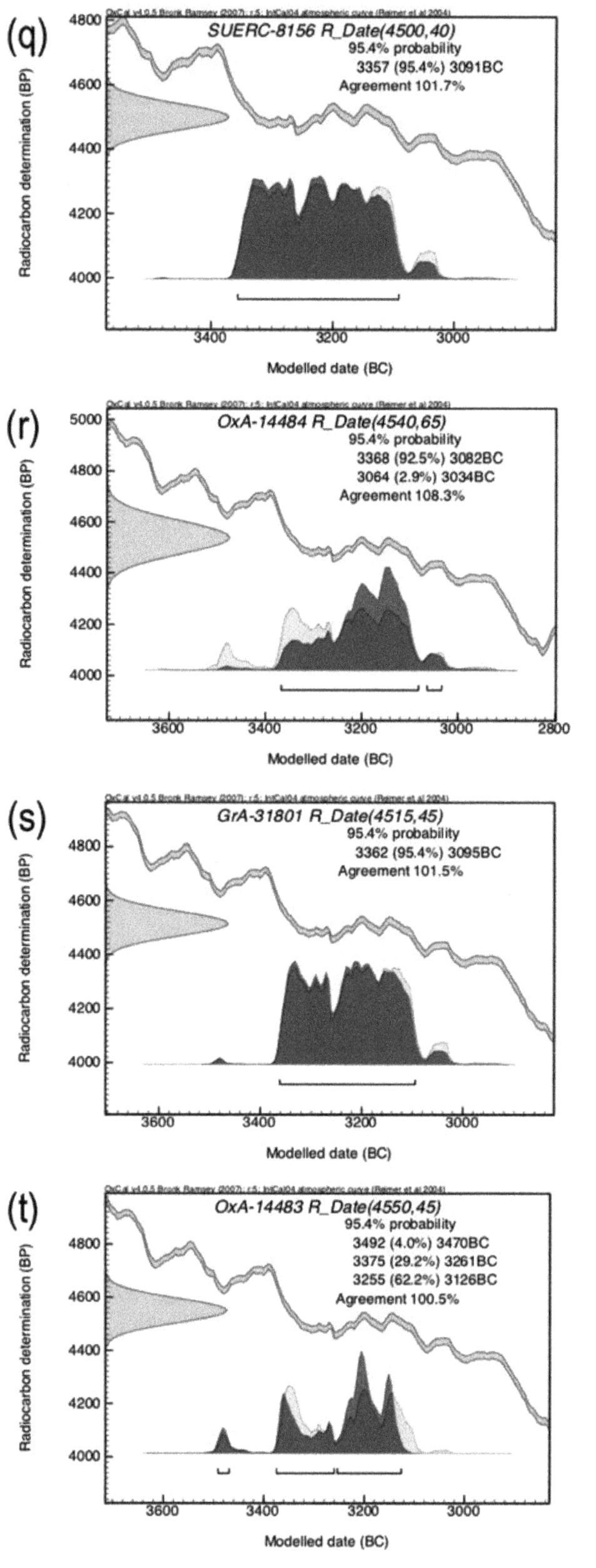

FIG 6.1 (A) TO (T) - OXCAL OUTPUT FOR THE MODELLED DATES REFERRED TO IN THE TEXT ABOVE. THE CURVE ON THE LEFT REPRESENTED THE UNCALIBRATED DATE; THE SINUOUS LINE THE CALIBRATION CURVE; THE LIGHT GREY CURVE AT THE BOTTOM REPRESENTS THE CALIBRATED DATE; AND THE DARK GREY CURVE REPRESENTS THE BAYESIAN MODELLED DATE.

Fire-clearance of the floodplain:

There was no evidence of occupation from the end of the Peterborough Ware phases until the establishment of Burnt Mound I, perhaps 800 years later. However, fire clearance of the woodland began in the Late Neolithic, indicating that the site retained an economic importance even if it was no longer considered favourable for occupation: contemporary occupation may have been present 1km to the north east (Beamish 2009: 145). This clearance represents an important change in the local landscape, as a necessary precursor to increased grazing and / or cultivation. Eight contexts directly relating to this clearance were dated (Fig 6.2a-l). Two samples (OxA-15898, GrA-31797) were taken from a charcoal-rich deposit that was immediately related to pockets of oxidised fire-reddened clay (which appeared to have been caused by fire-clearance). The charcoal had apparently been incorporated into the tree pit during felling, but the dates were not statistically consistent and were thus of different ages. GrA-31789 was taken from a charcoal-rich deposit again immediately related to pockets of fire-reddened clay and again apparently incorporated into the tree pit during felling. Two samples (OxA-15083, SUERC-7594) were taken from a deposit of fire-reddened charcoal and scorched clay, filling a probable tree pit. The two measurements were statistically consistent and could thus be of the same age (Marshall *et al.* 2009b: 4).

GrA-31800 was taken from a fragment of blackthorn from the base of a small burnt-out tree or shrub whilst GrA-31787 derived from an area of charcoal-rich soil with scorched red clay pockets. Two samples (OxA-15082, SUERC-7593) were taken from a reduced deposit of charcoal-rich clay, overlain by extensive deposits of reddened clay (the result of intense burning). These two samples were also statistically consistent and thus possibly of the same age. Two samples (OxA-15081, SUERC-7592) were taken from a deposit of charcoal-rich clay overlain by reddened, possibly scorched clay. Again, the two samples were statistically consistent and could be of the same age. Finally, GrA-31796 was taken from a single fragment of hazel charcoal from an area of scorched red clay with charcoal pockets. The overall interpretation of this suite of dates was that fire-clearance of the floodplain was concentrated in the mid-third to mid-second millennia cal BC (Marshall *et al.* 2009b: 4-5).

The excavator concluded that evidence for *in situ* burning of trees indicated that fire clearance began c.2500 cal BC and continued until c.1750 cal BC. He concluded that this was clearly not a one-off event, but took place over 500 to perhaps 1,000 years: clearance may have been sporadic and interleaved with regeneration (varying according to animal stock levels), or may have taken the form of ongoing slash and burn agriculture throughout this period. It is possible that the repeated burnings may have been of pastoral land in an attempt to promote fresh growth, but the presence of evidence for this within the profile of burnt out fallen trees makes this interpretation unlikely (i.e. trees were being burnt, not just scrub). It is also possible that the repeated burning might have resulted in the destruction of some material categories, such as charcoal and pottery sherds (Beamish 2009: 145-147).

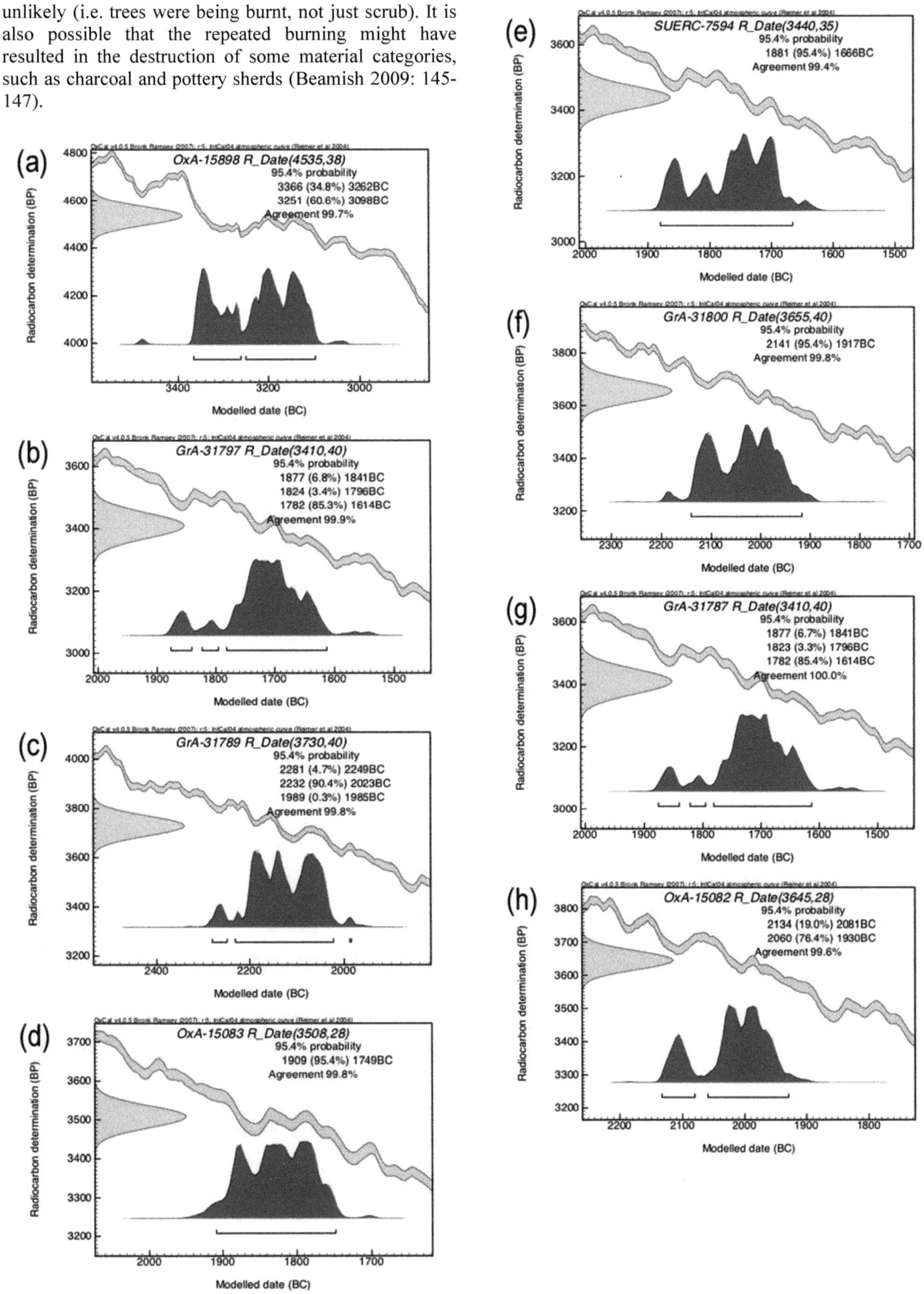

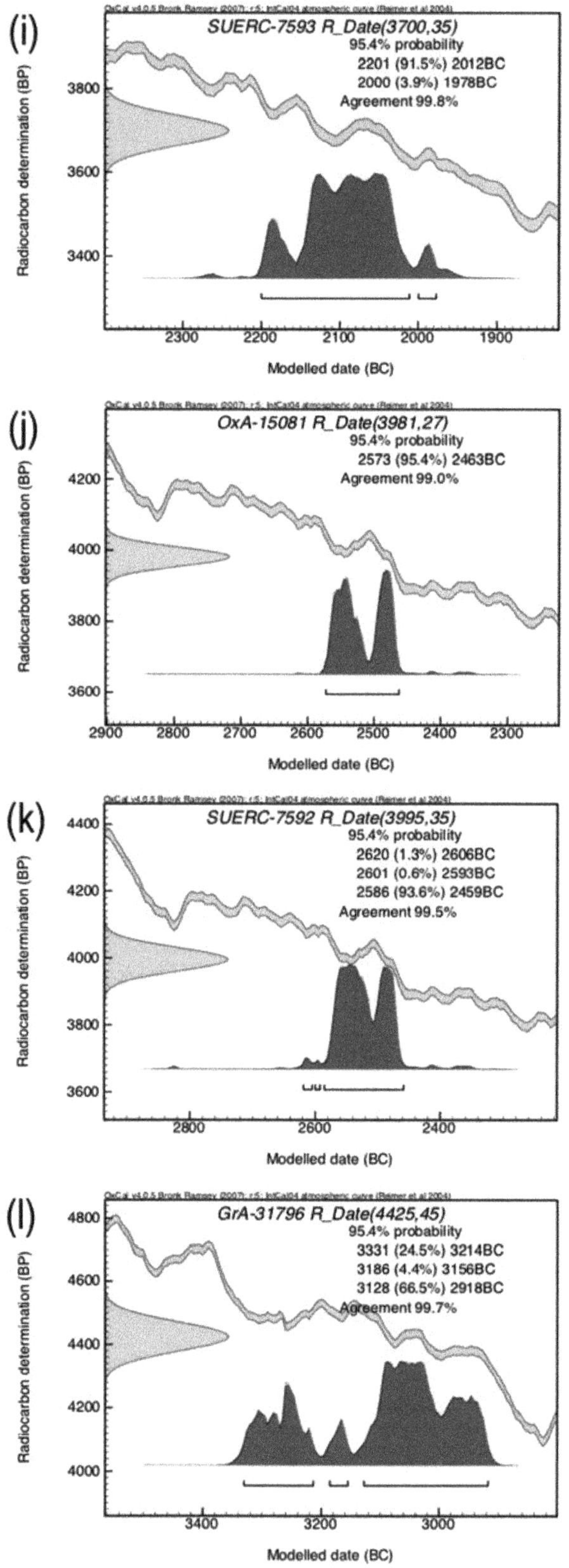

FIG 6.2 (A) TO (L) - OXCAL OUTPUT FOR THE MODELLED DATES REFERRED TO IN THE TEXT ABOVE.

Burnt Mound I and its context:

Burnt Mound I consisted of a layer of fire-cracked stone and charcoal with a series of central features, overlaying a spread of grey gravelly-clay which in turn sealed an earlier pit which cut remnants of more gravelly-clay (interpreted as a buried soil). Duplicated radiocarbon samples were taken from six contexts that were interpreted as part of the mound, or which came from stratigraphically-related layers (Fig 6.3a-m). Group 2550-01 contained one sample (OxA-14481), taken from residue adhering to a Neolithic sherd from a grey gravelly-clay that was sealed below all later layers and interpreted as a buried soil. Group 2550-02 contained two samples (OxA-15046, SUERC-7605), taken from a spread identified as representing earlier Neolithic activity pre-dating the mound. The two dates were statistically consistent (Marshall *et al.* 2009b: 5).

Group 2550-03 contained ten dated samples. OxA-15115 and SUERC-7606 were taken from a charcoal-rich deposit that post-dated the earliest spread. They were not statistically consistent and thus represented material of different age. OxA-15114 and SUERC-7604 were taken from the charcoal-rich fill of the central pit of the mound. They were statistically consistent and thus possibly of the same age. OxA-15113 and SUERC-7909 were taken from a charcoal and fire-cracked stone rich fill of an adjacent pit (interpreted as an oven or hearth). They were again statistically consistent and thus possibly of the same age. OxA-15112 and SUERC-7602 were taken from the primary fill of a substantial pit or tank adjacent to said hearth / oven and derived from burnt mound activity. Again, they were statistically consistent and possibly of the same age. Finally, OxA-15111 and SUERC-7598 came from a charcoal-rich layer derived from spent fuel and stone cleaned out of the central features. They were not statistically consistent and, as such, represented material of different ages (Marshall *et al.* 2009b: 5).

OxA-15046, SUERC-7605, and SUERC-7606 demonstrated that the base of the mound incorporated residual earlier Neolithic material. The three samples were statistically consistent and could be contemporary. The dating experts excluded the earlier Neolithic material due to its clearly residual nature. They concluded that their Bayesian model for Burnt Mound I showed good agreement between dates and sequence. It estimated that activity began between 2340-2060 cal BC and ended between 2120-1840 cal BC (both 95% confidence), during the Late Neolithic / Early Bronze Age (Marshall *et al.* 2009b: 5-6). Interpreting this evidence, the excavator noted that the mound overlay an existing Middle Neolithic spread, which itself contained Early Neolithic material: this suggests that this area had long remained a favoured location for activity. He favoured an interpretation of the mound as a cooking site (due to the presence of nutshell, sloe stones, a tooth, and a possible fragment of oven plate), perhaps taking place during late summer or early autumn (according to the charred plant remains). Yet he also noted that it was not possible to be certain in this conclusion, with residual material causing difficulties. Metalling in front of the mound suggested repeated, formal usage and the re-cut tank in front of the mound suggested some phasing to usage. The excavator concluded that he would favour an interpretation of the burnt mound as a location of the formal consumption of boiled and roasted food (Beamish 2009: 149-152, 157).

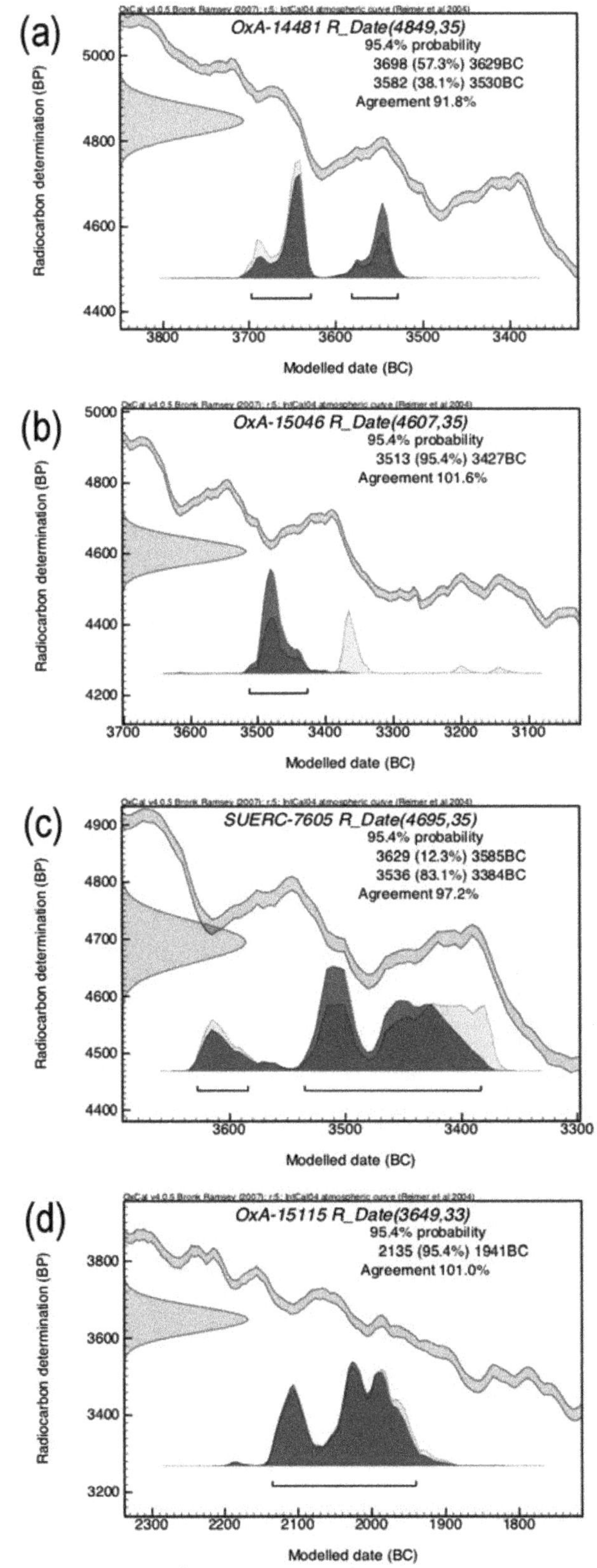

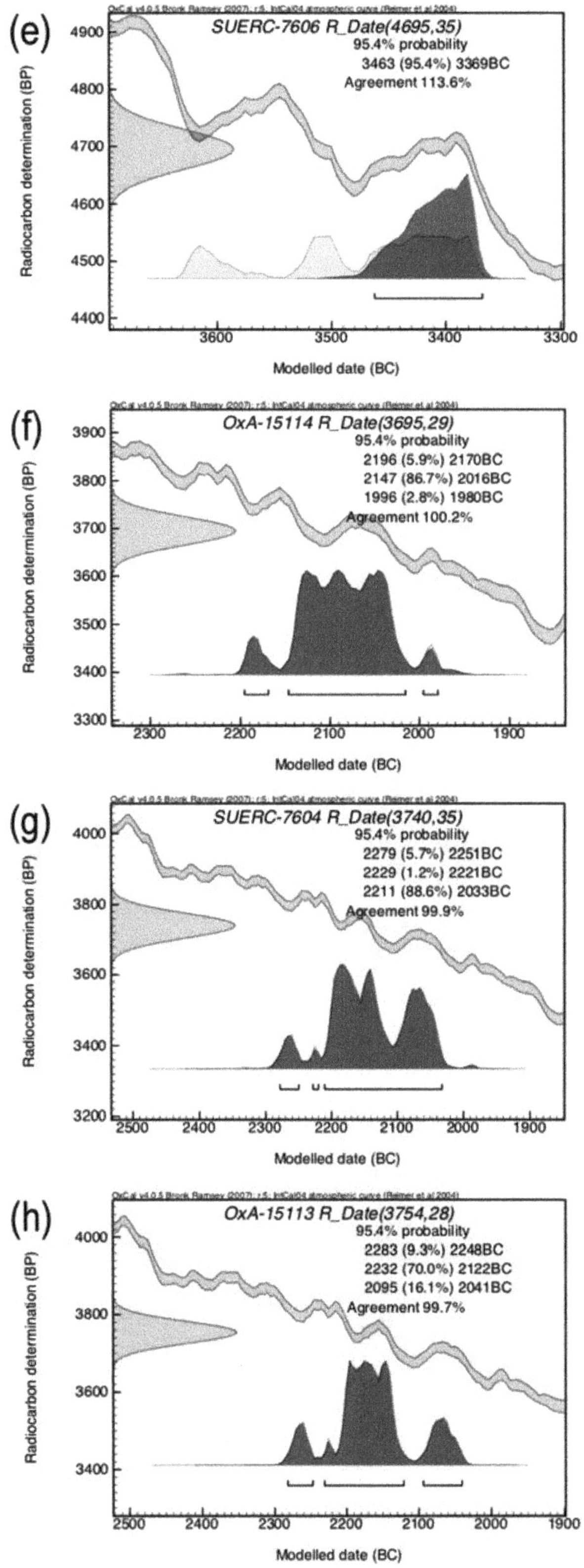

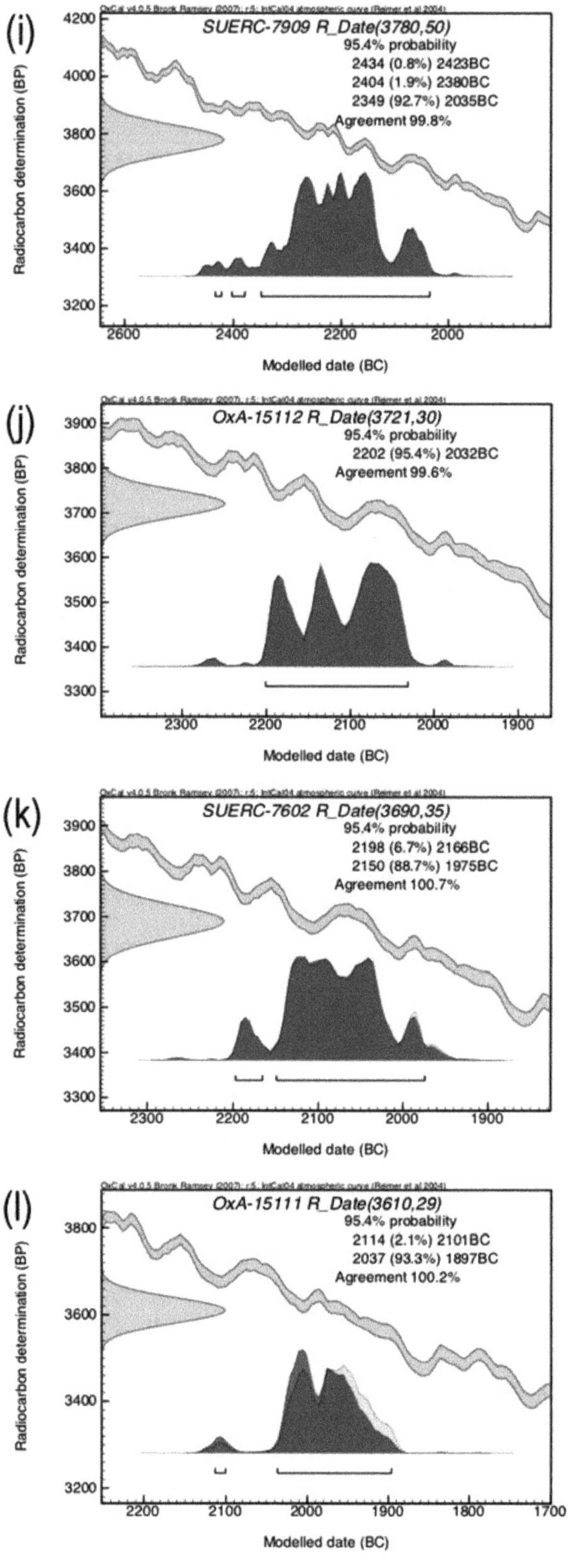

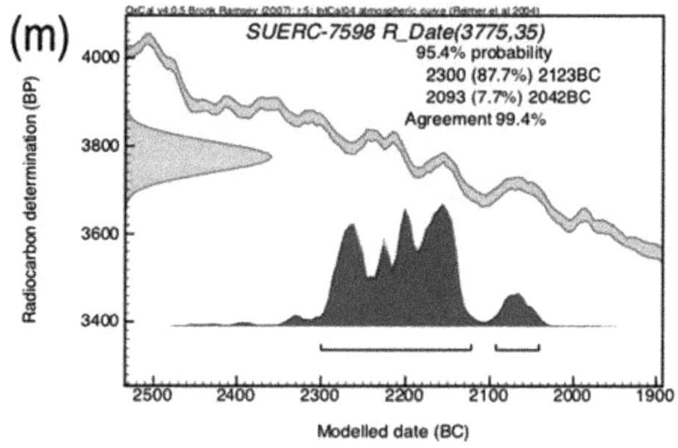

FIG 6.3 (A) TO (M) - OXCAL OUTPUT FOR THE MODELLED DATES REFERRED TO IN THE TEXT ABOVE.

Burnt Mound II:

Ten samples were dated from the second burnt mound discovered on the site (Burnt Mound II), and entered into a separate Bayesian model from the rest of the site (Fig 6.4a-k). Poz18029, Poz18009 and GrN30412 were a sequence of three samples from a silty peat layer beneath the trough. The timbers of the trough itself were dated via two samples (GrN30408, GrN30409), which were statistically consistent and thus could be of the same age. They were combined in the Bayesian model (labelled Comb_Trough). Poz18007 came from the bottom and Poz18006 from the top of the silty peat infill of the trough. Poz18010 came from a lens of leaf litter near the base of the adjacent palaeochannel, but had no direct stratigraphic relationship to the trough. GrN30410 and GrN30411 were taken from charcoal from a hearth-type feature that lay adjacent to the trough, and were statistically consistent and thus possibly of the same age (Marshall *et al.* 2009b: 6).

The dating experts concluded that their Bayesian model for Burnt Mound II showed a good correlation between dates and sequence. The trough construction was estimated to date to between 1290-1100 cal BC. The end of its use was estimated to date to 1170-1000 cal BC (both 95% confidence). The use-life of the trough was thus estimated to be between 20 and 210 years. The experts concluded that both burnt mounds were in use for a relatively short period of time, but that the gap between their use was between 640 and 960 years (Marshall *et al.* 2009b: 6), placing Burnt Mound II within the Middle Bronze Age. Again, the function of the mound could not be unequivocally diagnosed, but charred remains were present similar to those found in the first mound (hazelnut shell, sloe and fruit stones). The charred remains this time suggested an early to mid summer usage period, and the waterlogging of plant remains suggested that the mound was situated in a cleared pastoral landscape, specifically with damp grassland grading away into swamp. The cleared landscape situation, however, was at odds with the requirement for large amounts of coppiced woods to fuel the mound activity: it is possible that the situation was chosen to provide access to fuel supplies such as stands of alder along the marshy margin. Burnt Mound II showed clear signs of reuse over a period of time (Beamish 2009: 152-155, 158). Although previously thought to be a Late Bronze Age phenomenon, the two burnt mounds at Willington fit well into the recently recalculated schema for burnt mounds that has provided a broadly even temporal distribution between 2500-750 cal BC (Beamish 2009: 158).

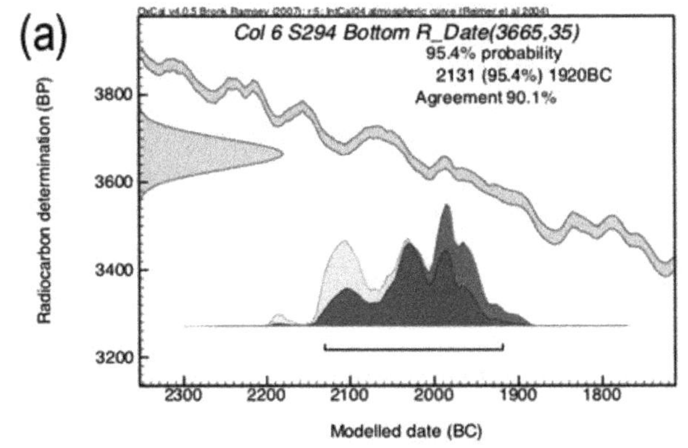

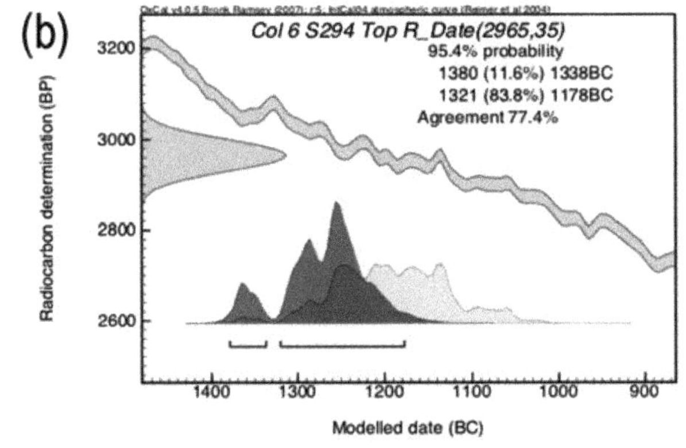

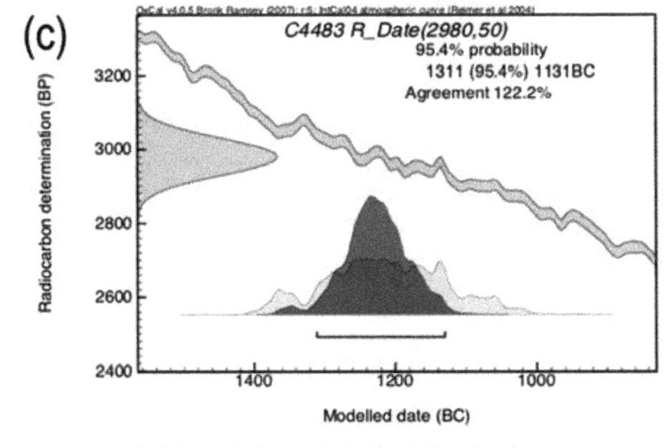

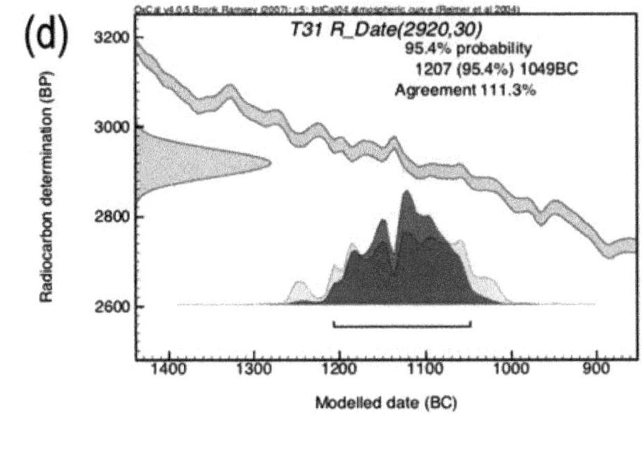

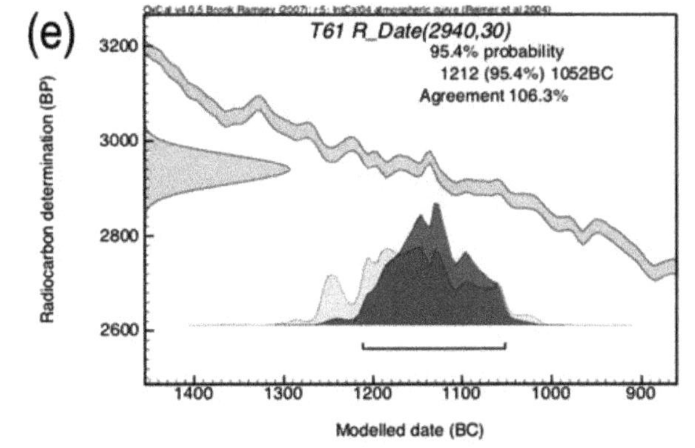

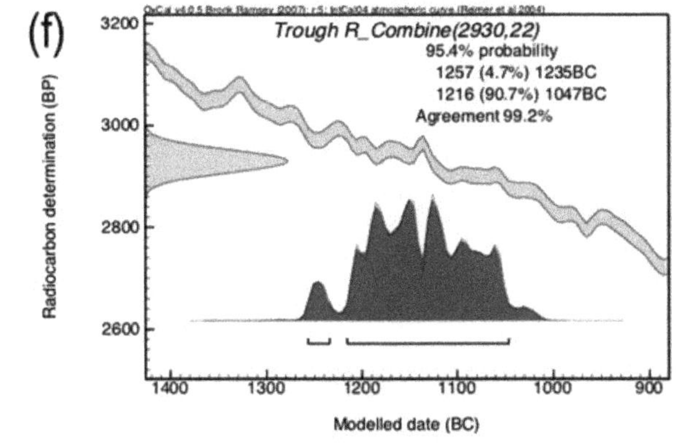

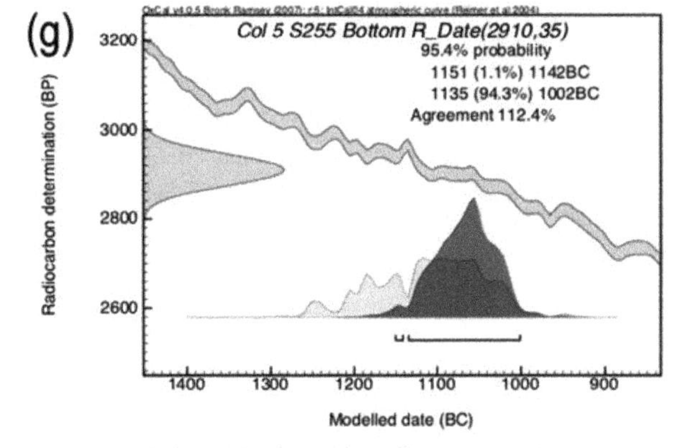

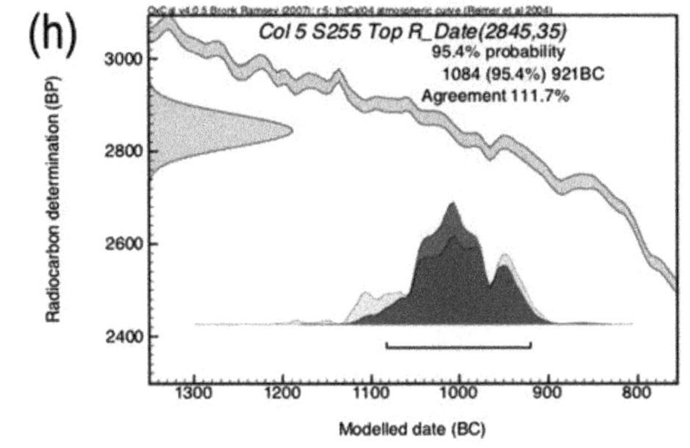

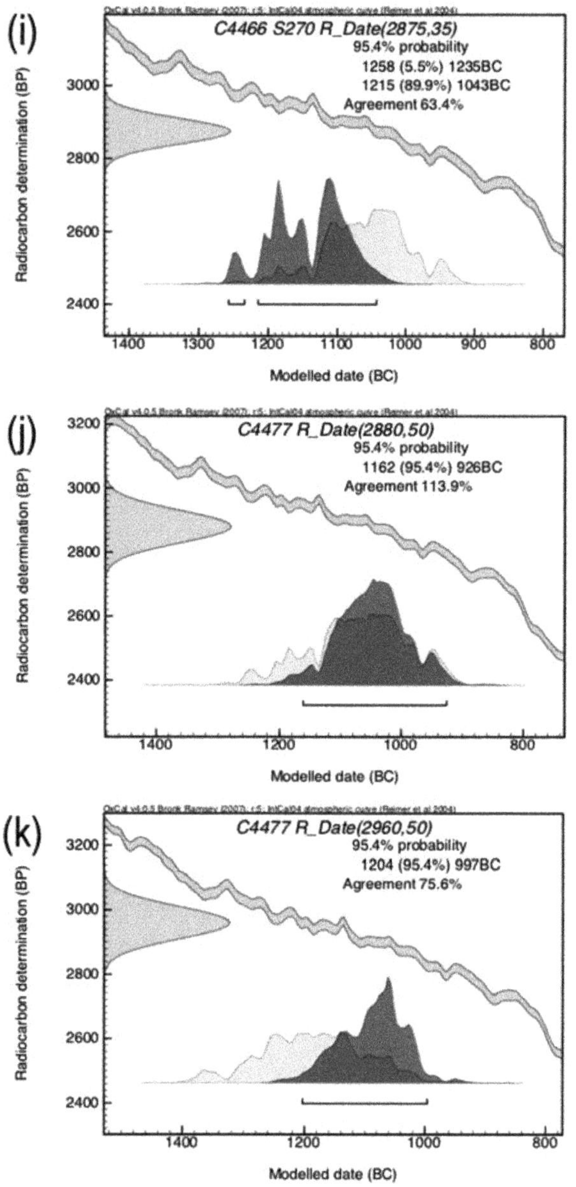

FIG 6.4 (A) TO (K) - OXCAL OUTPUT FOR THE MODELLED DATES REFERRED TO IN THE TEXT ABOVE. KEY: POZ18029 AND POZ18009 - COL 6 S294 BOTTOM AND TOP; GRN30412 - C4883; GRN30408 - T31; GRN30409 - T61; COMB_TROUGH - TROUGH; POZ18007 AND POZ18006 - COL 5 S255 BOTTOM AND TOP; POZ18010 - C4466 S270; GRN30410 - C4477 (2880, 50); GRN30411 - C4477 (2960, 50).

Alluviation:

Group 2530 - the base of a partly stone-lined feature that was interpreted as an oven - was dated to provide a *terminus post quem* for the alluviation event that buried the whole site in a silty clay (Fig 6.5)a-b. The charcoal dated from within the oven either represents its last usage or material deposited there after that usage. Following the collapse of the feature, part of the roof was washed down slope, demonstrating that the flooding that resulted in the build-up of alluvium had already begun before the feature was integrated into the surrounding soils. The two dates taken (OxA-15044, SUERC-7595) were, however, not statistically consistent and thus represent material of different ages. OxA-15044, the later, therefore provides a *terminus post quem* for the onset of alluvial conditions of 3490-3100 cal BC. The dating experts attempted to provide a more precise estimate by modelling the majority of the measurements on the site (excluding some that were residual or with an unclear relationship to the alluvium) on the basis of whether they lay above or below the alluvium, but the detailed recording of natural layers cut by features had not been undertaken during excavation as its significance had not been realised and, furthermore, the alluviation process was not a simple time-transgressive event. As such, the Bayesian model was unable to improve the estimate for the *terminus post quem*, but was able to provide an estimated *terminus ante quem* of 2200-1980 cal BC (95% confidence) (Marshall *et al.* 2009b: 6-7).

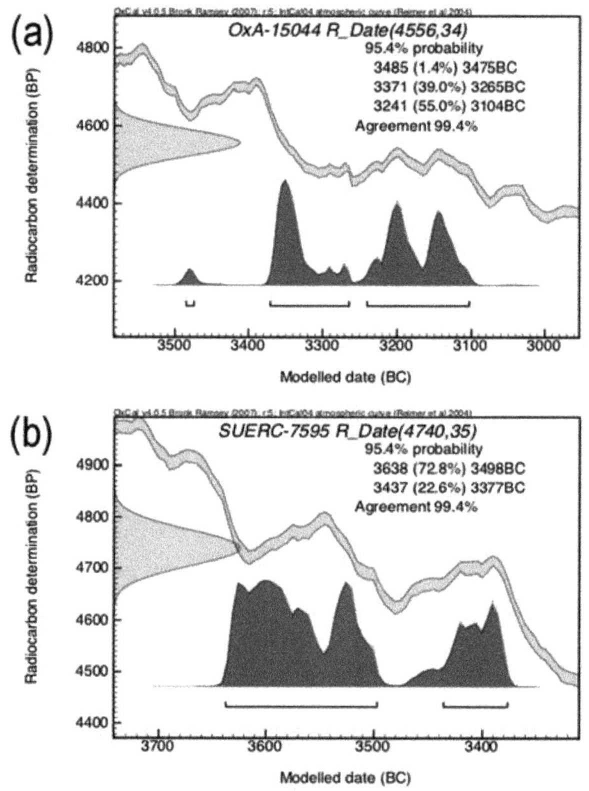

FIG 6.5 (A) TO (B) - OXCAL OUTPUT FOR THE MODELLED DATES REFERRED TO IN THE TEXT ABOVE.

Environmental column:

Four samples were dated from the column excavated for environmental sampling, the samples comprising seeds and twig fragments (Fig 6.6a-d). GrA-31468 and OxA-15897 came from within the 0-10cm section. They are not statistically consistent and thus only provide a *terminus post quem* for the column. SUERC-7350 came from 0.48-0.50cm depth and SUERC-7351 came from 0.96-0.98cm depth (Marshall *et al.* 2009b: 2-3).

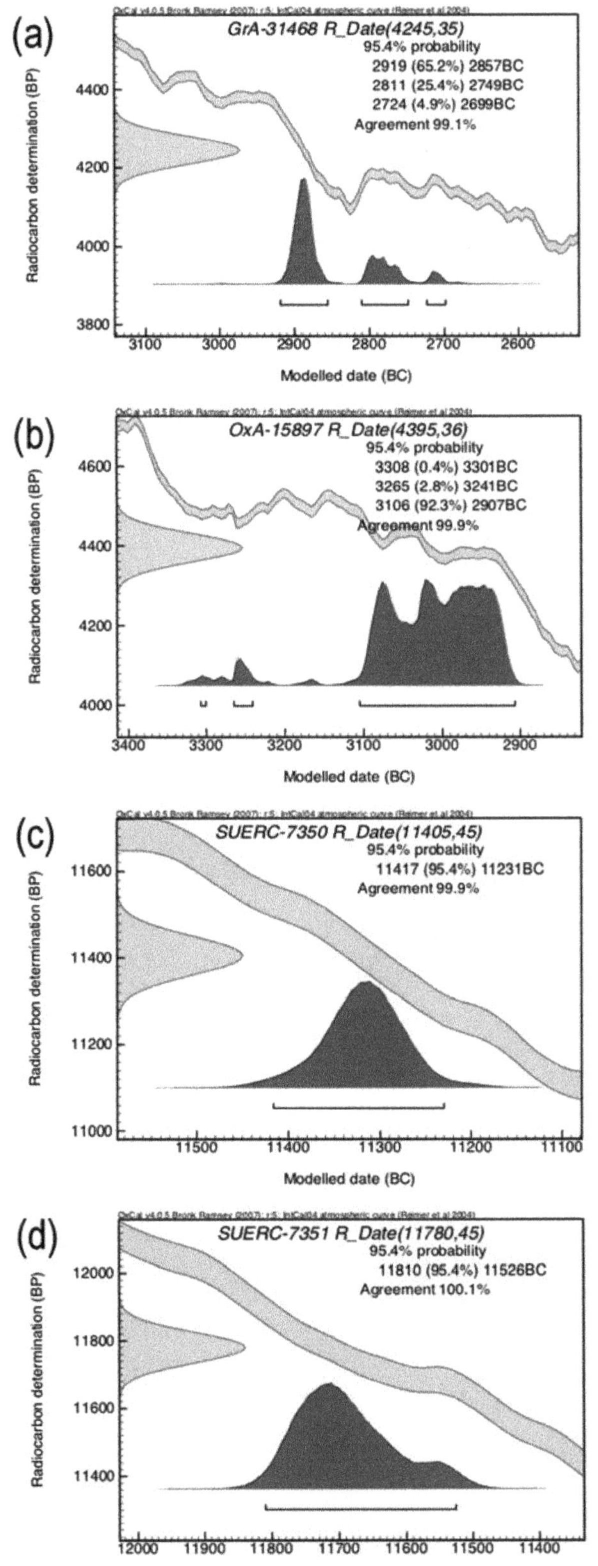

FIG 6.6 (A) TO (D) - OXCAL OUTPUT FOR THE MODELLED DATES REFERRED TO IN THE TEXT ABOVE.

Ceramic sequence:

The final element of the site that was radiocarbon dated was the ceramic sequence (Fig 6.7a-b). Six sherds with organic residues were dated (Neolithic bowl: OxA-14481; Peterborough Ware: OxA-14482 OxA-14483, OxA-14484, OxA-14485, OxA-14487). Three other contexts associated with significant amounts of Peterborough Ware were also dated (Context [458], a pit: GrA-31799, OxA-15899; Context [1499]: OxA-15084, SUERC-8158; Context [1477], a cooking pit: GrA-13801), each of which could only provide a *terminus post quem* for the ceramics' use either due to unclear relationships between ceramics and the dated charcoal or due to their being sealed under the ceramic-bearing context. The dating experts modelled these results alongside available Peterborough Ware dates from England to attempt to provide more precise estimates of the dates for the pottery style. They built three conservative models, the final one of which suggested that Peterborough Ware was used for a shorter period than previously believed (Marshall *et al.* 2009a: 68-81; 2009b: 7-9).

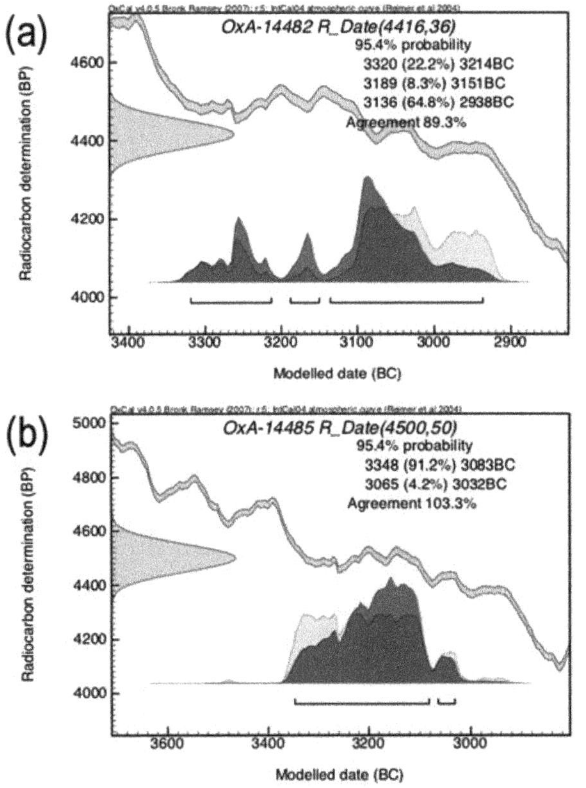

FIG 6.7 (A) TO (B) - OXCAL OUTPUT FOR THE MODELLED DATES REFERRED TO IN THE TEXT ABOVE. DATES NOT INCLUDED WHERE THEY HAVE ALREADY BEEN GIVEN ABOVE.

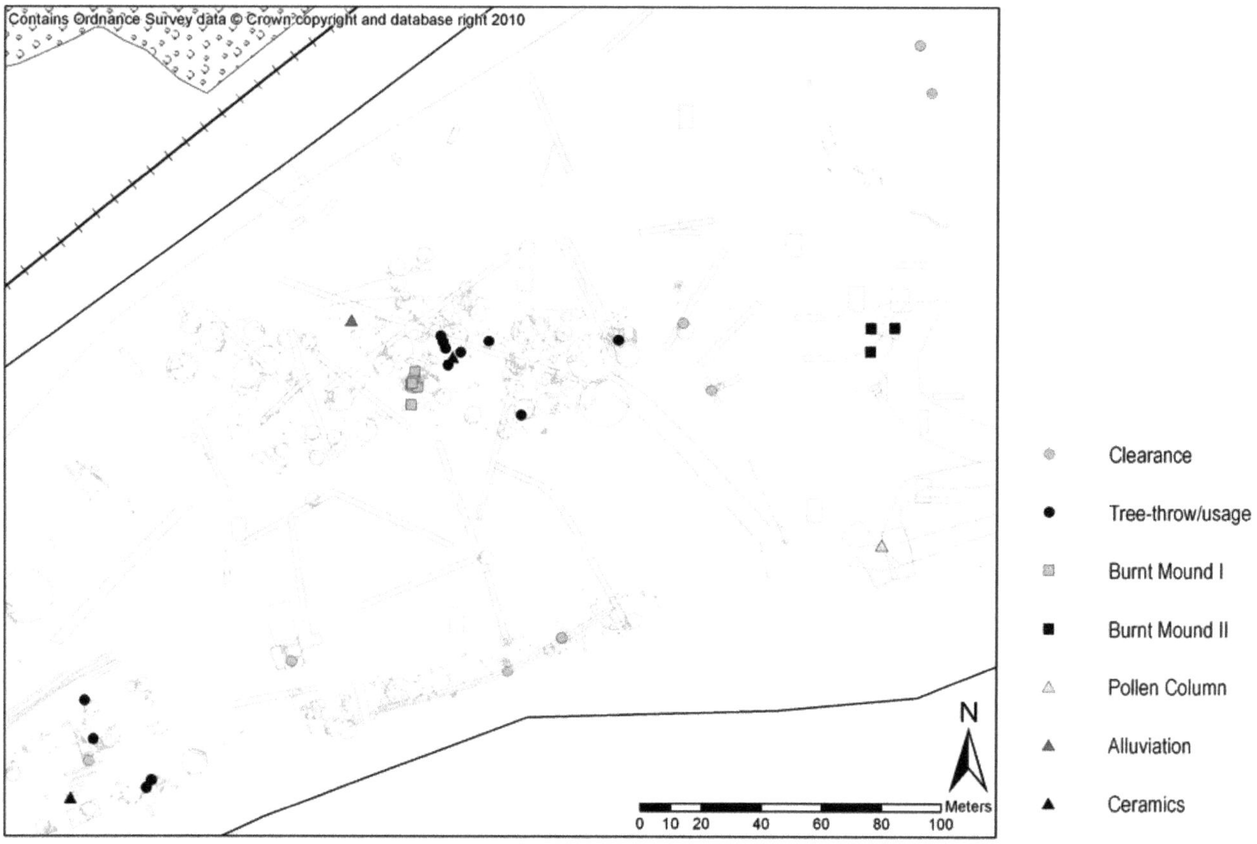

FIG 6.8 - WILLINGTON EXCAVATION PLAN, SHOWING DISTRIBUTION OF RADIOCARBON DATES CATEGORISED ACCORDING TO PRIMARY THEMATIC PHASE.

The TGIS approach

Having considered the dating evidence gathered at Willington and the conclusions reached by the excavators, the next stage was to import the data into the TGIS and attempt to spatialise and explore the key dating questions that had driven the radiocarbon program. The rest of this chapter will concentrate on presenting the results of the TGIS analysis, ending with a critical evaluation of the successes and problems arising from the TGIS method.

The implementation of radiocarbon evidence in the TGIS

The first stage in the implementation was to create a new map in ArcGIS based upon the TGIS template and then import the AutoCAD layers that provided the archaeological and geographic background to the radiocarbon dates. These layers were not subject to TGIS analysis, but provide the context within which the dated material studied was discovered and against which it must be analysed (Fig 6.8). The dates themselves were integrated into the TGIS as a series of layers. First was a layer containing all of the radiocarbon dates, their spatial locations, their calibrated (but un-modelled) minimum and maximum dates (at ±2 standard deviations), and an indication of the thematic phase to which they belonged. These phases were as follows: alluviation, Burnt Mound I, Burnt Mound II, ceramics, (fire) clearance, pollen column, and tree-throw / usage. Where a date belonged to more than one of these phases, the phase most likely to show spatial pattern was chosen. This mostly applied in situations where dates taken from ceramic residues related to both the ceramic and tree-throw / usage phases. In these cases, the tree-throw / usage phase was preferred as the ceramic phase was unlikely to have any meaningful spatial distribution within itself beyond where it provided an element of the spatial pattern within another phase, or overall. Specifically, this was because the ceramic dates were taken in an attempt to improve the dating of their typologies, rather than to consider any specific questions relevant to the site itself.

Another, almost identical, layer was then added to the map, except this time giving the minimum and maximum dates (at ±2 standard deviations) taken from the Bayesian models. A series of further layers were added containing the dates associated with each thematic phase: these extra layers thus included the full set of dates associated with a phase (i.e. all nine ceramic dates, as opposed to the two dates assigned to that grouping in the conglomerate layers) and were included to enable the spatial analysis of a single thematic group. In particular, it was hoped that it would prove possible to extract spatial pattern from the dates relating to the fire clearance - nevertheless, the

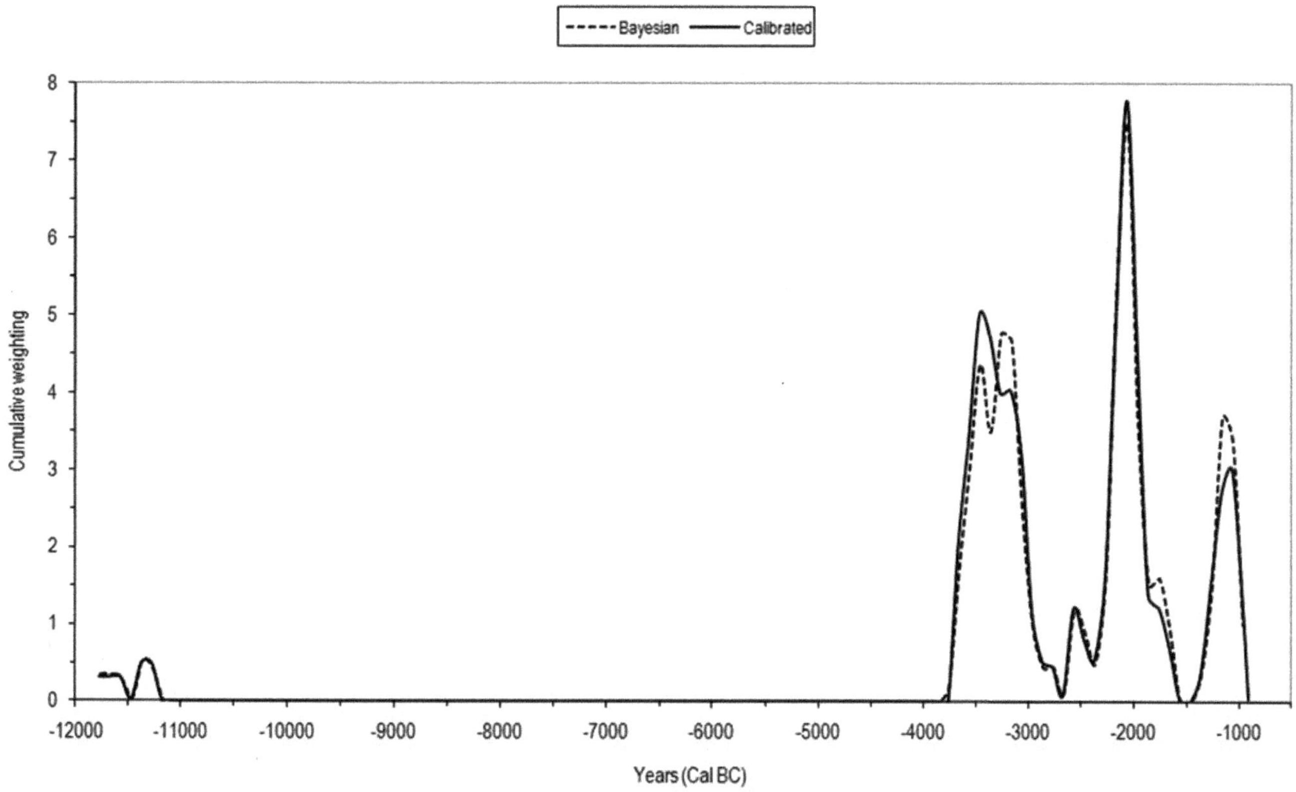

FIG 6.9 - GRAPH CONSTRUCTED FROM OUTPUT OF THE ANALYST TOOL IN THE TGIS COMPARING THE OVERALL PROBABILITY PROFILES FOR THE BAYESIAN MODELLED AND UN-MODELLED CALIBRATED DATE LAYERS.

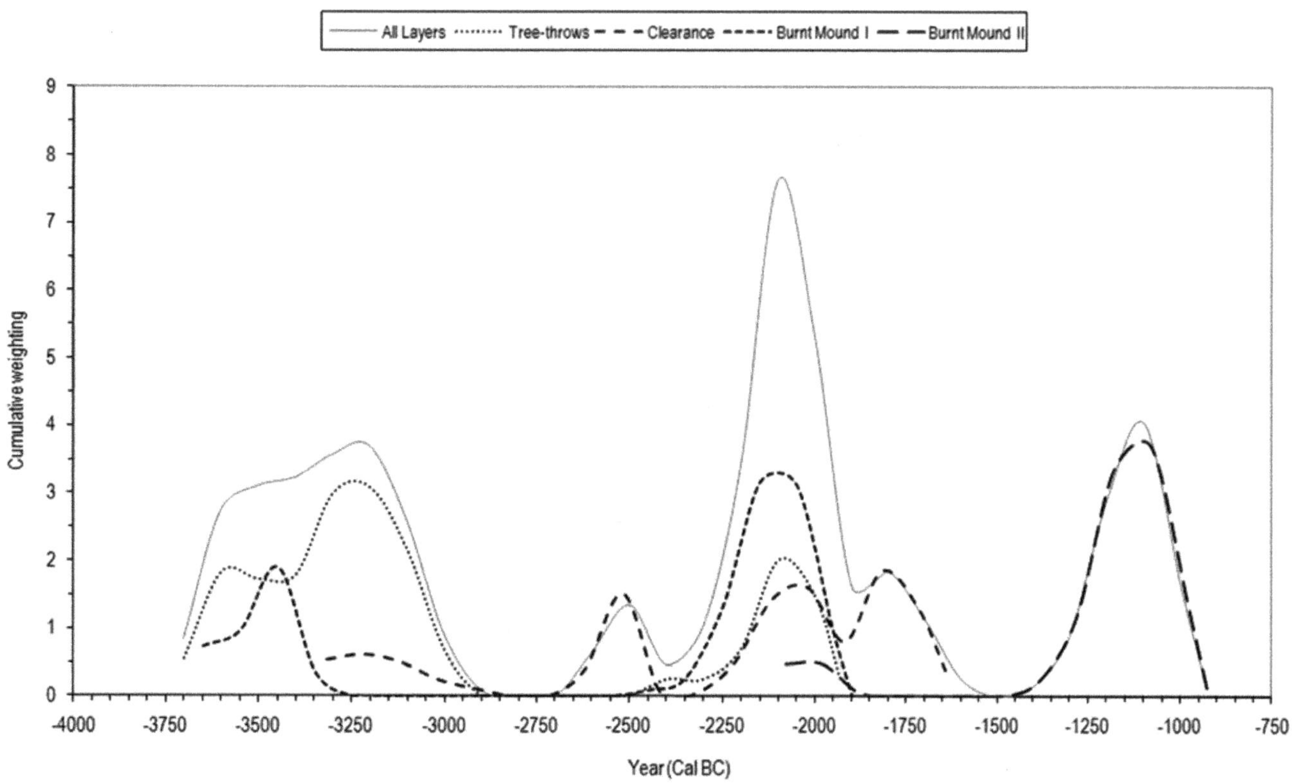

FIG 6.10 - GRAPH CONSTRUCTED FROM OUTPUT OF THE ANALYST TOOL IN THE TGIS COMPARING THE OVERALL PROBABILITY PROFILES FOR THE BAYESIAN MODELLED DATE LAYERS SEPARATED ACCORDING TO THEMATIC PHASE, COMPARED AGAINST THE PROFILE FOR ALL FOUR OF THESE PHASES.

other groups would also be variously subject to such analysis.

The next stage was to extract the detailed OxCal probabilities for each date. As noted above, the dating specialists had used OxCal v.3.10 to calibrate the dates and to construct their Bayesian models. However, OxCal v.3.10 cannot output the detailed probabilities needed by the TGIS. Therefore, the calibrations and the Bayesian models were re-run in the newer OxCal v.4.0. On the basis of a comparison between the ±2 standard deviation minima and maxima produced by the dating specialists and my own analysis, the results of the calibration and modelling were the same for both pieces of software. These detailed probabilities were entered into two large tables (one for the un-modelled output, one for the modelled) and added to the map document. The tables were then related to the relevant layers in the map, using the methodology discussed in Chapter 5. Following these preparations, the Willington radiocarbon data was ready for analysis in the TGIS.

Analysis and results

As discussed in Chapter 5, summing or multiplying the probabilities of radiocarbon is not a valid methodology when used to attempt to model **improved** probabilities for the radiocarbon dates on a site. However, so long as no great interpretative weight is placed on its output, the Analyst tool built into the TGIS can provide useful evidence of this type, as shown in fig 6.9. The graph was produced in Excel using data copied from the aforementioned tool: as such some of detail of the differences between the two curves plotted may be due to the algorithm used by Excel to produce smoothed curves. It illustrates the respective cumulative probability profiles for the un-modelled dates and for the modelled dates. The probabilities used were not the complex OxCal output, but simple percentage probabilities summed in 100 year time-slices. As noted in Chapter 5, the detailed probabilities are not usable in the Analyst tool as detailed consideration of these summed probabilities is not considered a valid exercise.

Two important conclusions can be drawn from looking at this graph. Firstly, there are five obvious temporal groupings in the radiocarbon data: 12000-11000 cal BC; 11000-4000 cal BC; 4000-c.2750 cal BC; c.2750-1500 cal BC; and 1500-c.750 cal BC. This compares somewhat favourably to the site phasing constructed by the excavator (in turn: pollen column, no data, tree-throws / usage, clearance / Burnt Mound I, Burnt Mound II - the alluviation phase is not obvious due to its lack of dating evidence). The second thread to draw out from reading this graph is the various differences in probability between the two sets of data. Specifically, the Bayesian model produces a different probability emphasis for the sequence of the tree-throw / usage phase. The greater peak in the Bayesian model for the Burnt Mound II phase is due to the extra date in that grouping (Comb_trough). Bearing this in mind, it is likely that the different probability emphasis for the tree-throw / usage phase is also partly due to the two combined dates in that grouping. Overall, however, the graphs illustrate the relatively minor differences between the modelled and un-modelled dates on a broad brush basis.

In this section, I will begin by working through and discussing the overall spatial distribution and the radiocarbon probabilities for a series of time-slices through the site as a whole, specifically 11820 to 4000 cal BC (the period prior to major activity on the site), 4000 to 3000 cal BC (the period of activity associated with fallen trees), 3000 to 1500 cal BC (the period of fire clearances and Burnt Mound I), then finally 1500 to 910 cal BC (the period of Burnt Mound II). Then, we shall look in more detail at the specific spatial distributions of the following thematic phases: tree-throws / usage, fire clearance, Burnt Mound I, and Burnt Mound II. The environmental column and alluviation phases will not be considered individually, as they are each present only on a single location and, as such, cannot show any wider spatial trends. The ceramic phase will likewise also not be considered individually as any spatial distribution ought only to be relevant where the dates relate to other phases.

Before moving on to consider the spatial distribution of the radiocarbon dates, it is worth looking in a little more detail at the cumulative probabilities of the four major thematic phases that will be discussed below. Fig 6.10, to which the same issues and caveats apply as Fig 6.9, compares these cumulative probabilities. The thin line shows the overall trend for the four phases in combination. Looking at the tree-throw / usage curve, it becomes obvious that the later (3rd millennium BC) peak appears concurrently with peaks in the Burnt Mound I and clearance phases: if the date sampless within this later peak are spatially close to those of those other phases, then this would add weight to the suggestion that this material was probably intrusive. Looking at the Burnt Mound I curve, the two peaks accord well with the picture painted by the excavator of a Late Neolithic / Early Bronze Age mound overlaying earlier Neolithic layers.

Looking at the clearance curve, four peaks are obvious: the first short peak is contemporary to the main peak in the tree-throw / usage curve. Again, therefore, if these samples have a close spatial relationship to the tree-throw / usage dates of that age, then the material will be shown to be probably residual. The other issue to consider with this thematic phase is whether the three later peaks are a true indication of variation in activity, or whether they simply reflect the choices made in selection of dating samples: again, this ought to be apparent from the dates' spatial proximity. Finally, looking at the Burnt Mound II curve, it is apparent that this forms the entire radiocarbon assemblage for the post-1500 cal BC period: the slight difference in the peak is due to a combination of the Excel smoothing algorithm and the specific positions of the 100 year time-slice divisions chosen by the Analyst tool. The various issues outlined here will be considered further below.

11820 to 4000 cal BC:

Two maps (Figs 6.11 and 6.12) demonstrate that the only radiocarbon dates on the site that pre-date 4000 cal BC are the dates taken from the environmental column. As these two dates fall on the same spatial location, they cannot show any spatial pattern. Thus, the TGIS can add little to the account written by the excavator for this particular time period.

4000 to 3000 cal BC:

Four maps (Figs 6.13 to 6.16) illustrate the probabilities of modelled and un-modelled radiocarbon dates falling within the periods 4000-3500 cal BC and 3500-3000 cal BC.

Beginning with the earlier period (Figs 6.13 and 6.14), several points arise. First, it is apparent that there is a significant probability that activity associated with tree-throws / usage began prior to 3500 cal BC, in particular towards the south-western corner of the site. These particular dates fall within Groups 802, 803 and 809 as described above, and will be considered in more detail below. Secondly, there is also a probability of several dates towards the centre of the site falling before 3500 cal BC. Of these, the alluviation date is the earlier date taken to provide a *terminus post quem* for that process and, as such, should be discounted (as per the dating experts' account). The other dates are of lesser probability (particularly those based upon the Bayesian model) and relate to the earlier Neolithic layers below Burnt Mound I.

Turning to the second period (Figs 6.15 and 6.16), the large number of red probability hotspots show that this is indeed the primary period associated with tree-throw / usage activity. For this period, there is very little difference between the modelled and un-modelled dates at this temporal scale of analysis. Furthermore, the Early Neolithic layers below Burnt Mound I have again manifested themselves. However, it should also be noted that two dates associated with the supposedly later clearance phase have a high probability of falling within this earlier time period: it is also apparent that these two dates are not of close spatial proximity to the contemporary dates of the tree-throw / usage phase, decreasing (but not removing) the possibility that they might derive from residual material. The later pollen column dates also may fall within this period, but their probability overlaps into the next time-slice. Finally, a problem with simple visual examination of the TGIS results is present in these two maps: the later of the alluviation dates (which has a probability of 1.0) is hidden behind the symbol for the less likely earlier date. Whilst it is obvious that there is a probability of a date on that location falling within the selected period, the apparent level of probability illustrated is misleading. This demonstrates that it is essential to look behind the map and consider the probability outputs themselves.

3000 to 1500 cal BC:

Four maps (Figs 6.17 to 6.20) illustrate the period 3000-1500 cal BC.

The first two maps (Figs 6.17 and 6.18) show what is essentially the transitional period between the tree-throw / usage and clearance phases of the site. The various dates are split between the two thematic phases and are of rather low probability. The only exceptions of higher probability are a clearance date towards the north-east of the site and one of the dates in the pollen column (especially in the un-modelled data). The second two maps (Figs 6.19 and 6.20) show the period identified by the excavator as that of primary clearance activity (2500-1500 cal BC). The main features of Burnt Mound I also date to within this time period. The interpretation reached is clearly fairly strong, as all of the clearance dates shown during this time-slice are of high probability. However, it should be noted that several dates associated with the tree-throw / usage thematic phase are also of relatively high probability on this map. Bearing in mind the nature of tree-throw deposit formation as discussed above, it is possible that the samples used to produce these dates were intrusive: this is somewhat supported by the close spatial proximity of Burnt Mound I. In the alternative, it is also possible that there is some temporal concurrence between the two phases of activity: after all, if clearance was not complete on the site, then the remaining standing trees would continue to die and fall. It should also be noted that the earliest date associated with Burnt Mound II also has a high probability of falling within this period.

1500 to 910 cal BC:

Two maps relate to the final period of activity discovered on the site (Figs 6.21 and 6.22). The only thematic phase associated with this period is that of Burnt Mound II and it is apparent from the maps that all of the dates shown relate to that phase. As such, any wider spatial patterns during this period are not possible to reconstruct from the radiocarbon evidence.

Tree-throws / usage:

Windthrow would have been the most common type of natural gap in temperate deciduous woodland. Furthermore, floodplain environments such as at Willington were probably more affected by gap-forming processes (beavers, river erosion, greater wind exposure, etc.) and, as such, may have been more attractive for human activity. That activity could be agricultural, but could also carry many other meanings (nodal points on path networks, areas of increased visibility / light, etc.). Islands in particular, again as at Willington, may have carried especial ritual significance due to the contradiction between their proximity and inaccessibility. It is apparent to me that the use of these natural forest

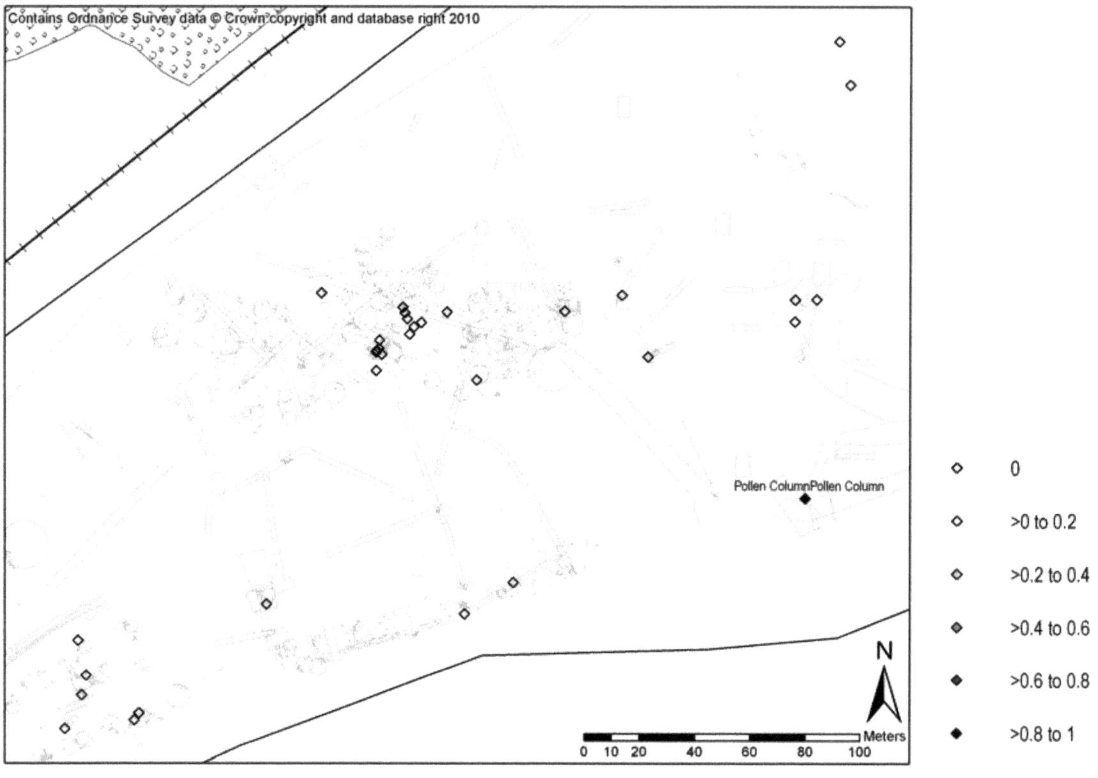

FIG 6.11 - SPATIAL DISTRIBUTION OF MODELLED RADIOCARBON DATES, SHADED ACCORDING TO PROBABILITY OF THEM FALLING WITHIN THE PERIOD 11810-4000 CAL BC. DATES OF GREATER THAN 0.0 PROBABILITY LABELLED ACCORDING TO THEMATIC PHASE.

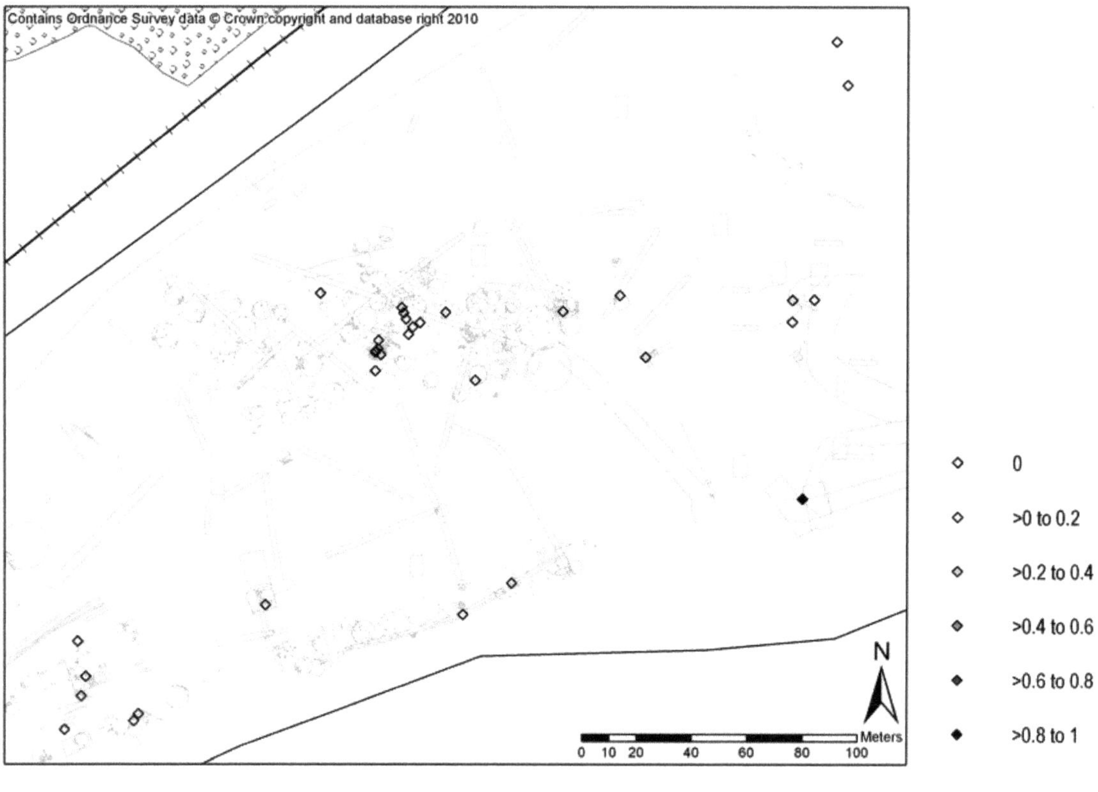

FIG 6.12 - SPATIAL DISTRIBUTION OF UN-MODELLED RADIOCARBON DATES, SHADED ACCORDING TO PROBABILITY OF THEM FALLING WITHIN THE PERIOD 11820-4000 CAL BC.

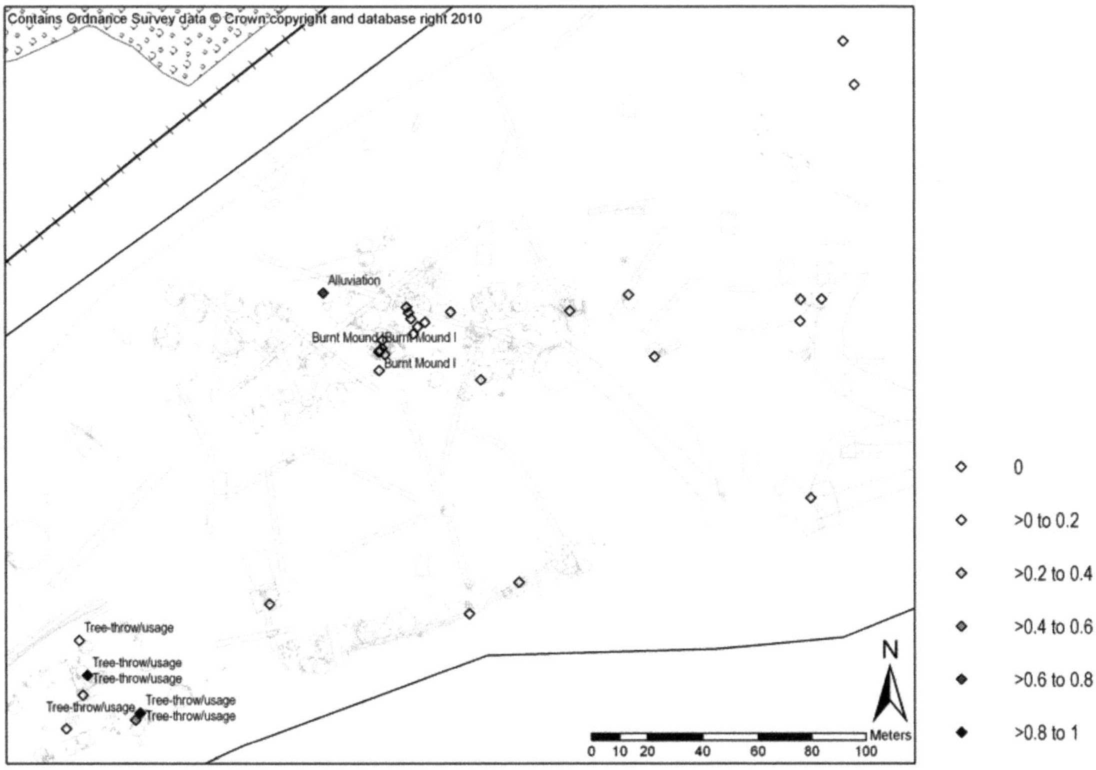

FIG 6.13 - SPATIAL DISTRIBUTION OF MODELLED RADIOCARBON DATES, SHADED ACCORDING TO PROBABILITY OF THEM FALLING WITHIN THE PERIOD 4000-3500 CAL BC. DATES OF GREATER THAN 0.0 PROBABILITY LABELLED ACCORDING TO THEMATIC PHASE.

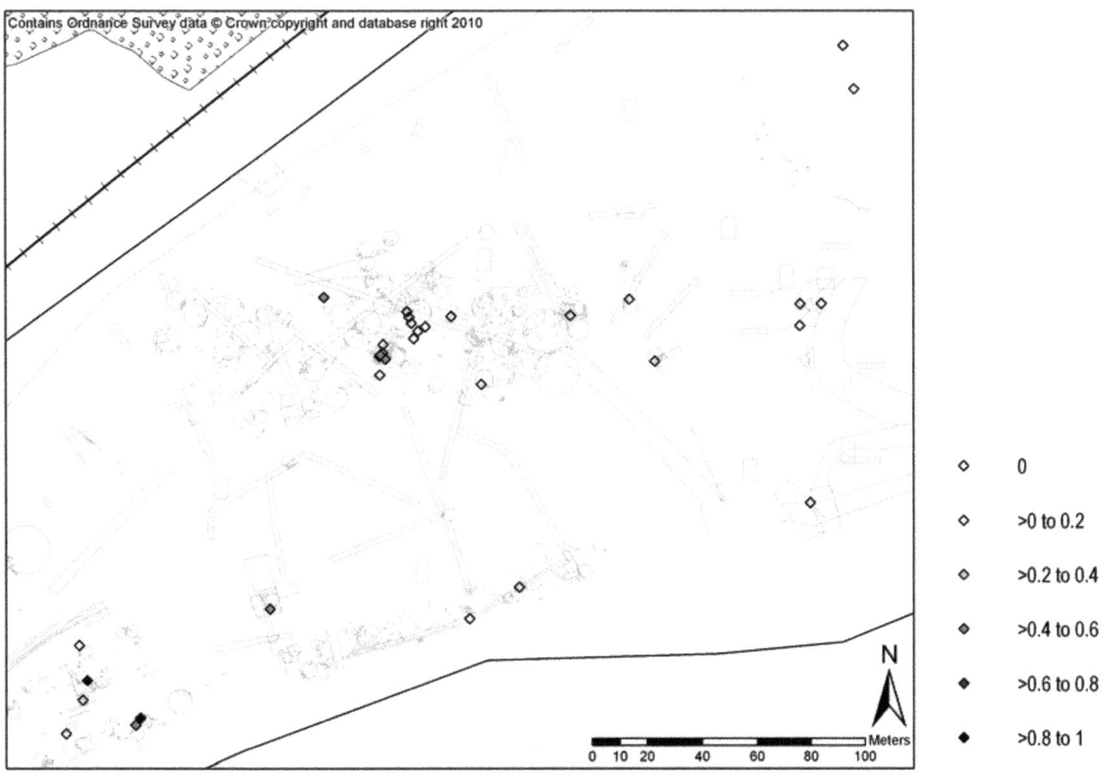

FIG 6.14 - SPATIAL DISTRIBUTION OF UN-MODELLED RADIOCARBON DATES, SHADED ACCORDING TO PROBABILITY OF THEM FALLING WITHIN THE PERIOD 4000-3500 CAL BC.

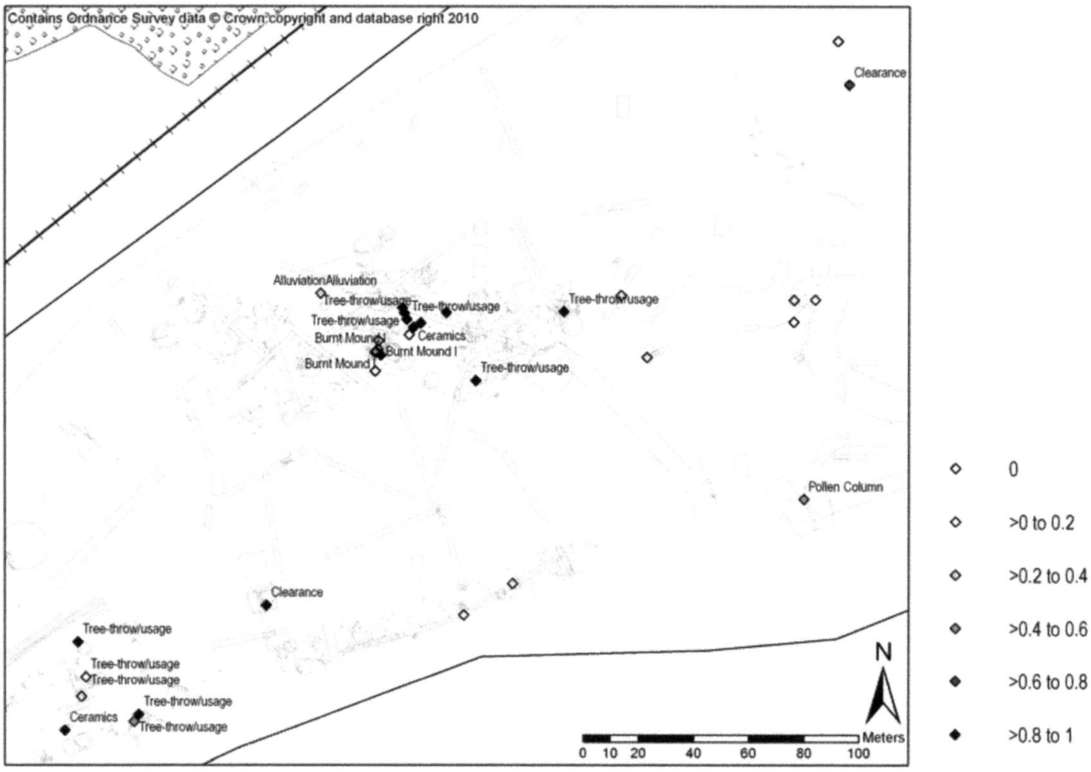

FIG 6.15 - SPATIAL DISTRIBUTION OF MODELLED RADIOCARBON DATES, SHADED ACCORDING TO PROBABILITY OF THEM FALLING WITHIN THE PERIOD 3500-3000 CAL BC. DATES OF GREATER THAN 0.0 PROBABILITY LABELLED ACCORDING TO THEMATIC PHASE.

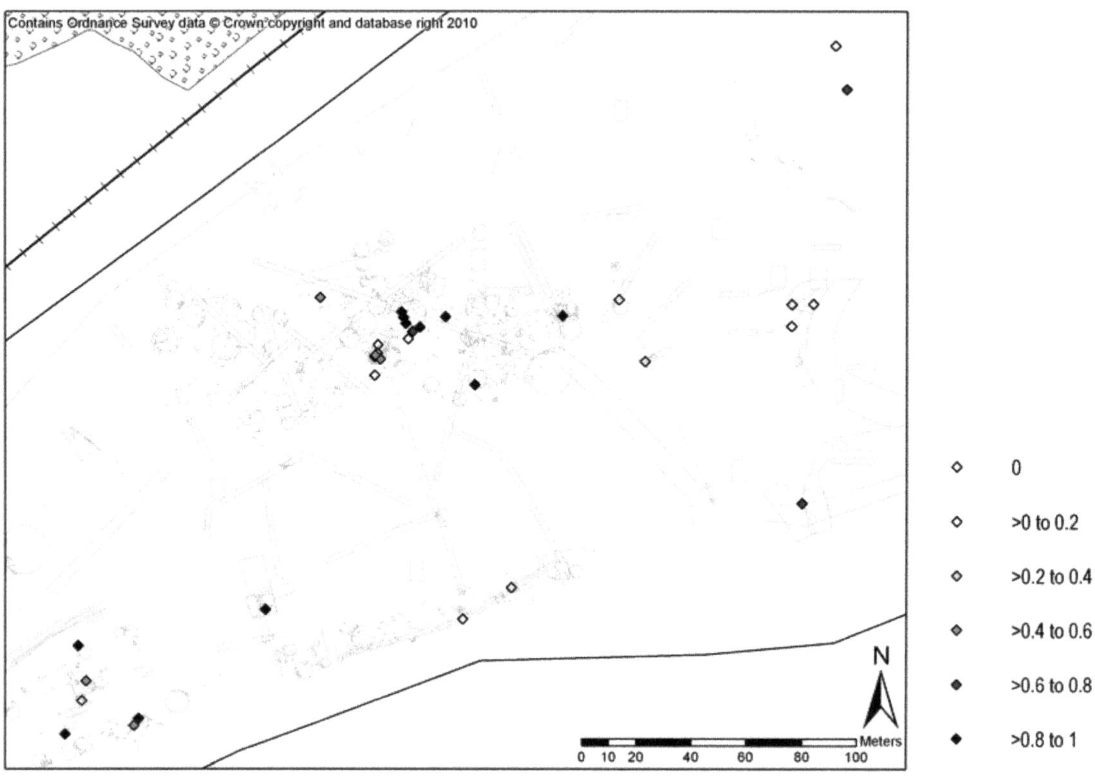

FIG 6.16 - SPATIAL DISTRIBUTION OF UN-MODELLED RADIOCARBON DATES, SHADED ACCORDING TO PROBABILITY OF THEM FALLING WITHIN THE PERIOD 3500-3000 CAL BC.

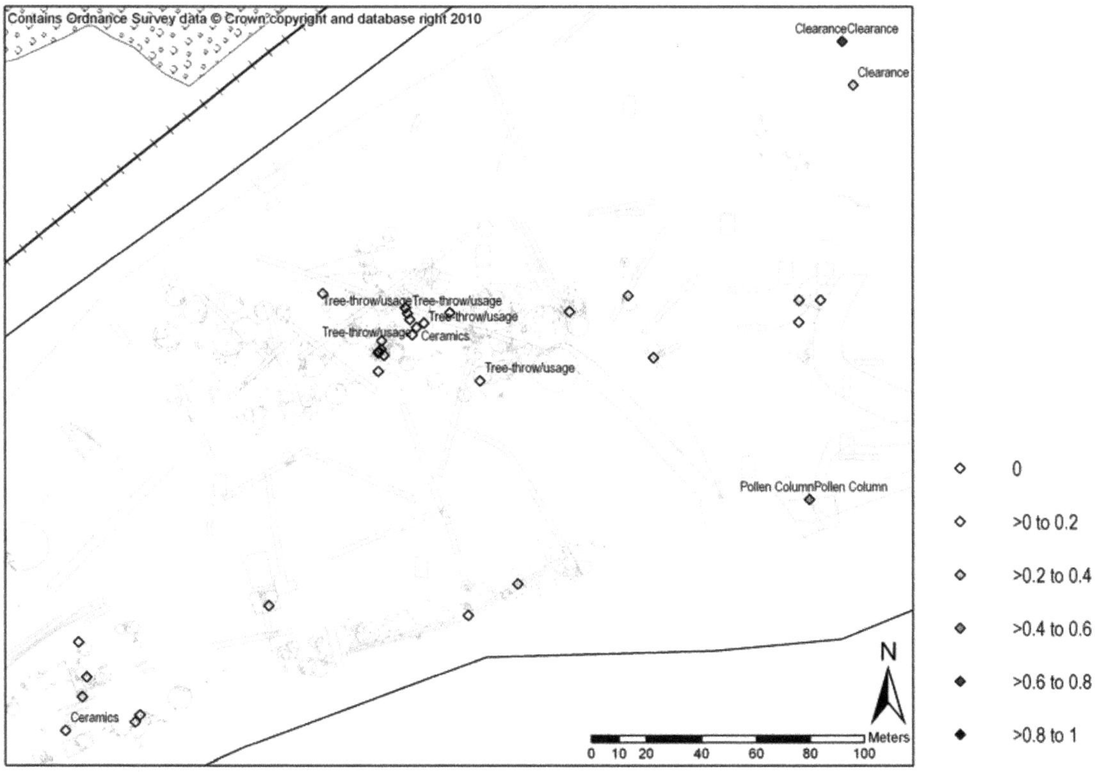

FIG 6.17 - SPATIAL DISTRIBUTION OF MODELLED RADIOCARBON DATES, SHADED ACCORDING TO PROBABILITY OF THEM FALLING WITHIN THE PERIOD 3000-2500 CAL BC. DATES OF GREATER THAN 0.0 PROBABILITY LABELLED ACCORDING TO THEMATIC PHASE.

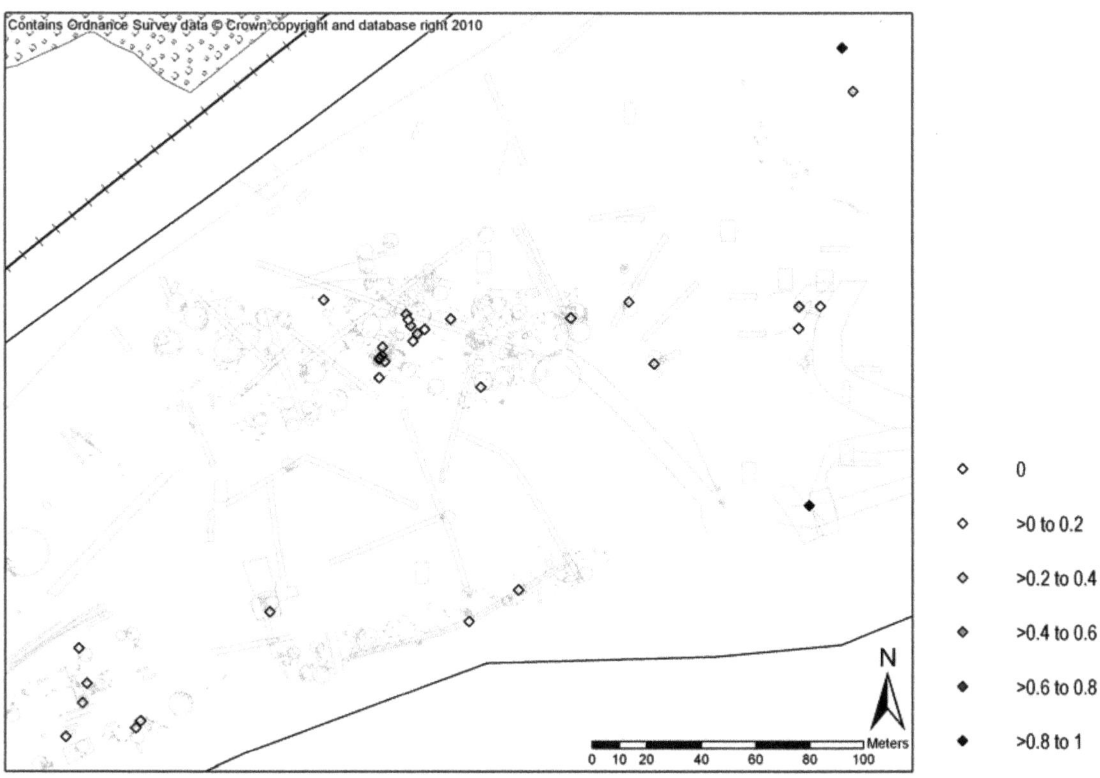

FIG 6.18 - SPATIAL DISTRIBUTION OF UN-MODELLED RADIOCARBON DATES, SHADED ACCORDING TO PROBABILITY OF THEM FALLING WITHIN THE PERIOD 3000-2500 CAL BC.

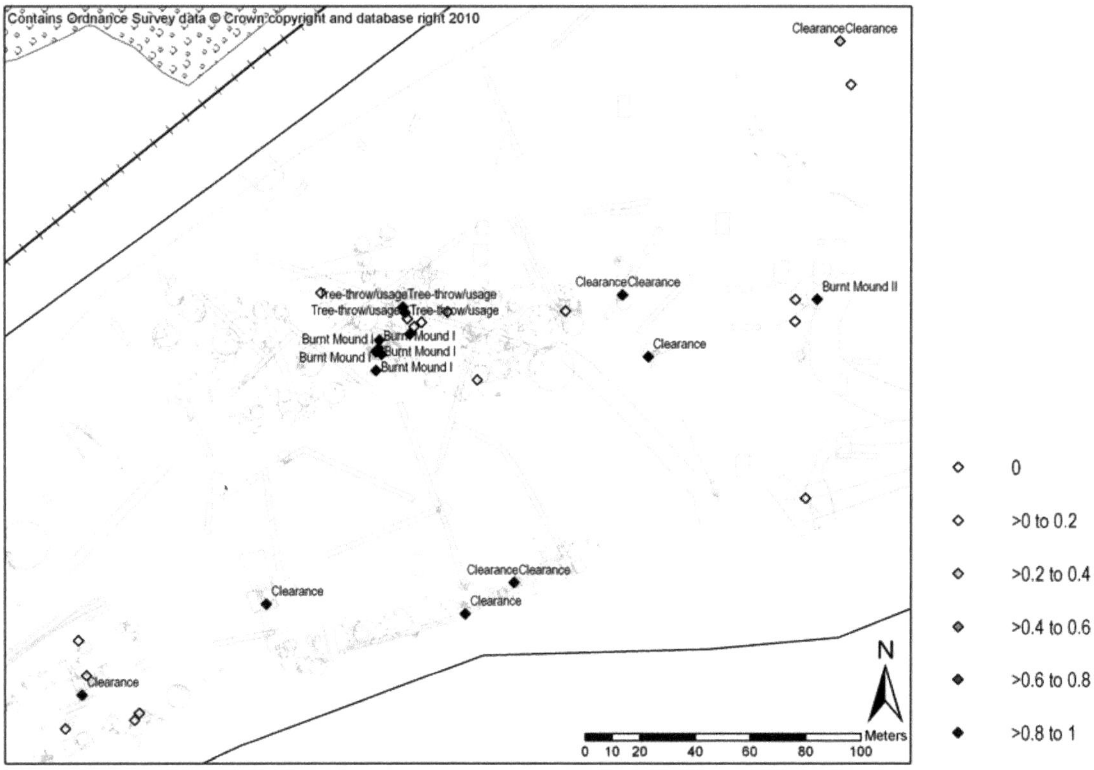

FIG 6.19 - SPATIAL DISTRIBUTION OF MODELLED RADIOCARBON DATES, SHADED ACCORDING TO PROBABILITY OF THEM FALLING WITHIN THE PERIOD 2500-1500 CAL BC. DATES OF GREATER THAN 0.0 PROBABILITY LABELLED ACCORDING TO THEMATIC PHASE.

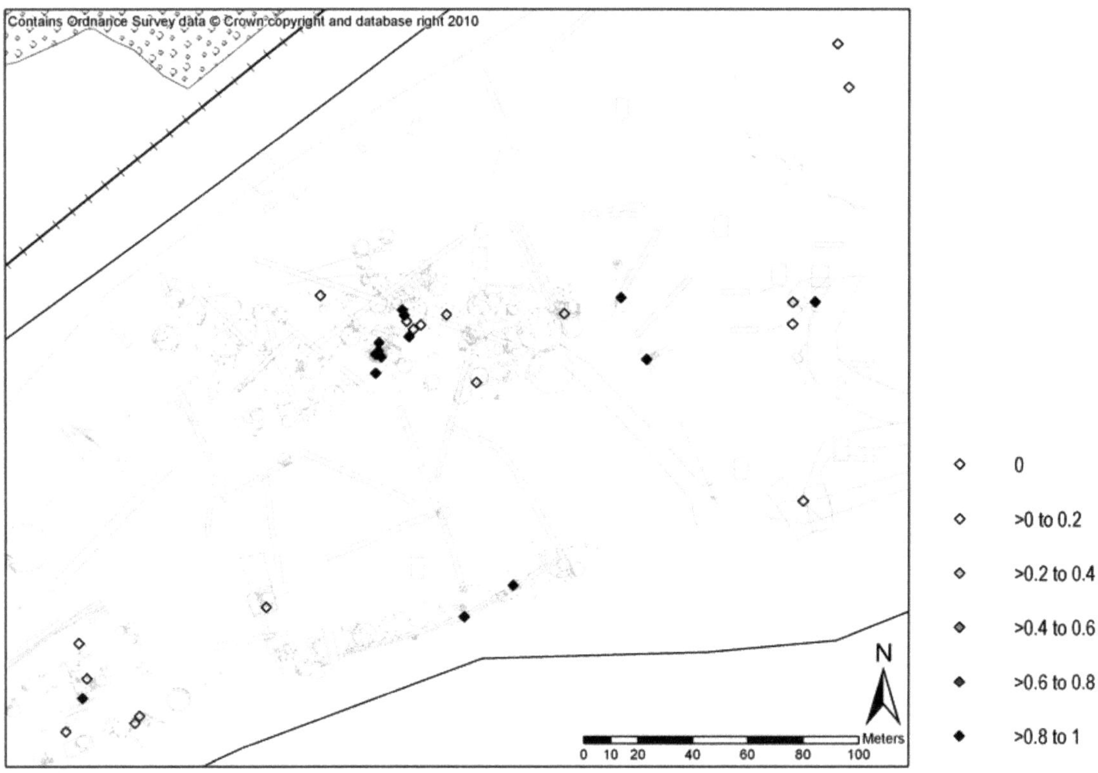

FIG 6.20 - SPATIAL DISTRIBUTION OF UN-MODELLED RADIOCARBON DATES, SHADED ACCORDING TO PROBABILITY OF THEM FALLING WITHIN THE PERIOD 2500-1500 CAL BC.

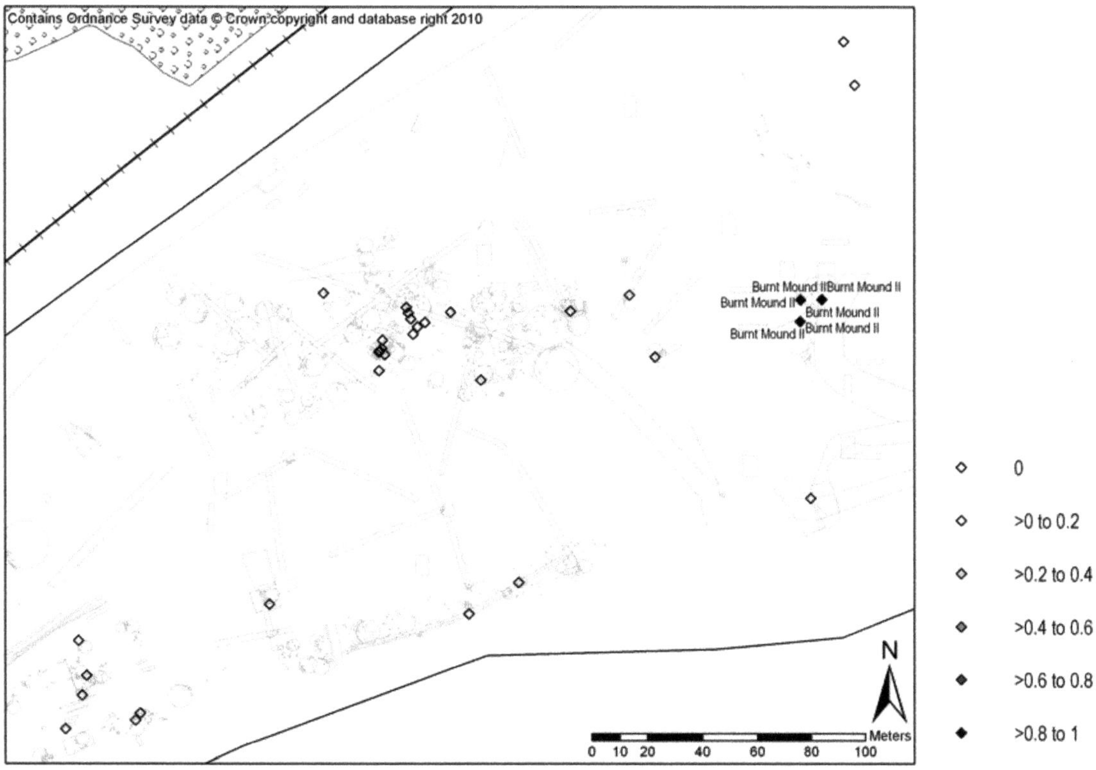

FIG 6.21 - SPATIAL DISTRIBUTION OF MODELLED RADIOCARBON DATES, SHADED ACCORDING TO PROBABILITY OF THEM FALLING WITHIN THE PERIOD 1500-921 CAL BC. DATES OF GREATER THAN 0.0 PROBABILITY LABELLED ACCORDING TO THEMATIC PHASE.

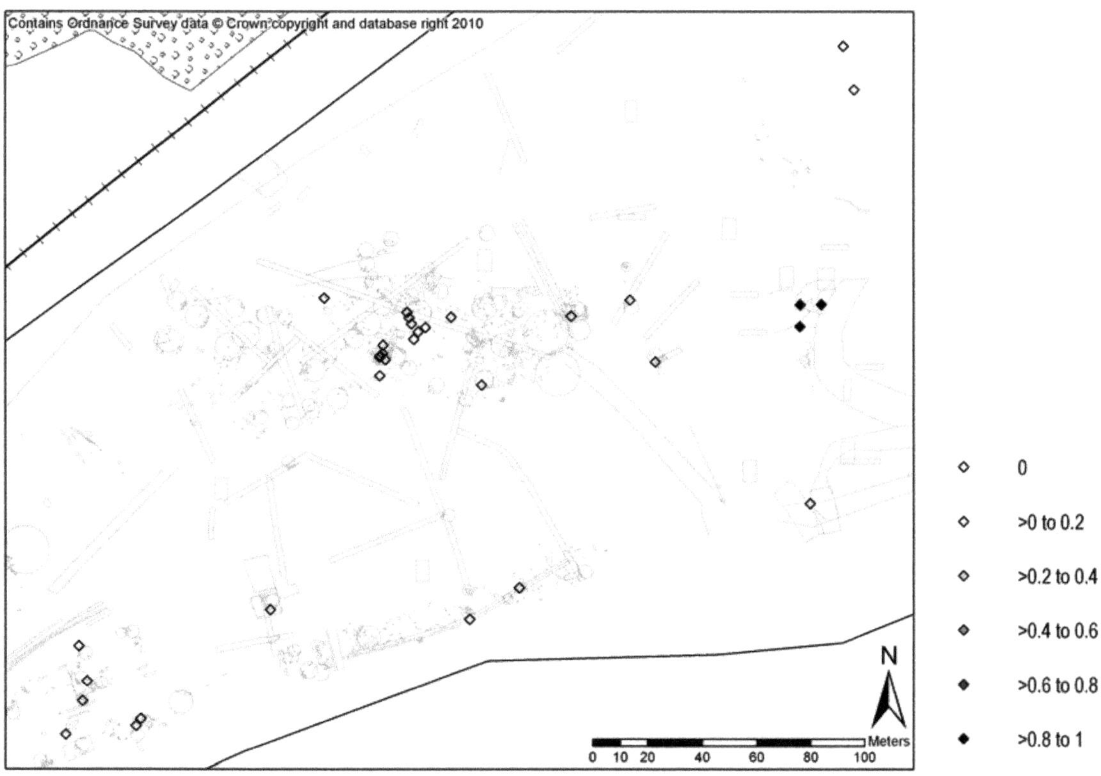

FIG 6.22 - SPATIAL DISTRIBUTION OF UN-MODELLED RADIOCARBON DATES, SHADED ACCORDING TO PROBABILITY OF THEM FALLING WITHIN THE PERIOD 1500-910 CAL BC.

gaps would inevitably have a spatial character, albeit one primarily determined by natural tree-fall processes. Nevertheless, human choices about which gaps to use for activity could have had many influencing factors, both functional and social (Brown 2000: 49-50; 61). As such, I decided to next look for spatio-temporal patterns within the tree-throws / usage thematic phase, based upon the Bayesian model output. This phase, when taken in isolation, has dates with probabilities that cover the span between 3750 and 1915 cal BC. A table (Table 6.1, page 147) gives the probabilities of the dates assigned to the tree-throw / usage phase falling within a series of c.250 year time-slices. The maps show the spatial distribution of these probabilities (Fig 6.23).

These maps show a very clear spatial trend: on the basis of the dating evidence analysed, activity appears to begin in the south-western part of the site at some time between c.3750-3500 cal BC (Fig 6.23a): this is shown by the very high probability (98%) of OxA-15899 falling within this period, especially when taken in conjunction with the high probability (76%) for GrA-31799. As noted above, the dates in this corner of the site relate to Groups 802, 803 and 809: specifically, these two dates both came from Group 803 and were rated as likely to be of the same age by the dating experts. Remaining in this corner of the site, the other associated dates all span various portions of this time-slice and the two following: there is no probability of any of these dates being later than 3000 cal BC (Table 6.1, page 147). We can thus be confident that activity associated with tree-throws in the south-western corner of the site dates to between 3750-3000 cal BC, with a fairly even spread of probabilities across that period.

Moving to the centre of the site (towards the northern edge of the maps in Fig 6.23), there is no probability of any of the dates assigned to this thematic phase pre-dating 3500 cal BC. The main phase of tree-throw / usage activity in this area of the site is largely over by 3000 cal BC, although there is a very small probability (1-4%) of a few of the dates falling within the 3000-2750 cal BC time-slice (Fig 6.23d). It is, therefore, probable that the majority of tree-throw / usage activity in this area of the site began slightly later than in the south-west, but peaked over the same 3500-3000 cal BC period (Fig 6.23b-c). However, five of the dates (including one combined date) fall within the three time-slices from 2500-1915 cal BC (Fig 6.23f-h), with their greatest peak within the middle 2250-2000 cal BC time-slice (Fig 6.23g). These dates (Comb_[1328]<64>, GrA-31785, GrA-31786, GrA-31803, and OxA-15110) all came from Group 2503 and principally contexts (with the exception of GrA-31786) that also contained material dated to the earlier peak of tree-throw / usage activity (specifically the dates SUERC-7596 and -7597). The fact that these dates came from temporally-mixed contexts means that a degree of residuality or intrusion was a factor in this Group: it is very close spatially to the broadly contemporary Burnt Mound I activity and it is possible that the later material came from deposition associated with activity during that thematic phase.

The TGIS analysis of this thematic phase thus broadly agrees with the results produced by the excavator. However, it has added some specific detail to that picture: specifically, the strong evidence for an earlier start to activity on the site within the south-western grouping and the evidence for residual / intrusive material in the central grouping. Unfortunately, any spatial analysis beyond this, whilst possible, would not be prudent: results could be produced, but the strong spatial grouping of the dates associated with this thematic phase would make them questionable. The two closely associated spatial groups are an interesting spatial pattern in themselves, but there is not enough evidence for the areas between the groups to confidently interpolate further based upon these dates alone. Studies in the Chilterns in 1987 suggested that natural gaps in deciduous woodland have an average diameter of 80m, and there is no reason to assume that this was an atypical situation (Brown 2000: 50). The two areas discussed above are roughly 100-150m apart (depending on edge definition). If we assume that the openness of a clearing edge increases the likelihood of windfall (Brown 2000: 61), then it is just possible that the two areas represent a single clearing opening in the south-west and spreading north-eastwards. On the other hand, the two groups could equally represent two clearings, the south-western forming and used first, with activity then moving towards the central group as the previous began to close up. Without further evidence for the intervening area, however, these contrasting conclusions must remain speculative.

Fire clearance:

Of all the thematic phases, fire clearance could be expected to have the most obvious human-influenced spatial character. In particular, it could be expected to either be systemic, deliberate and directional, or more piecemeal and opportunistic. As such, in this section, I shall look for spatio-temporal patterns within the fire clearance thematic phase, based upon the Bayesian model output. This phase, when taken in isolation, has dates with probabilities that cover the span between 3366 and 1614 cal BC. A table (Table 6.2, page 148) gives the probabilities of the dates assigned to the clearance phase falling within a series of c.250 year time-slices. The maps show the spatial distribution of these probabilities (Fig 6.24).

Visual examination of these time-slice maps shows four obvious peaks where the time-slices have dates of higher probability: 3250-3000 cal BC (Fig 6.24b); 2750-2500 cal BC (Fig 6.24d); 2250-2000 cal BC (Fig 6.24f); and 1750-1614 cal BC (Fig 6.24h). However, there is no obvious spatial trend, with dates from different periods being fairly well spread across the site at each peak of activity. Looking at the spread of probabilities for each date, there appear to be four phases corresponding to the four peaks already mentioned: 3366-2750 cal BC (Fig 6.24a-c); 2750-2250 cal BC (Fig 6.24d-e); 2250-2000 cal BC (Fig 6.24f); 2000-1614 cal BC (Fig 6.24g-h). The excavator interpreted the clearance as taking place

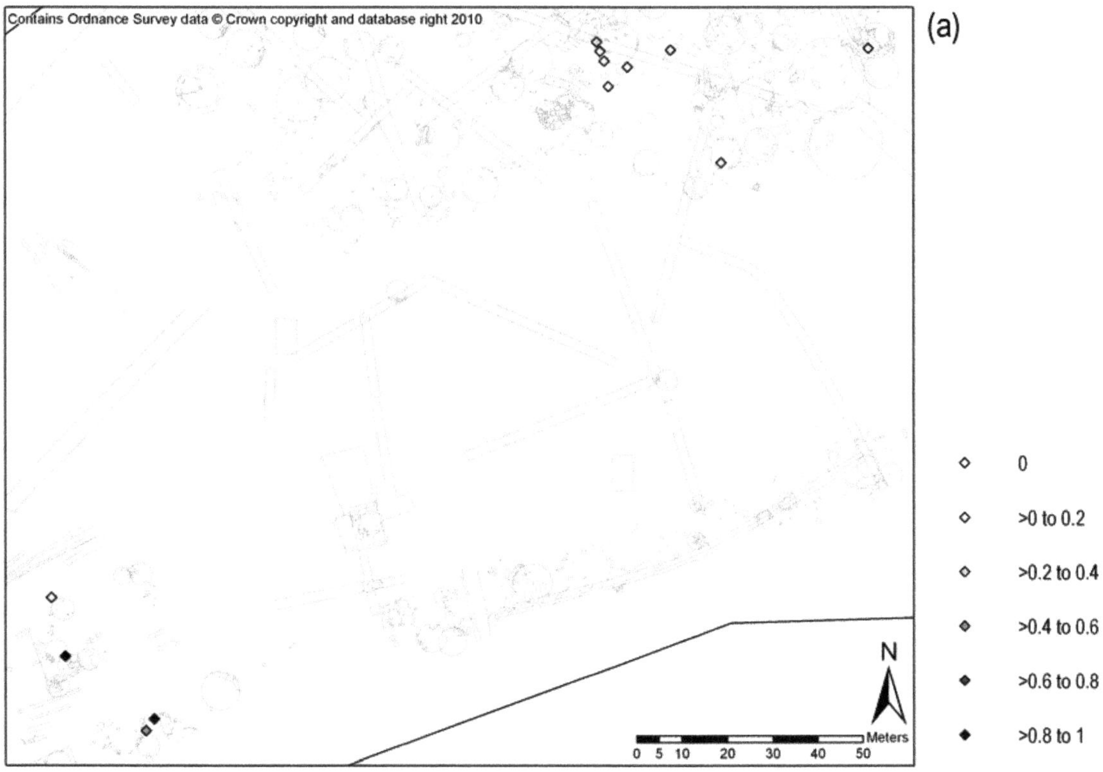

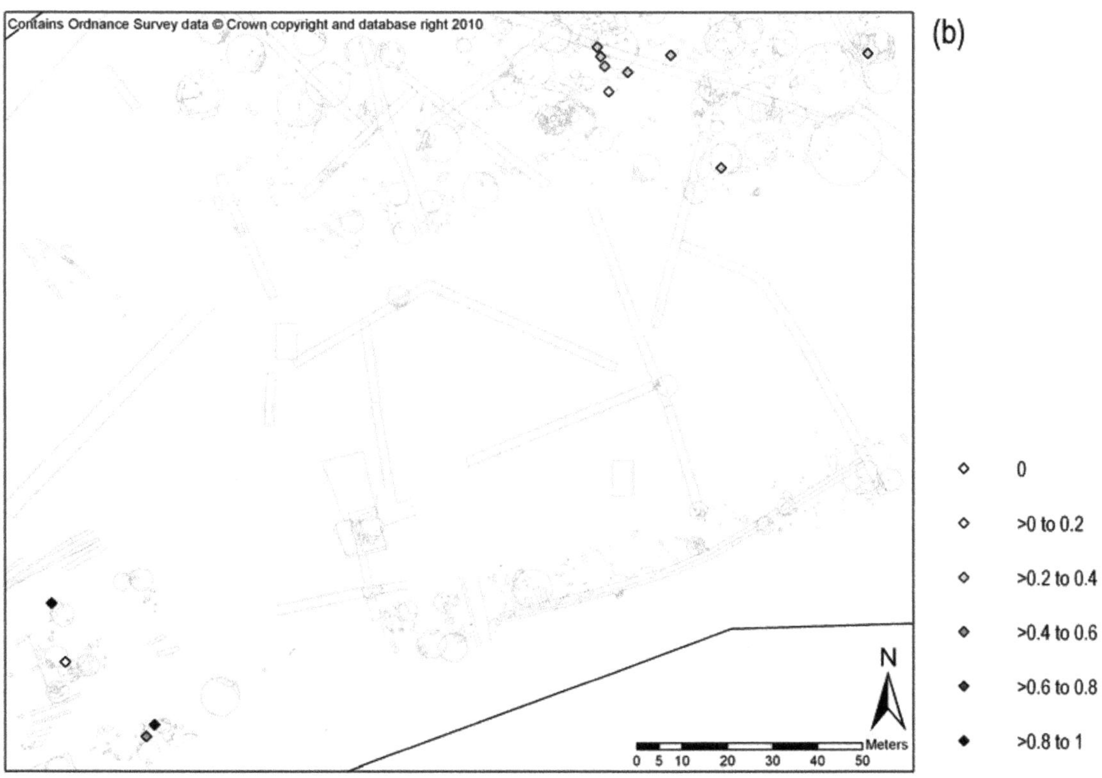

FIG 6.23 (A) TO (B) - PROBABILITIES OF TREE-THROW / USAGE DATES FROM 3750-3500 CAL BC AND 3500-3250 CAL BC.

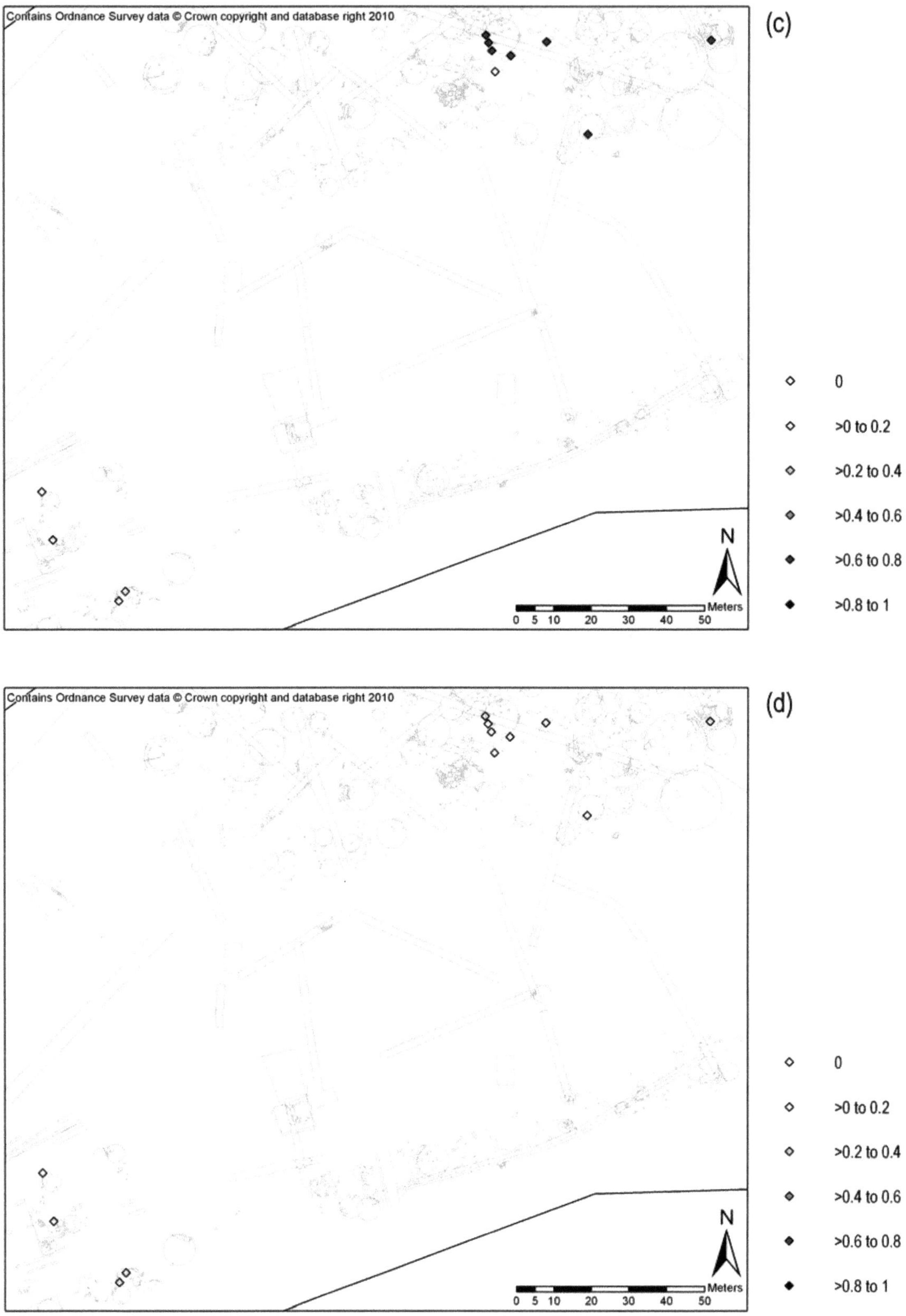

FIG 6.23 (C) TO (D) - PROBABILITIES OF TREE-THROW / USAGE DATES FROM 3250-3000 CAL BC AND 3000-2750 CAL BC.

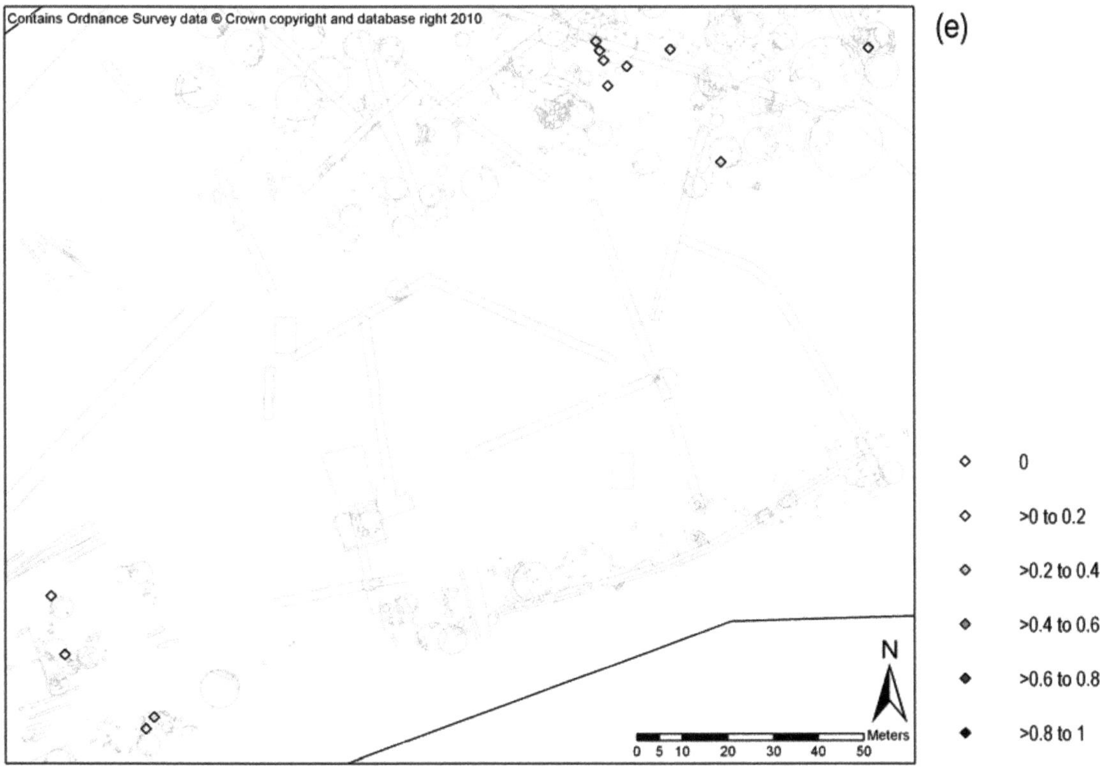

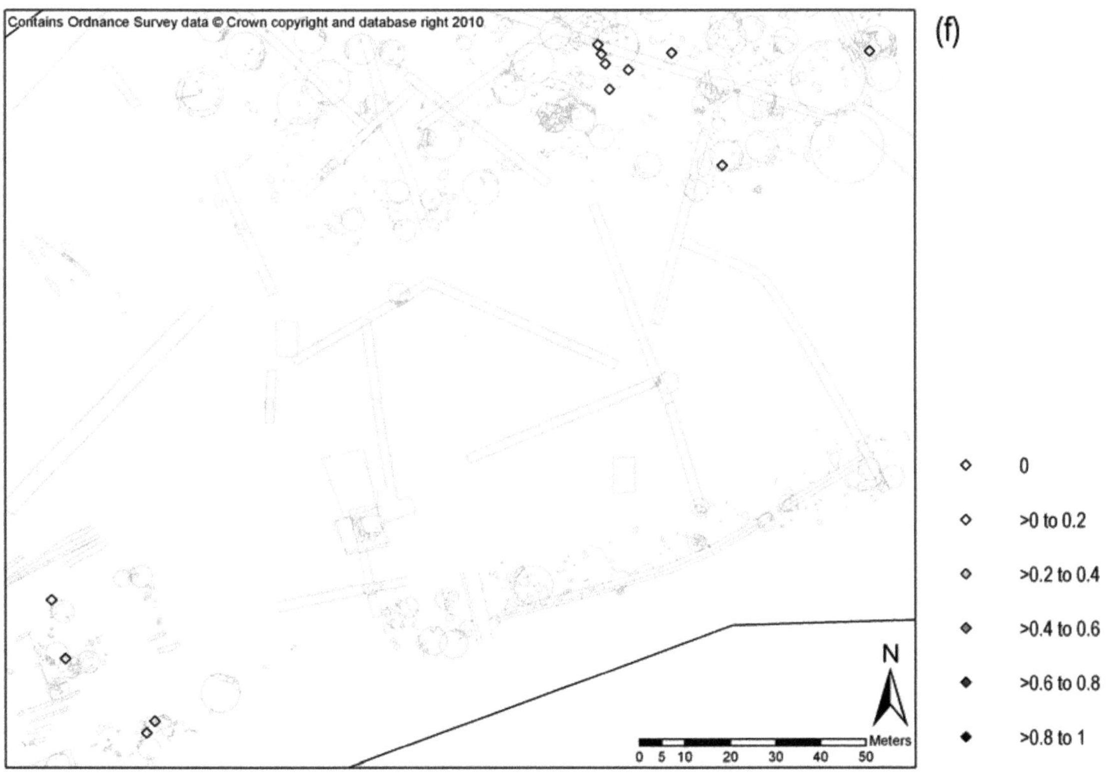

FIG 6.23 (E) TO (F) - PROBABILITIES OF TREE-THROW / USAGE DATES FROM 2750-2500 CAL BC AND 2500-2250 CAL BC.

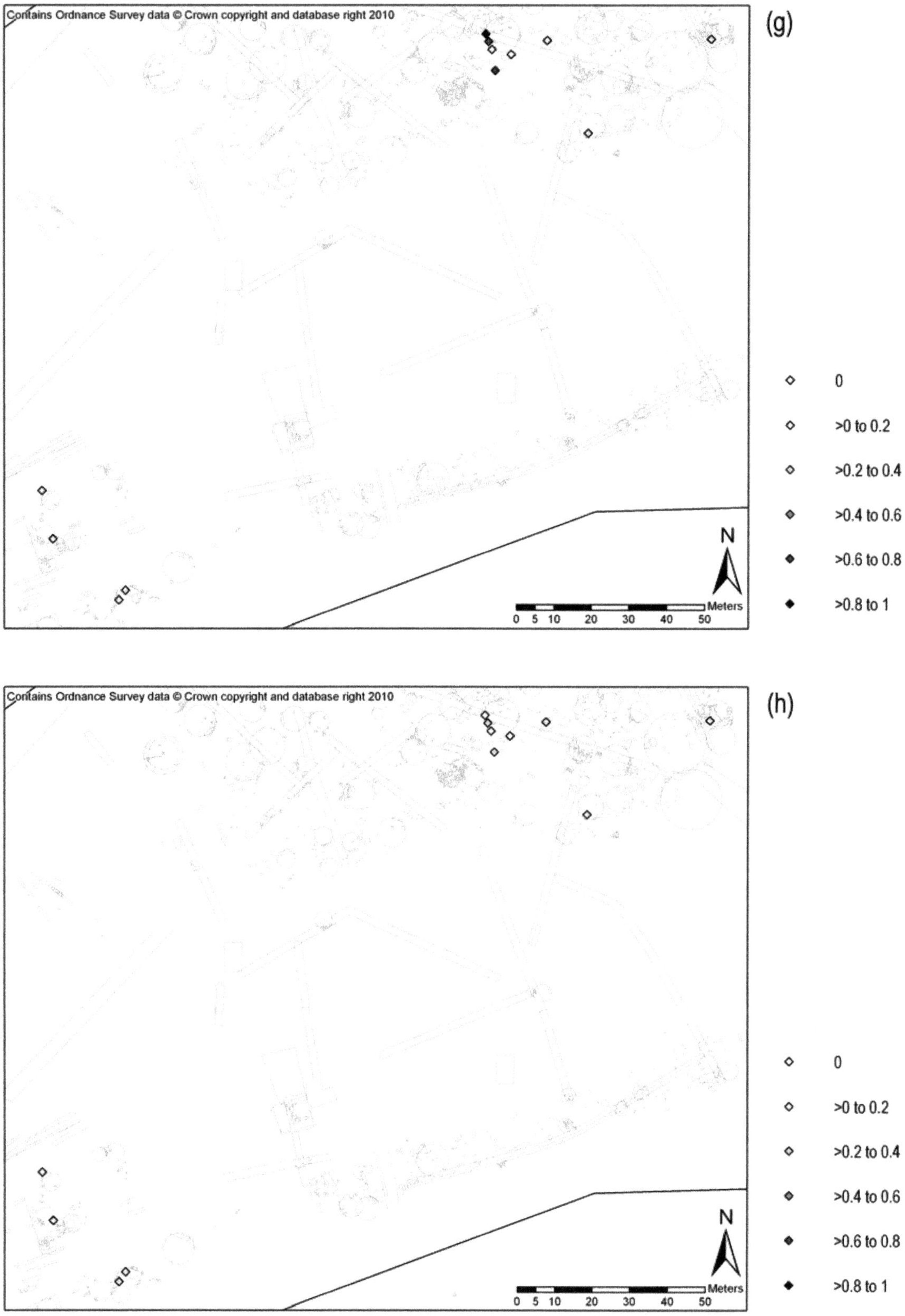

FIG 6.23 (G) TO (H) - PROBABILITIES OF TREE-THROW / USAGE DATES FROM 2250-2000 CAL BC AND 2000-1915 CAL BC.

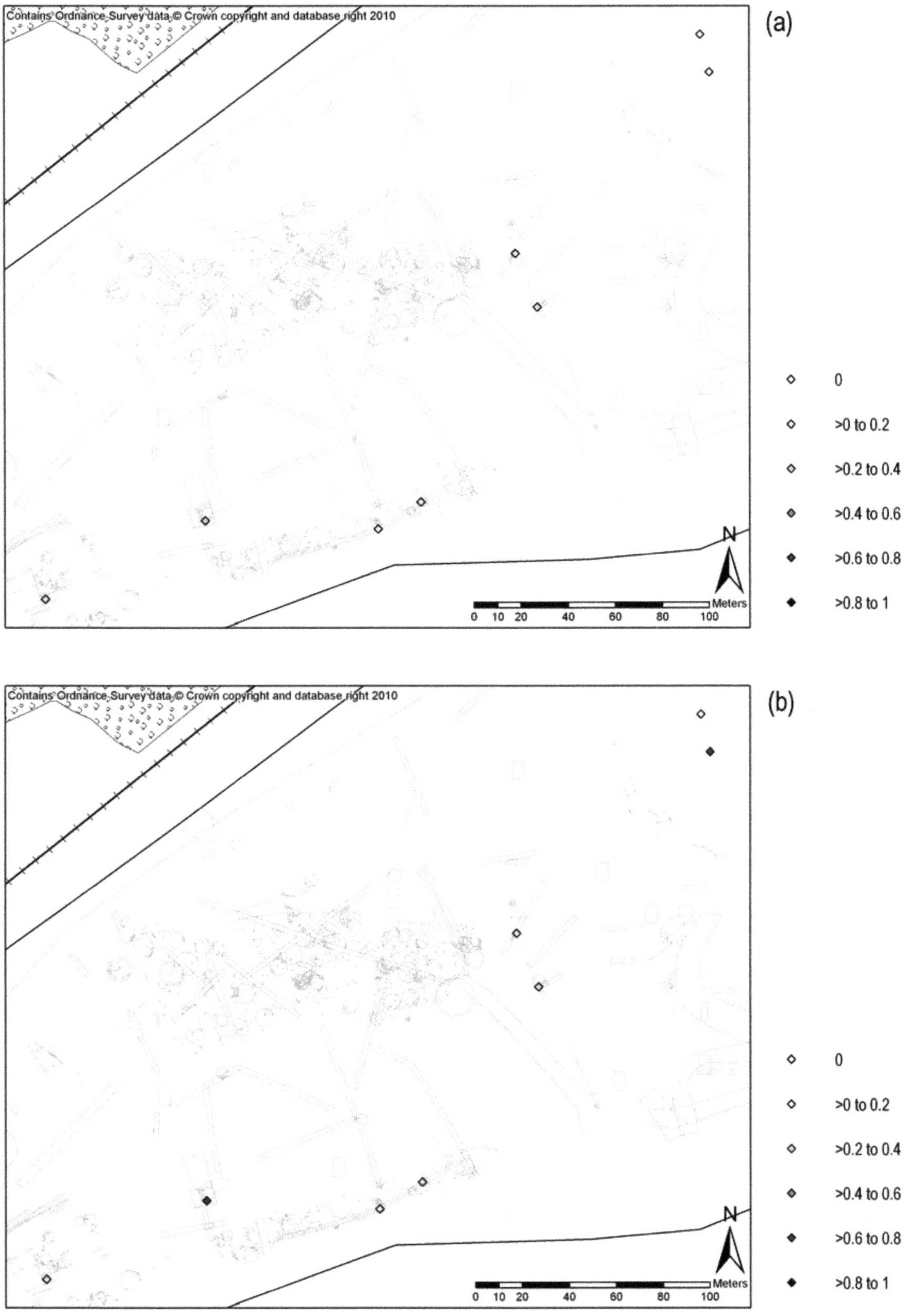

FIG 6.24 (A) TO (B) - PROBABILITIES OF FIRE CLEARANCE DATES FROM 3366-3250 CAL BC AND 3250-3000 CAL BC.

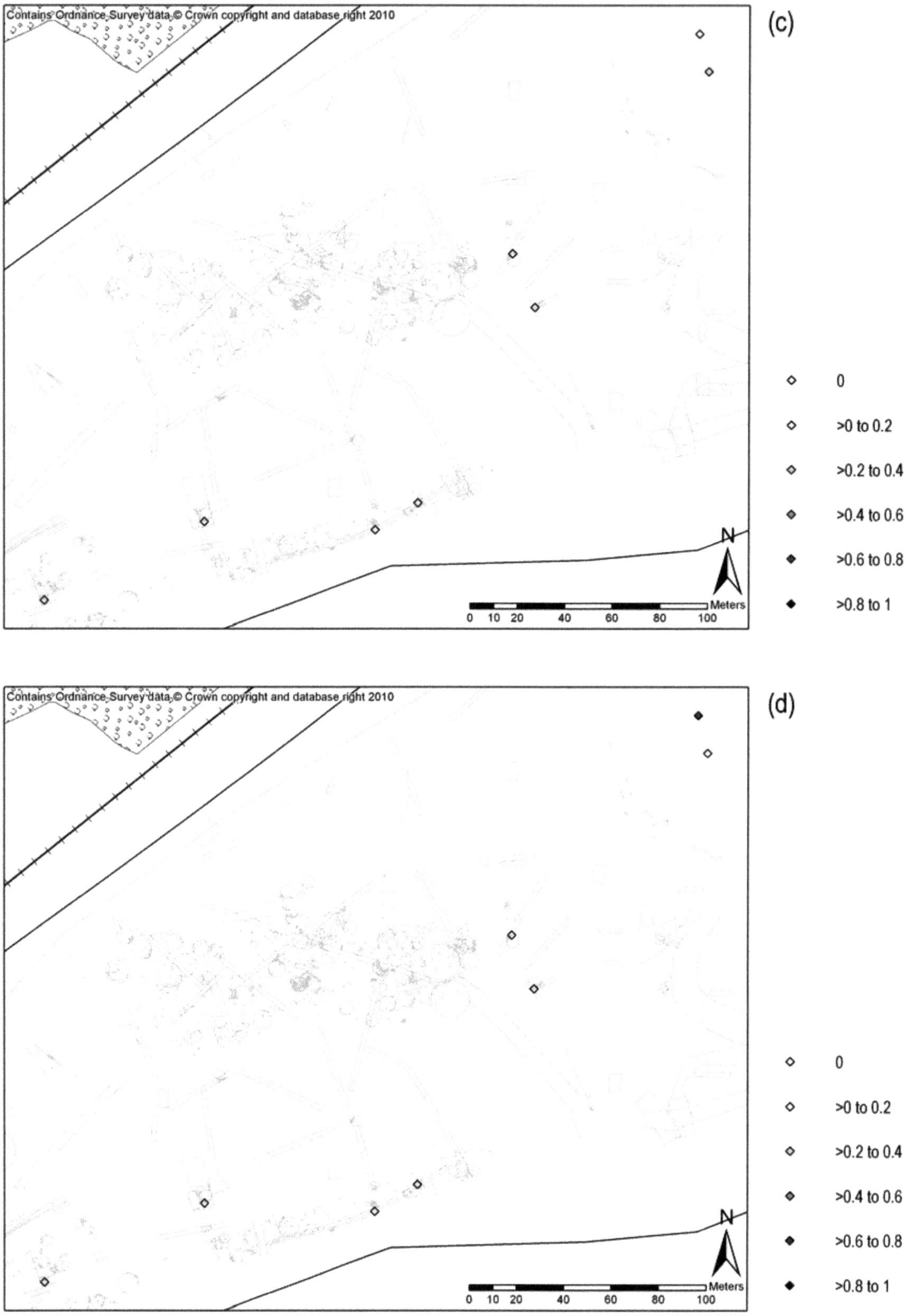

FIG 6.24 (C) TO (D) - PROBABILITIES OF FIRE CLEARANCE DATES FROM 3000-2750 CAL BC AND 2750-2500 CAL BC.

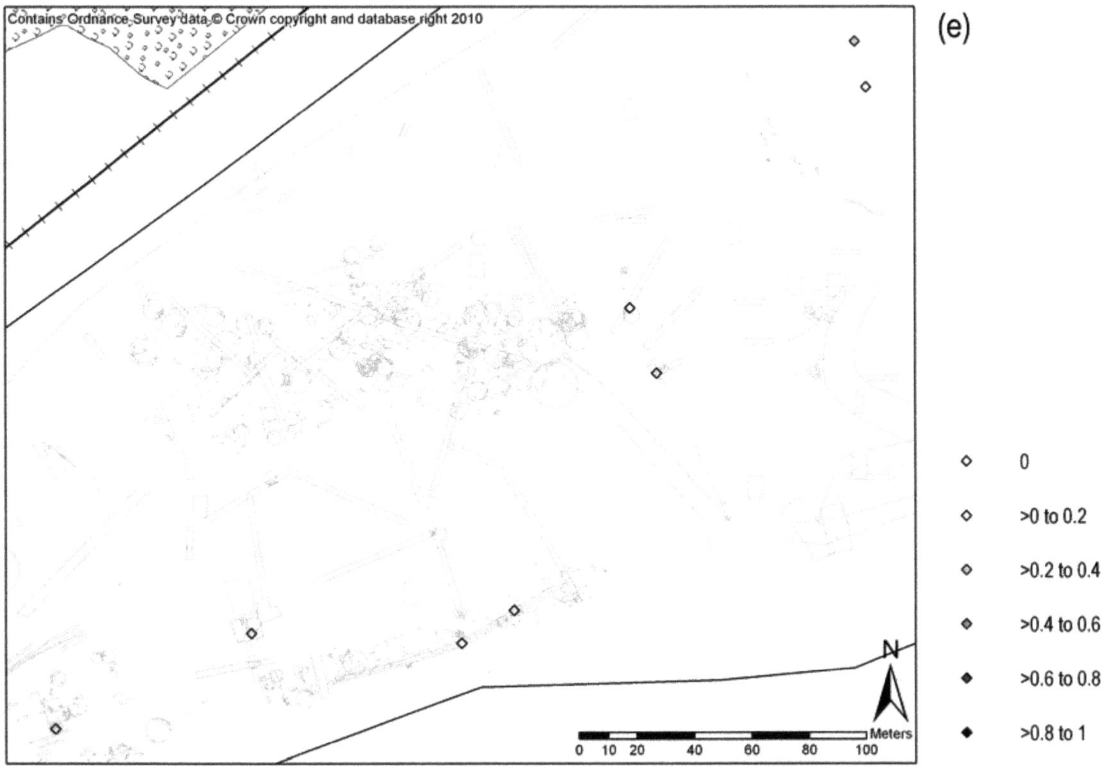

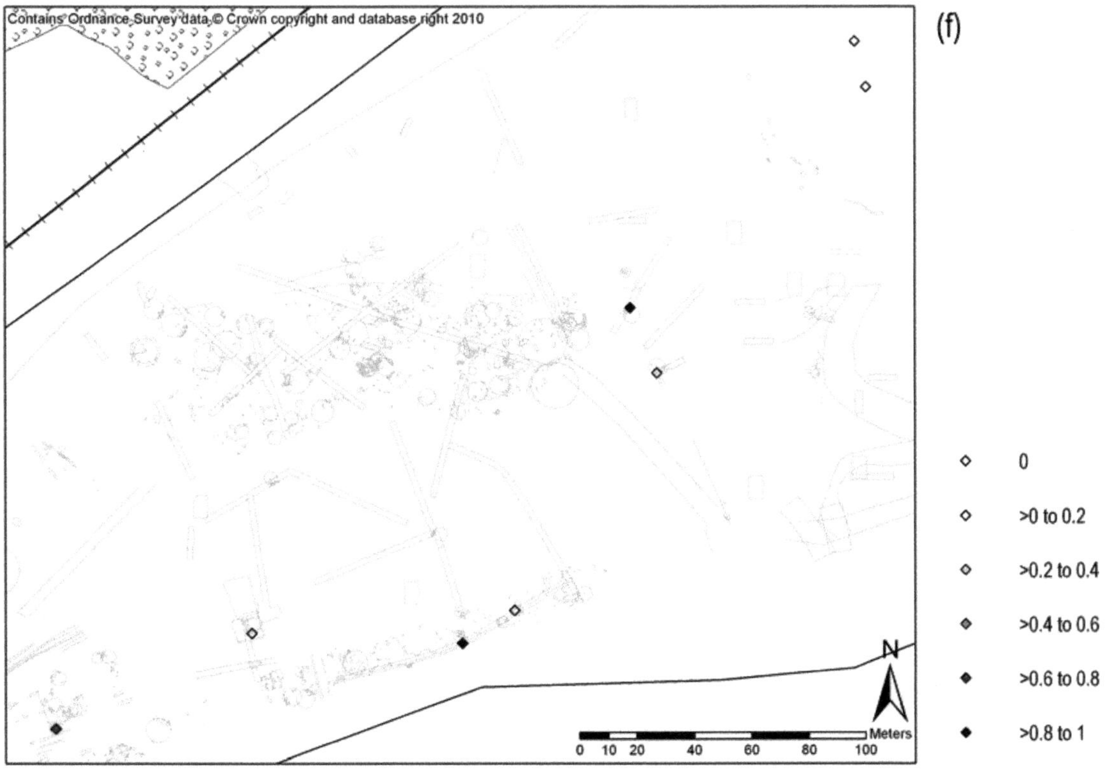

FIG 6.24 (E) TO (F) - PROBABILITIES OF FIRE CLEARANCE DATES FROM 2500-2250 CAL BC AND 2250-2000 CAL BC.

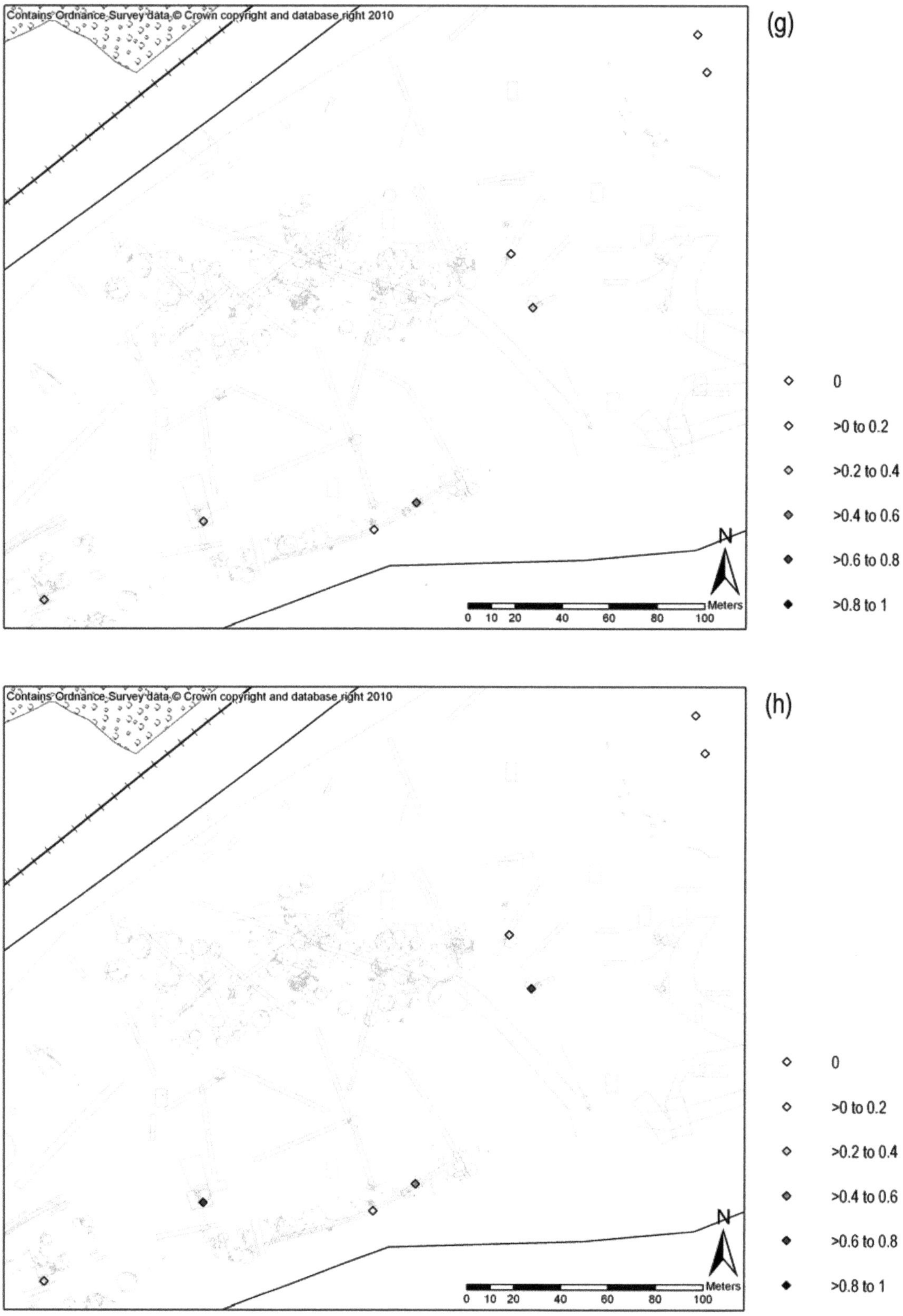

FIG 6.24 (G) TO (H) - PROBABILITIES OF FIRE CLEARANCE DATES FROM 2000-1750 CAL BC AND 1750-1614 CAL BC.

between c.2500 cal BC and c.1750-1500 cal BC, in a series of indeterminate phases. Two dates in particular fall before 2500 cal BC: OxA-15898 and GrA-31796. It is possible that these two samples were taken from residual material (especially in the case of OxA-15898 as it came from the same deposit as the much later GrA-31797), but it is also possible that fire clearance began up to c.800 years prior to 2500 cal BC.

Within this thematic phase, the dates form a vague line across the site and, as such, it would appear to be possible to perform further spatial analysis upon the data (although, ideally, there would be more points). However, in practice, when using the probability output as the z-field for an interpolation algorithm, the fact that only a few points have a value of greater than zero for each phase removes the possibility of undertaking any quantitative spatial analysis based upon this thematic layer in isolation. Nevertheless, the fact that the Willington clearance material shows little obvious directional or systematic spatio-temporal patterning suggests that perhaps it was of a more opportunistic nature: natural woodland gaps form an obvious and convenient starting point for human clearance (Brown 2000: 49-50). However, there are too many issues over possible residuality / intrusion to attach any great weight to this speculation.

Burnt Mound I:

In this section, I shall look for spatio-temporal patterns within the Burnt Mound I thematic phase, based upon the Bayesian model output. Note that this is at a much closer spatial scale: in effect, looking at events within a feature rather than the site as a whole. This phase, when taken in isolation, has dates with probabilities that cover the span between 3698 and 1897 cal BC. A table (Table 6.3, page 149) gives the probabilities of the dates assigned to the Burnt Mound I phase falling within a series of c.250 year time-slices (3250-2500 cal BC forms one slice as no dates fall within this period). The maps show the spatial distribution of these probabilities (Fig 6.25).

The dates associated with the burnt mounds are, by their very nature, closely grouped spatially. The first two time-slices show the Early Neolithic samples taken from beneath Burnt Mound I, which form a fairly tight spatial grouping (Fig 6.25a-b). There then follows a conglomerate time-slice which shows the gap between the earlier and the later Neolithic / Bronze Age activity: a gap of at least 750 years (Fig 6.25c). The final three time-slices show the main phase of the mound (Fig 6.25d-f). However, it is clear that the 2500-2250 (Fig 6.25d) and 2000-1897 cal BC (Fig 6.25f) time-slices only carry fairly low probabilities within the mound itself (no more than 35% - the 64% probability within the final time-slice [OxA-15111], is the date towards the south of the maps in question, beyond the boundaries of the mound): the dating specialists' Bayesian model placed the main phase of usage of Burnt Mound I within the time period 2340-1840 cal BC (see above). Thus, finally, it is useful to ask how this time period is reflected within the wider radiocarbon dataset. The map produced for this purpose (Fig 6.26) shows quite clearly that there is a high probability of Burnt Mound I usage being contemporary with at least some of the fire clearance. The large spatial distance between the mound samples and the clearance samples means that this is not likely to be the result of residual or intrusive material and is probably a true relationship.

Burnt Mound II:

Finally, I shall look for spatio-temporal patterns within the Burnt Mound II thematic phase, based upon the Bayesian model output. This phase, when taken in isolation, has dates with probabilities that cover the span between 2131 and 921 cal BC. A table (Table 6.4, page 150) gives the probabilities of the dates assigned to the Burnt Mound II phase falling within a series of c.250 year time-slices. The maps show the spatial distribution of these probabilities (Fig 6.26).

The table shows that the date which appears in the first two time-slices here is PoZ18029 (Fig 6.27a-b). That sample was taken as the first of a sequence of three dates (together with Poz18009 and GrN30412) from a silty peat layer beneath the trough (see above). As such, it is logical that it is earliest and, presumably, it came from material that pre-dated the construction of the mound. The rest of the dates all fall within the 1500-921 cal BC time-slices and show no particular spatial pattern (Fig 6.27d-f). This is perhaps not too surprising given the paucity of samples. The mound consisted of a large amount of preserved timber and, as such, the spatial locations of the samples taken were presumably determined according to a sampling strategy designed to obtain a date from each of four quadrants. As such, the TGIS can add little meaningful spatial detail to the excavator's interpretation. Finally, as with Burnt Mound I, it is again useful to compare the modelled usage period for Burnt Mound II to the wider radiocarbon dataset. In this case, the dating experts' model placed the mound within the period 1290-1000 cal BC (see above). Looking at the map produced (Fig 6.28), it is clear that none of the other radiocarbon evidence is contemporary to Burnt Mound II.

Probability interpolation

Interpolation is a GIS process used to determine the value of an attribute at locations intermediate to known points, and is often necessary to obtain more information. As a process, it involves relationships (understanding which points are appropriate neighbours) and axioms (assumptions about the surface between measured locations). Some methods of interpolation apply a global model (e.g. trend surfaces - see Chapter 7), but most work more locally (Chrisman 1997: 210; Laurini and Thompson 1992: 258-259; Burrough and McDonnell 1998: 98-99). The main problem with interpolation is that it is difficult to work out how accurate the models produced are without obtaining further data, and yet

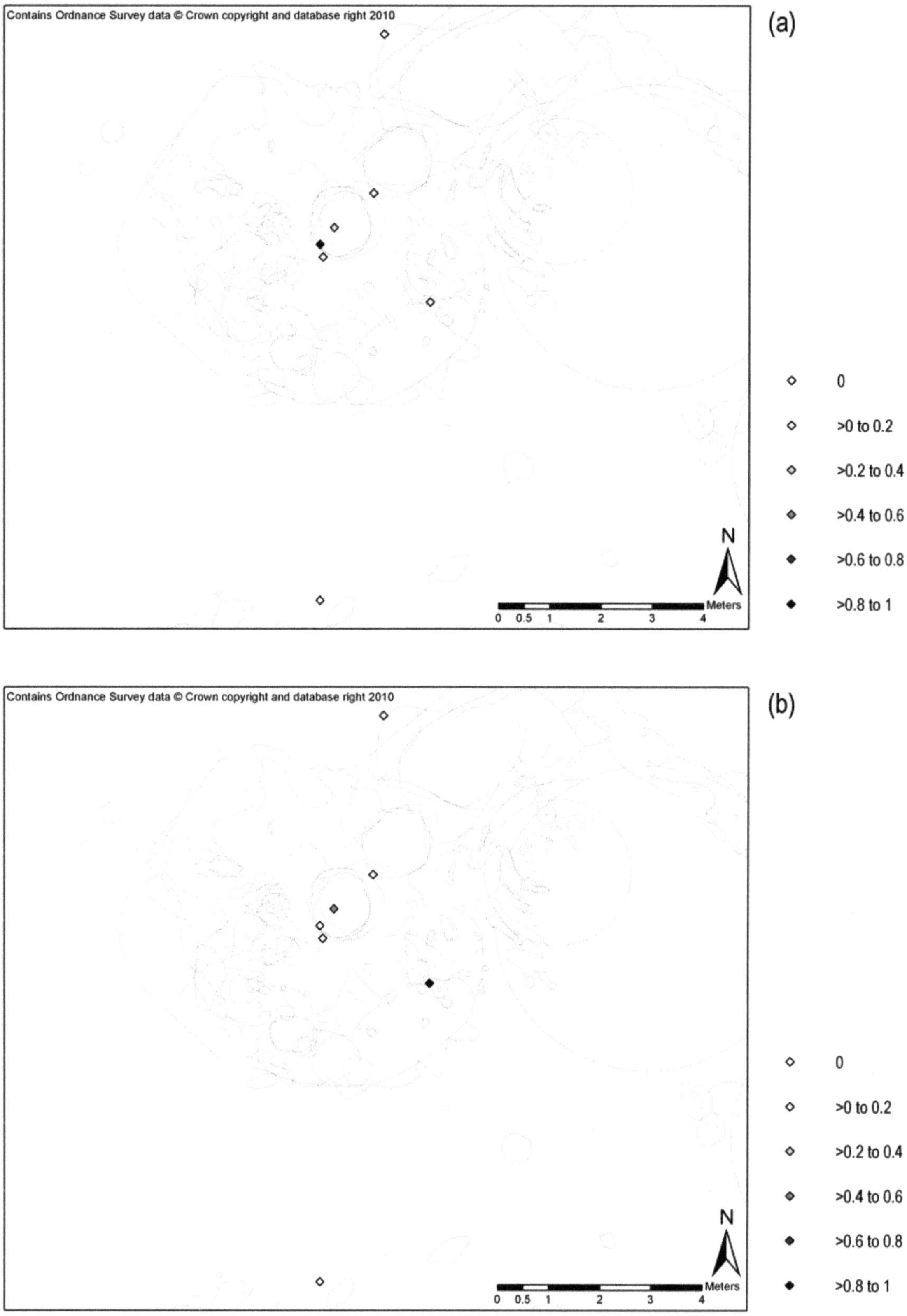

FIG 6.25 (A) TO (B) - PROBABILITIES OF BURNT MOUND I DATES FROM 3698-3500 CAL BC AND 3500-3250 CAL BC.

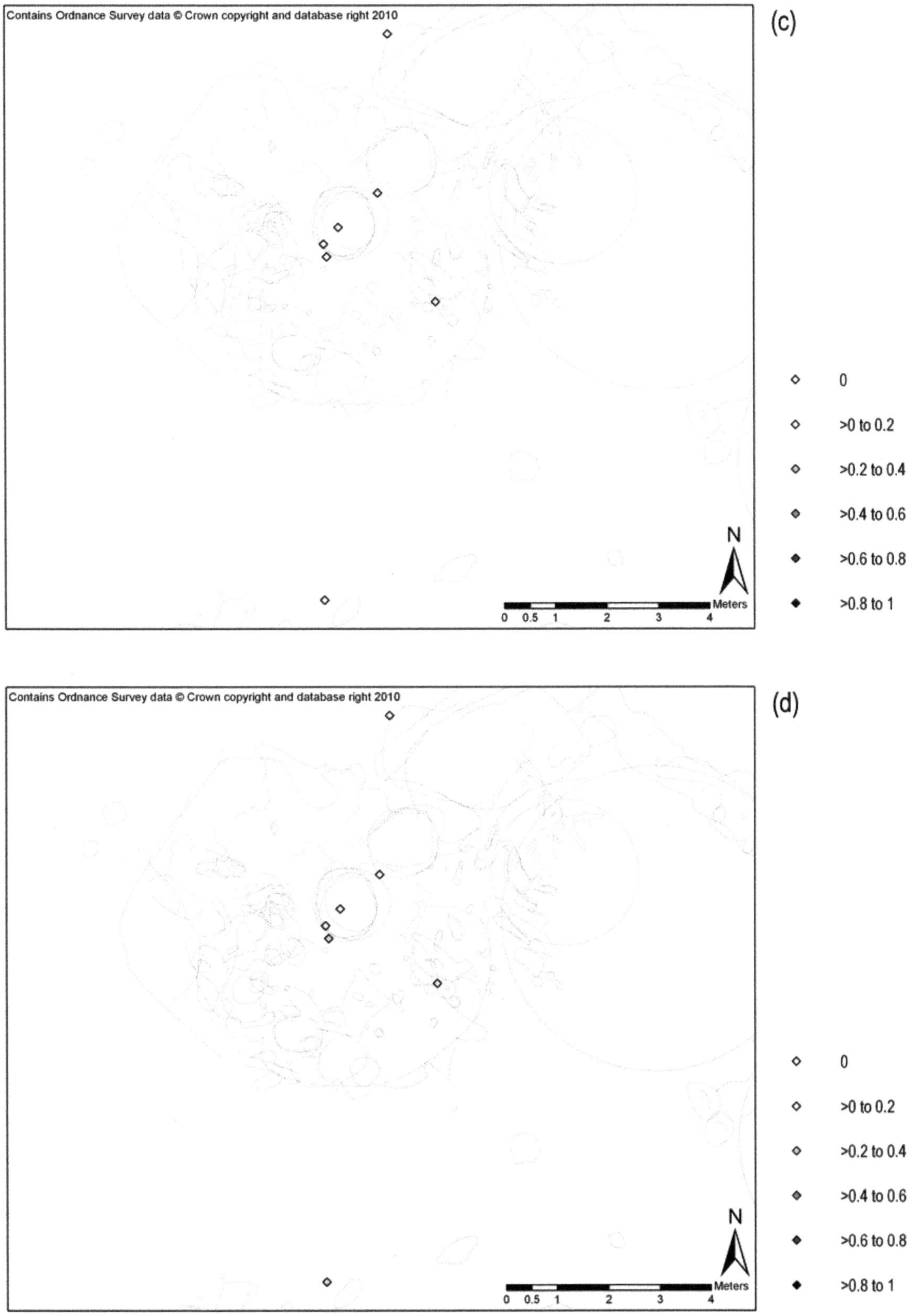

FIG 6.25 (C) TO (D) - PROBABILITIES OF BURNT MOUND I DATES FROM 3250-2500 CAL BC AND 2500-2250 CAL BC.

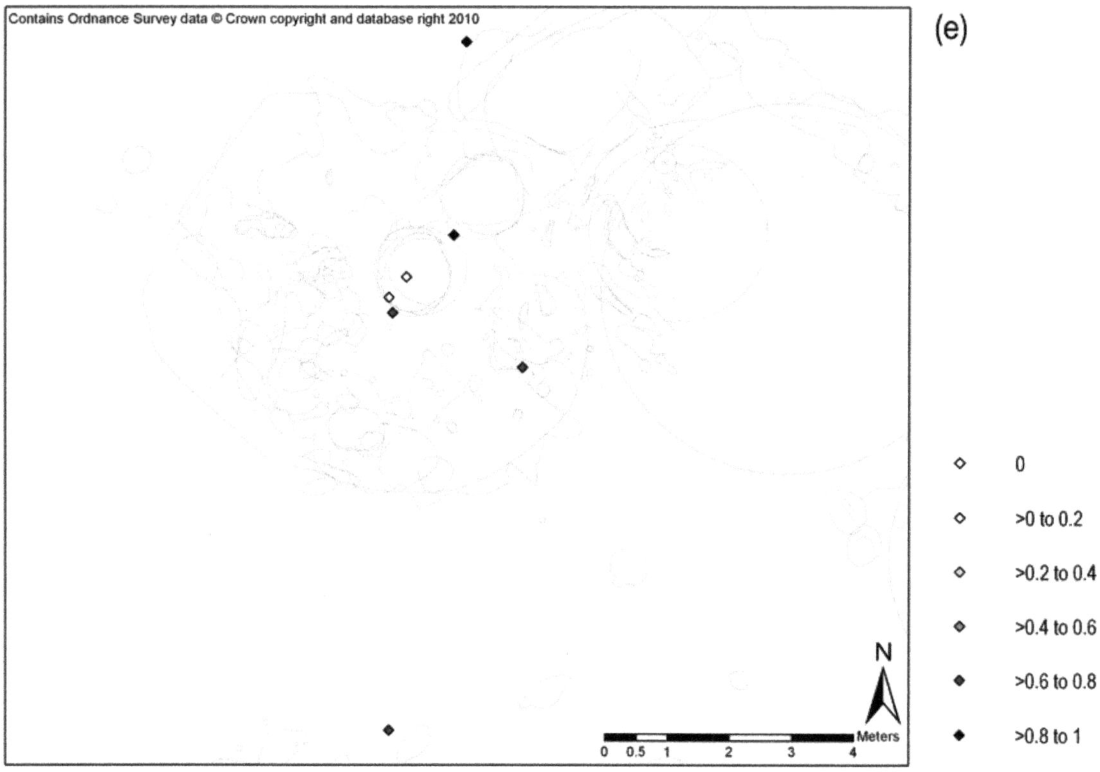

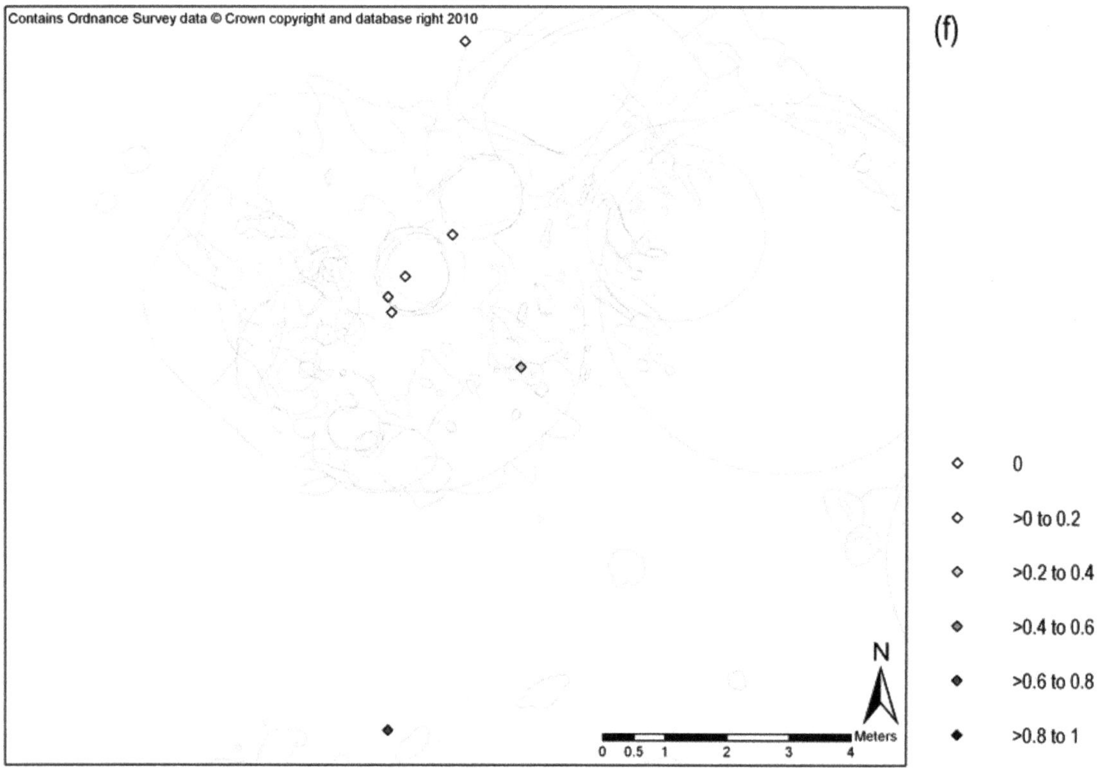

FIG 6.25 (E) TO (F) - PROBABILITIES OF BURNT MOUND I DATES FROM 2250-2000 CAL BC AND 2000-1897 CAL BC.

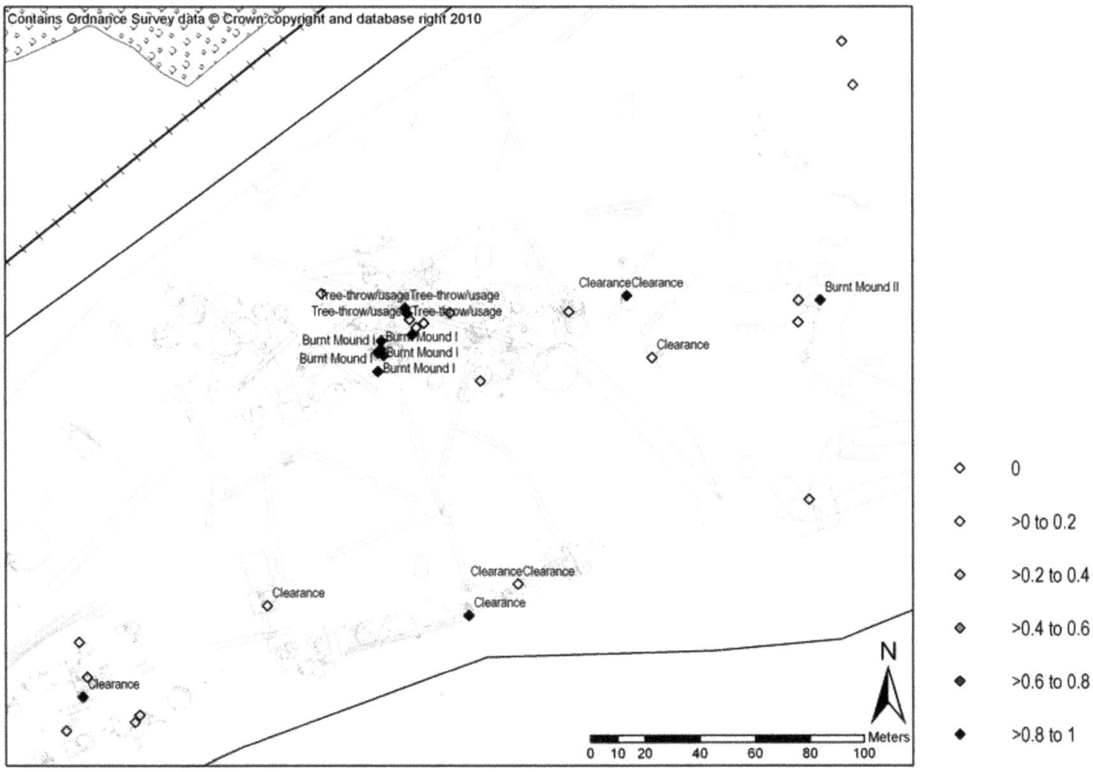

FIG 6.26 - SPATIAL DISTRIBUTION OF MODELLED RADIOCARBON DATES, SHADED ACCORDING TO PROBABILITY OF THEM FALLING WITHIN THE PERIOD 2340-1840 CAL BC. DATES OF GREATER THAN 0.0 PROBABILITY LABELLED ACCORDING TO THEMATIC PHASE.

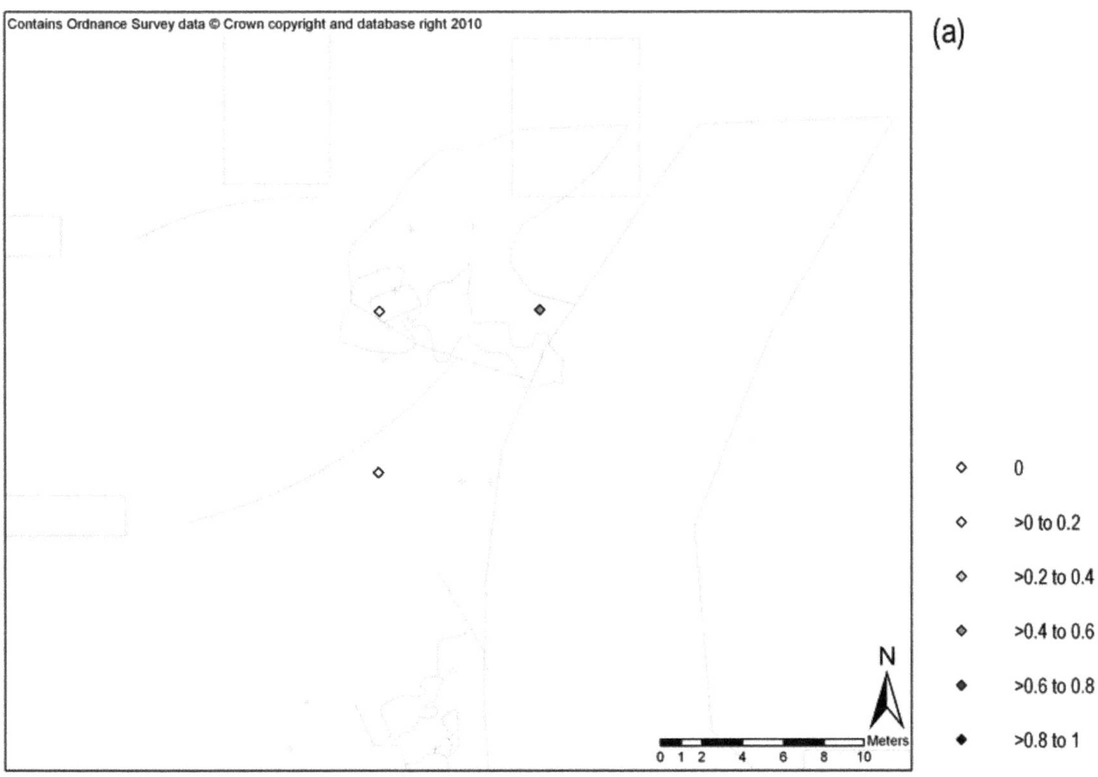

FIG 6.27 (A) - PROBABILITIES OF BURNT MOUND II DATES FROM 2131-2000 CAL BC.

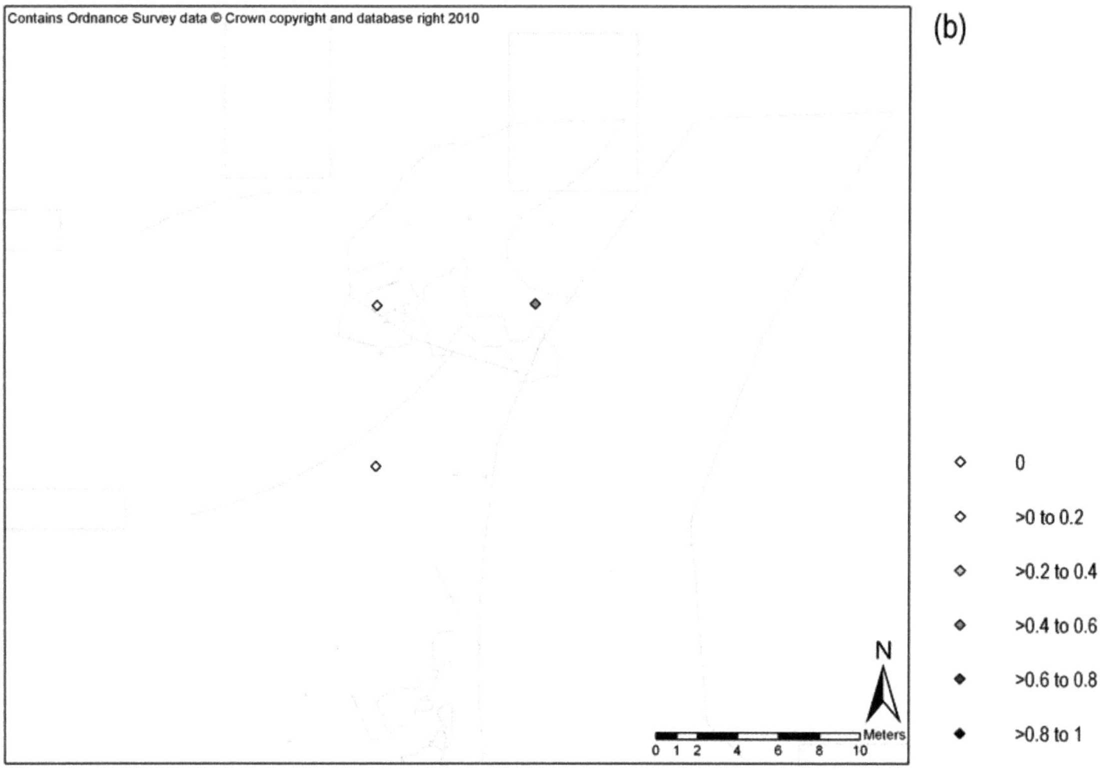

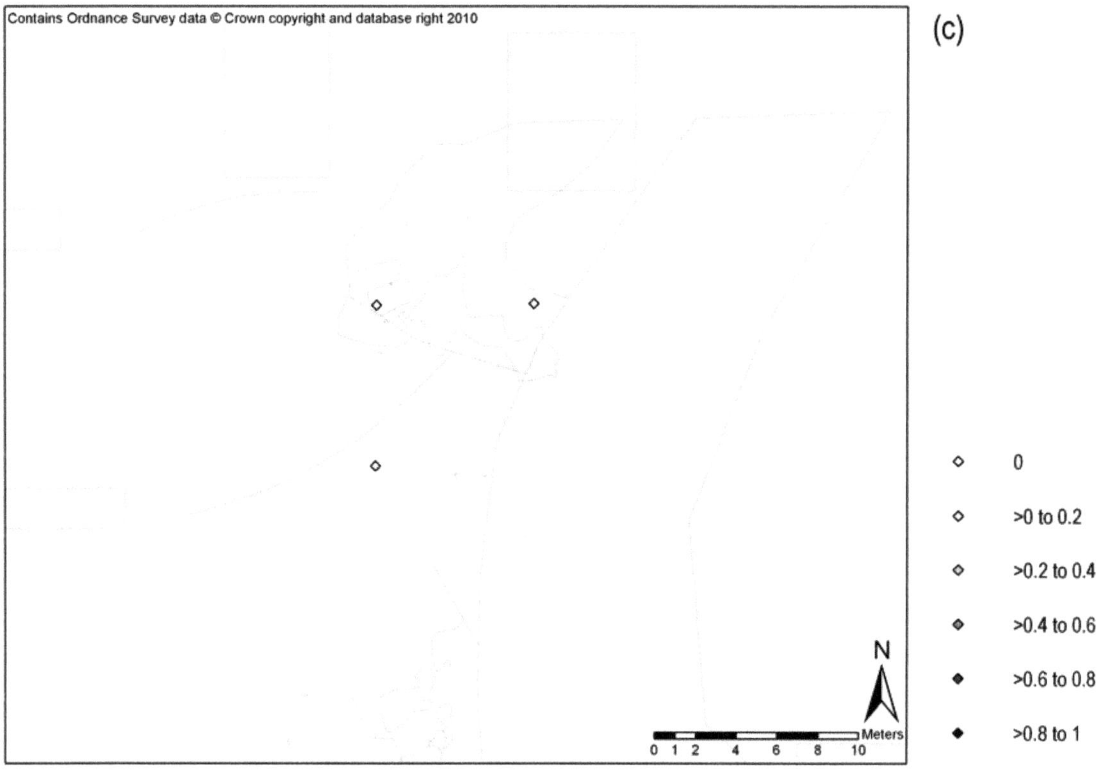

FIG 6.27 (B) TO (C) - PROBABILITIES OF BURNT MOUND II DATES FROM 2000-1750 CAL BC AND 1750-1500 CAL BC.

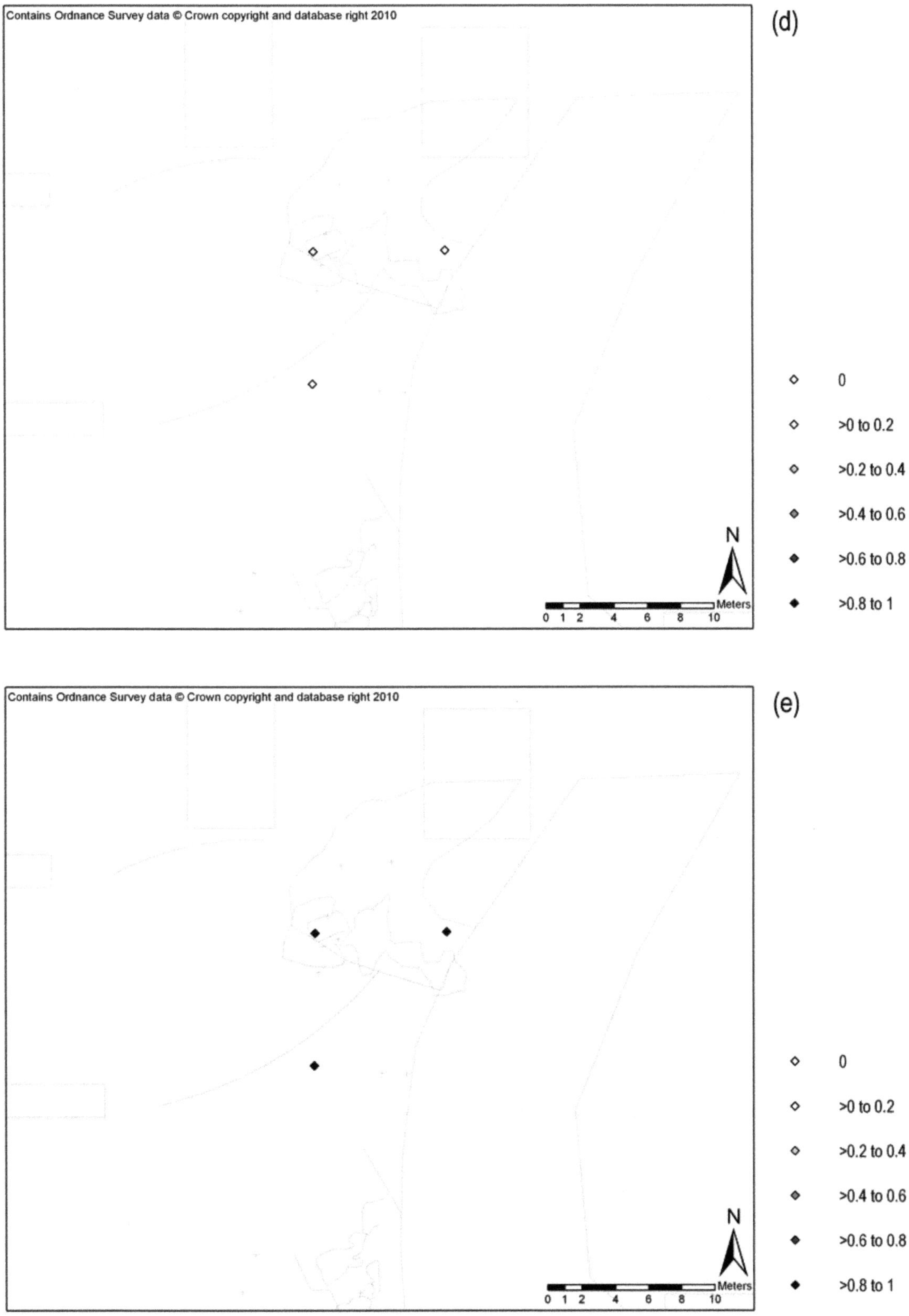

FIG 6.27 (D) TO (E) - PROBABILITIES OF BURNT MOUND II DATES FROM 1500-1250 CAL BC AND 1250-1000 CAL BC.

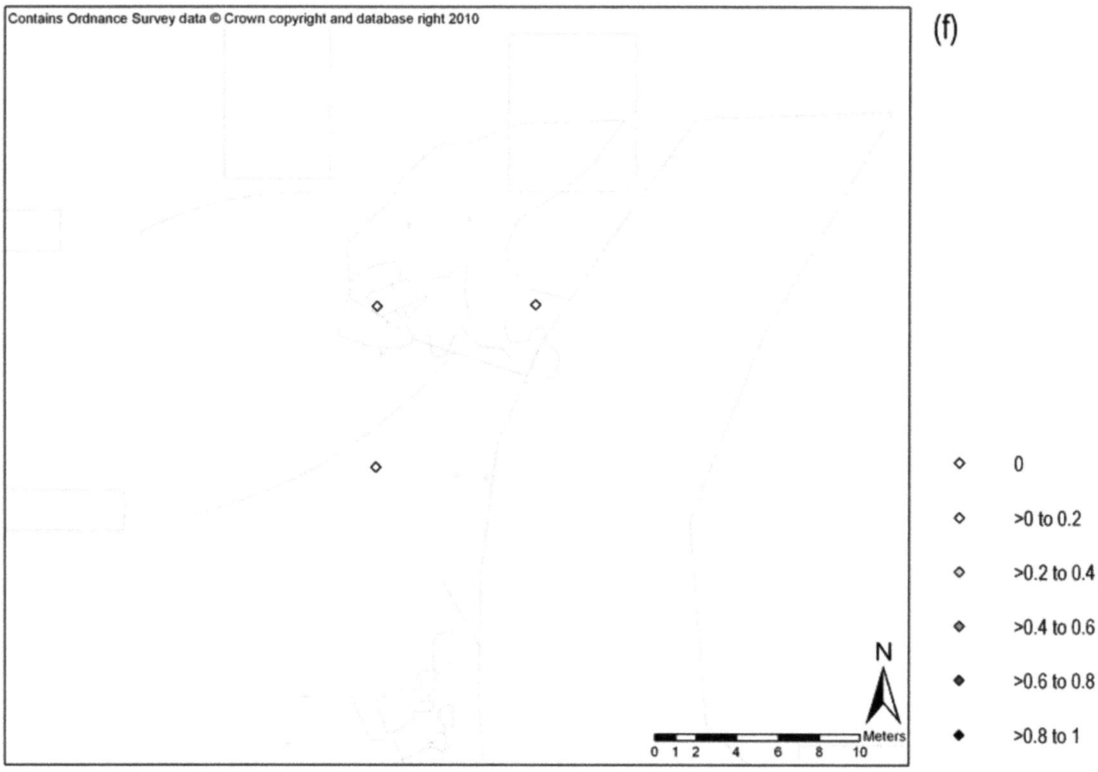

FIG 6.27 (F) - PROBABILITIES OF BURNT MOUND II DATES FROM 1000-921 CAL BC.

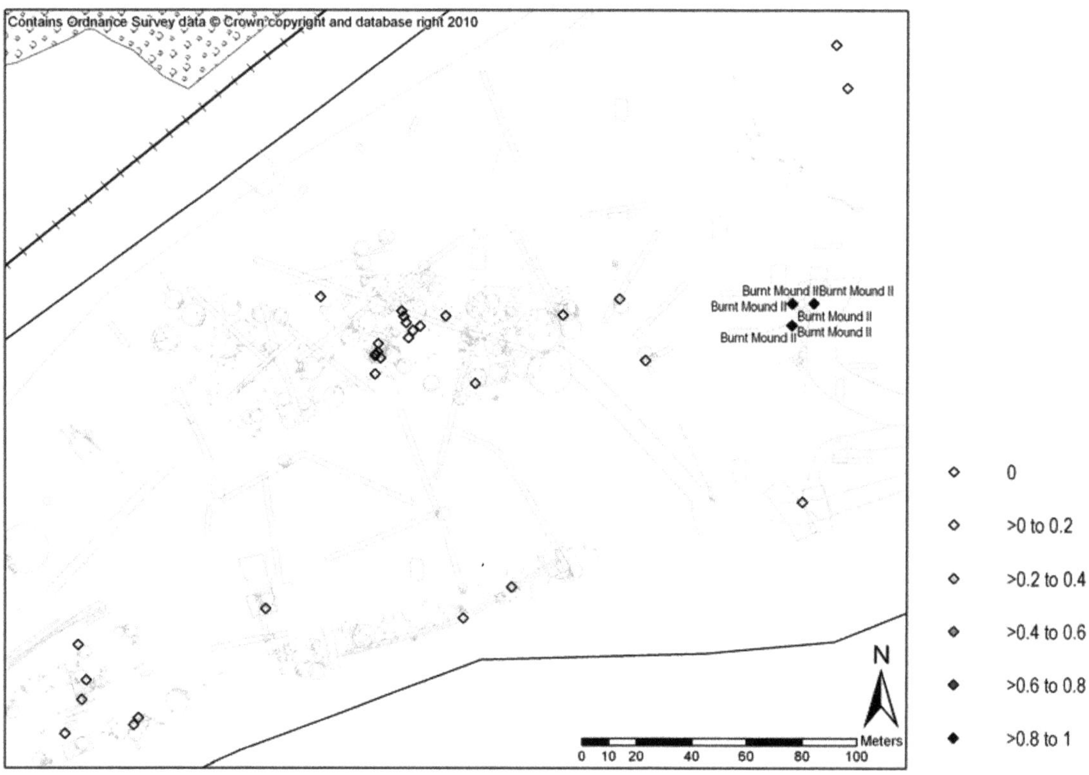

FIG 6.28 - SPATIAL DISTRIBUTION OF MODELLED RADIOCARBON DATES, SHADED ACCORDING TO PROBABILITY OF THEM FALLING WITHIN THE PERIOD 1290-1000 CAL BC. DATES OF GREATER THAN 0.0 PROBABILITY LABELLED ACCORDING TO THEMATIC PHASE.

obtaining further data reduces the need to interpolate in the first place (Wheatley 1996: 285).

Due to the lack of data points within the reduced datasets (see above), it was not deemed prudent to produce any interpolations of the probability maps produced for each thematic phase. Instead, it was decided to produce a series of interpolated probability maps for the site as a whole. These give an indication of the probability of dates falling within any area of the site and help to demonstrate the power of the TGIS in this regard. As with any interpolations, they should be interpreted with care. Firstly, less weight should be given to the probabilities as they move away from dense concentrations of sample points. Furthermore, any low probability areas should only be seen as representing a low probability of radiocarbon dates falling within that spatio-temporal neighbourhood, rather than being taken as a certain indication of low probability of contemporary activity. The following periods were chosen for interpolation, based upon the discussions above:

3750-3500 cal BC - to consider tree-throws pre-dating 3500 cal BC;

3500-3000 cal BC - to consider the main period of tree-throw usage as per the excavator's conclusions;

2500-1500 cal BC - to consider the main period of fire clearance as per the excavator's conclusions;

3366-2750 cal BC - to consider the first of four peaks of fire clearance as discussed above;

2750-2250 cal BC - to consider the second of four peaks of fire clearance;

2250-2000 cal BC - to consider the third of four peaks of fire clearance;

2000-1614 cal BC - to consider the fourth of four peaks of fire clearance;

The interpolation algorithm used was the Inverse Distance Weighted tool in ArcGIS's Spatial Analyst toolbar. This is the simplest interpolation methodology, estimating the interpolated cell values by averaging the values of data points nearby each processing cell: the closer a point is to the cell being estimated, the greater the weight it is given in the process.[1] A no data value of zero probability was set to discount these points due to the old adage that absence of evidence is not evidence of absence. More complex interpolation methods (e.g. kriging, spline interpolation) did not function correctly with the Willington dataset due to concurrent points at the same spatial location and due to the limited numbers of points overall of greater than zero probability.

A general point applies to all of these maps: any patterns in the areas towards the north-west and south-east of the site should be treated with caution due to their lack of sample points. With the tree-throw / usage interpolations (Figs 6.29 and 6.30), the picture painted is one of intensification in activity after 3500 cal BC particularly around the centre of the site. However, the data in the earlier period actually possessed a mild degree of negative spatial autocorrelation, so the result is not robust. For the fire clearance interpolations, there is a clear contrast between the 2500-1500 cal BC time-slice (Fig 6.31) and the four smaller time-slices (Fig 6.32 to 6.35). These four shorter phases do appear to show a changing spatial pattern in the radiocarbon probabilities for tree clearances, with clearance beginning in the centre and east of the site, retreating to the north east, then down to the centre and south, and finally taking place in three clusters in the west, south and east. It is tempting to suggest that this pattern is a reflection of the true pattern of tree clearance in the past, but that would perhaps be an interpretative leap too far. In particular, the data presented in Figs 6.33 and 6.34 showed little positive spatial autocorrelation. In any event, they certainly do provide an indication of how fire clearance **might** have progressed in the past: with more data, a truer picture could probably be claimed.

Conclusions

Working with real world data, produced according to a sampling strategy that did not envisage integration into a TGIS, introduces complications. With an idealised dataset, the TGIS could produce all sorts of handsome maps and patterns, but such a dataset would not reflect conditions at the archaeological coal face. Nevertheless, the application of the TGIS to the Willington material has produced some interesting results. Beginning with the negatives, it is apparent that interpolation of probability data would ideally need more sample points and over a wider spatial area: the interpolations do paint some interesting pictures, particularly in regard to possible change in fire clearance patterns, but are not based on enough evidence to be robust. The lack of sample points also made TGIS study of the alluviation thematic phase impossible, and the Burnt Mound II thematic phase problematic. Furthermore, great care is needed when considering whether the TGIS results contain any elements caused by sample residuality / intrusion. This is equally true of any chronological work, but the nature of tree-throw deposit formation (see above) does create particular difficulties in this regard at Willington.

It is apparent that archaeologists are beginning to appreciate the value of obtaining many radiocarbon samples (naturally, subject to cost considerations) and integrating them with stratigraphic information in order to reveal and improve temporal pattern (e.g. Bayliss *et al.* 2007; Whittle *et al.* 2007). However, there is not yet a general appreciation of the importance of the (horizontal) spatial dimension of these samples. As a result, if not chosen with spatial considerations in mind, the full analytical potential of taking many radiocarbon samples may not be achieved. Hopefully, as the possibilities of TGIS start to become recognised, the importance of dense

[1] http://webhelp.esri.com/arcgisdesktop/9.2/index.cfm?TopicName=Implementing_Inverse_Distance_Weighted_(IDW)

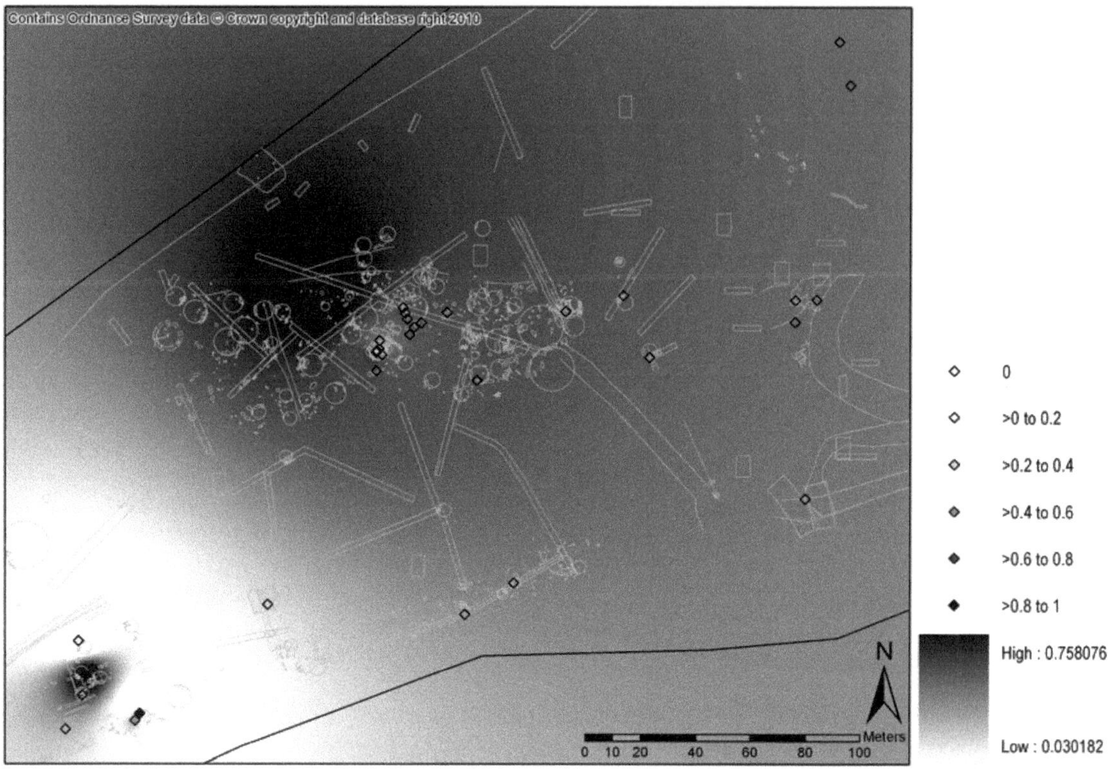

FIG 6.29 - INVERSE DISTANCE WEIGHTED INTERPOLATION OF BAYESIAN PROBABILITIES: 3750-3500 CAL BC.

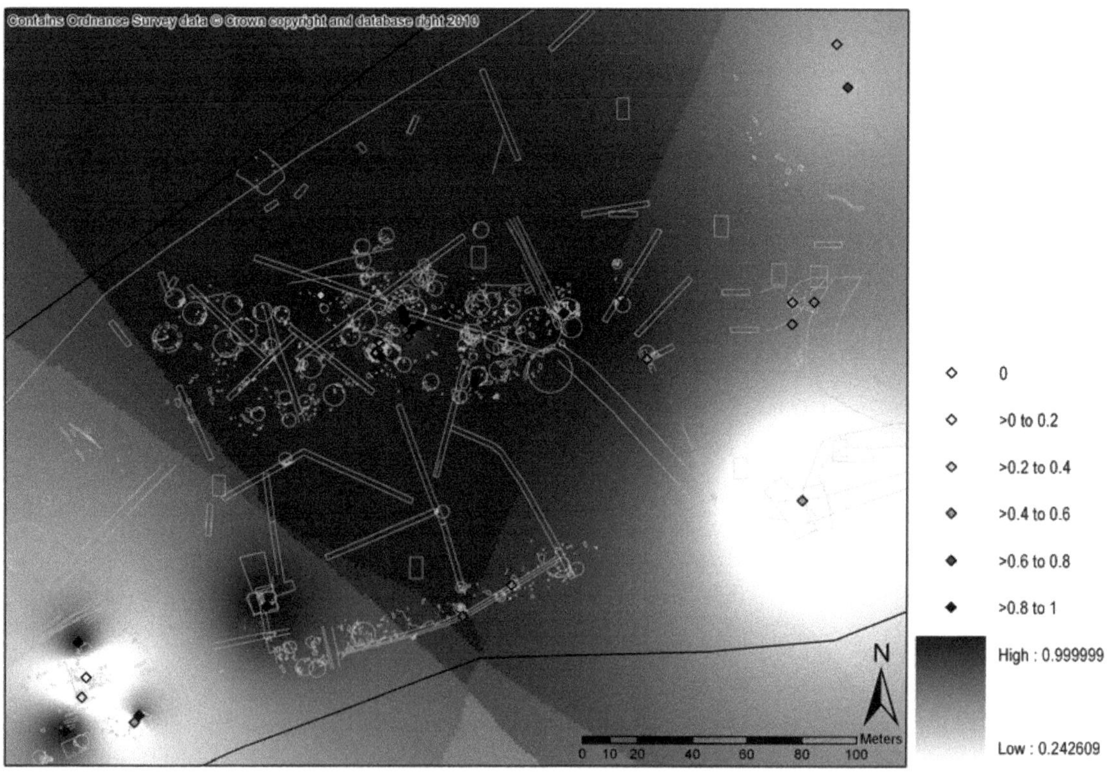

FIG 6.30 - INVERSE DISTANCE WEIGHTED INTERPOLATION OF BAYESIAN PROBABILITIES: 3500-3000 CAL BC.

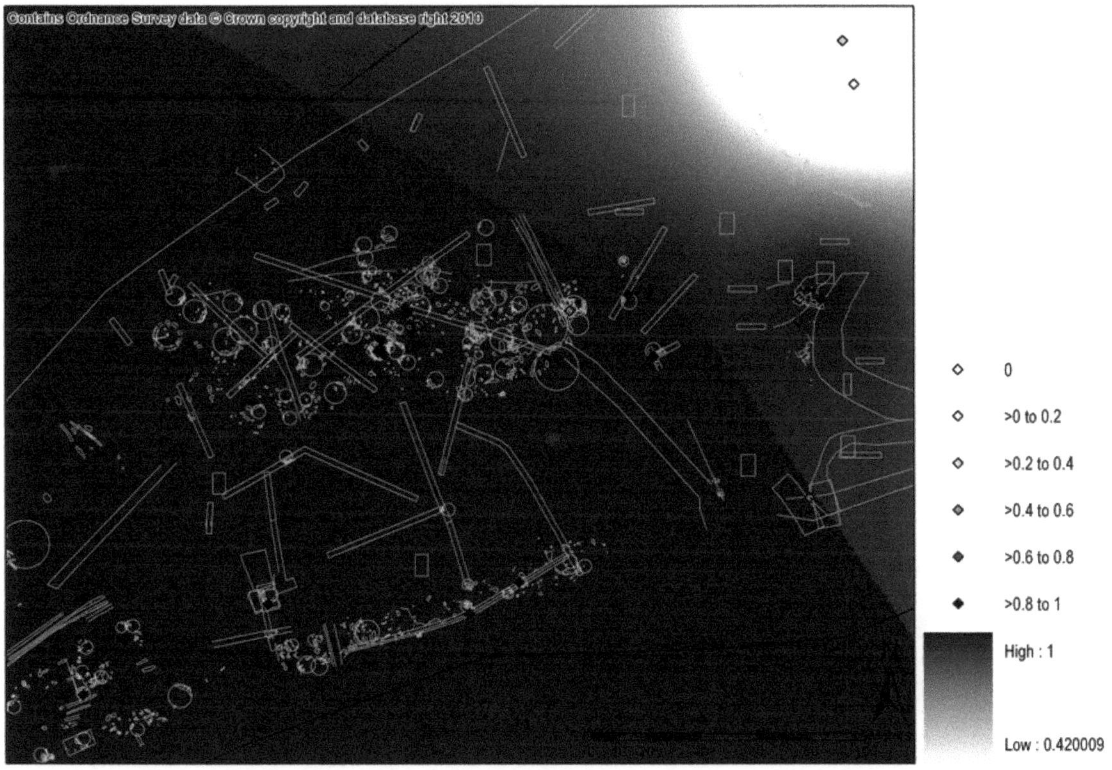

FIG 6.31 - INVERSE DISTANCE WEIGHTED INTERPOLATION OF BAYESIAN PROBABILITIES: 2500-1500 CAL BC.

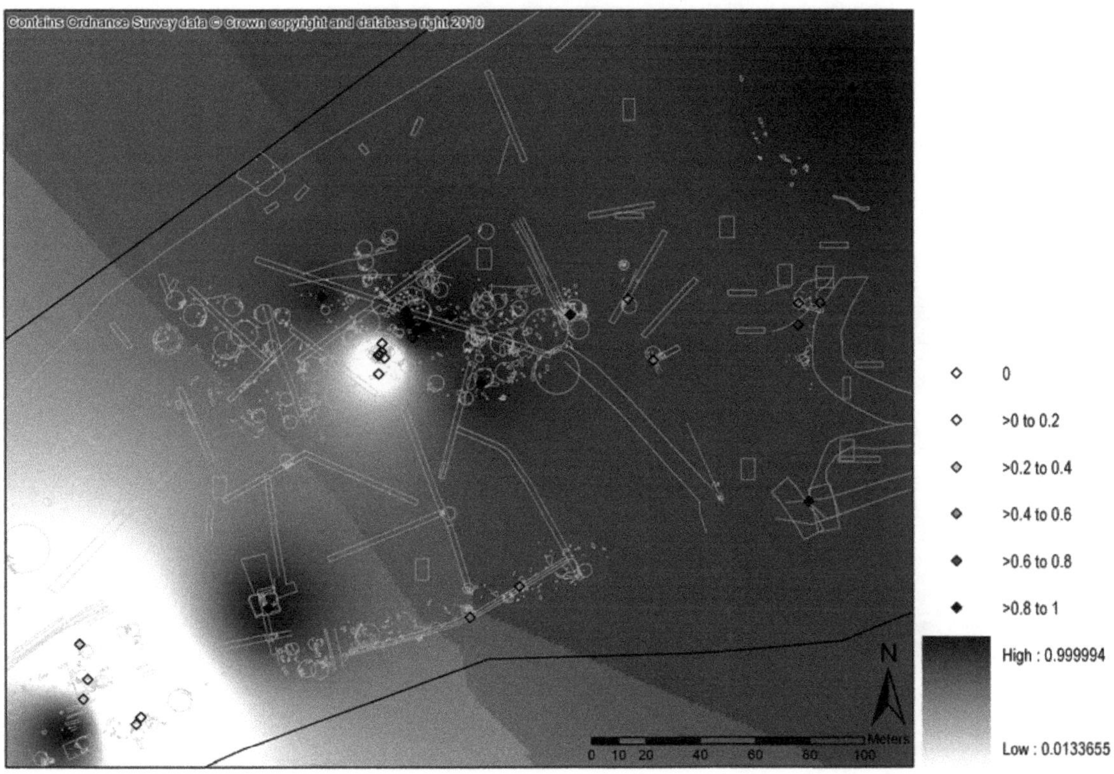

FIG 6.32 - INVERSE DISTANCE WEIGHTED INTERPOLATION OF BAYESIAN PROBABILITIES: 3366-2750 CAL BC.

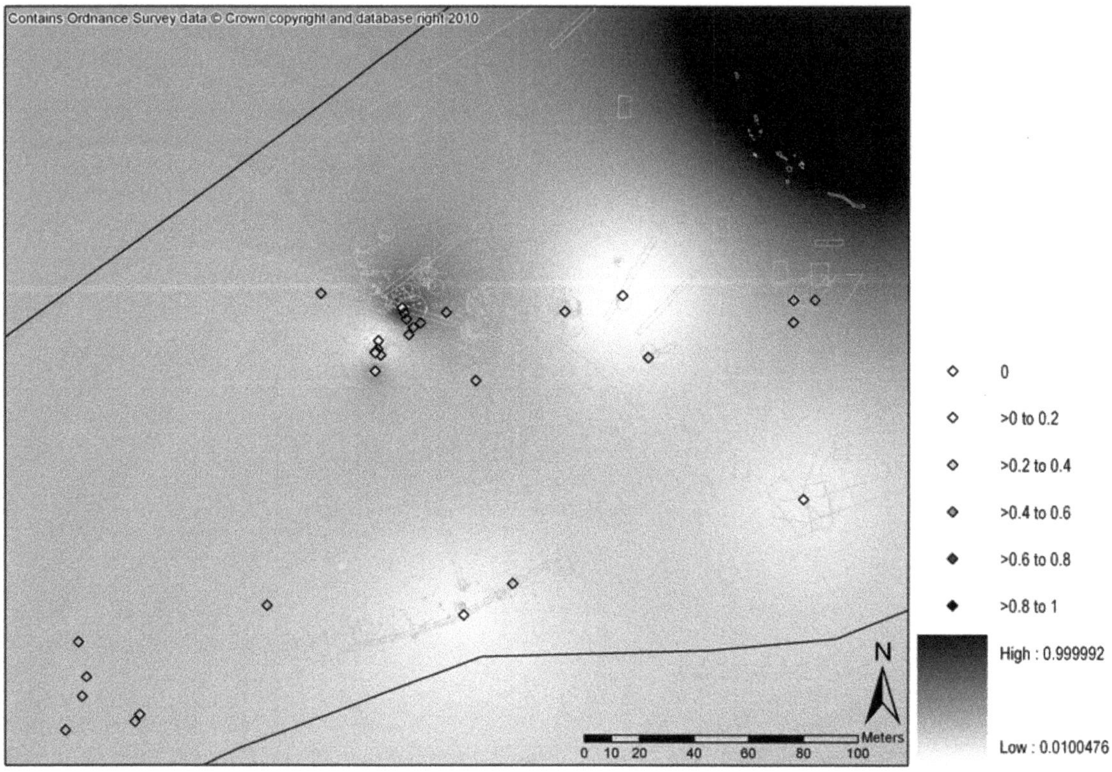

FIG 6.33 - INVERSE DISTANCE WEIGHTED INTERPOLATION OF BAYESIAN PROBABILITIES: 2750-2250 CAL BC.

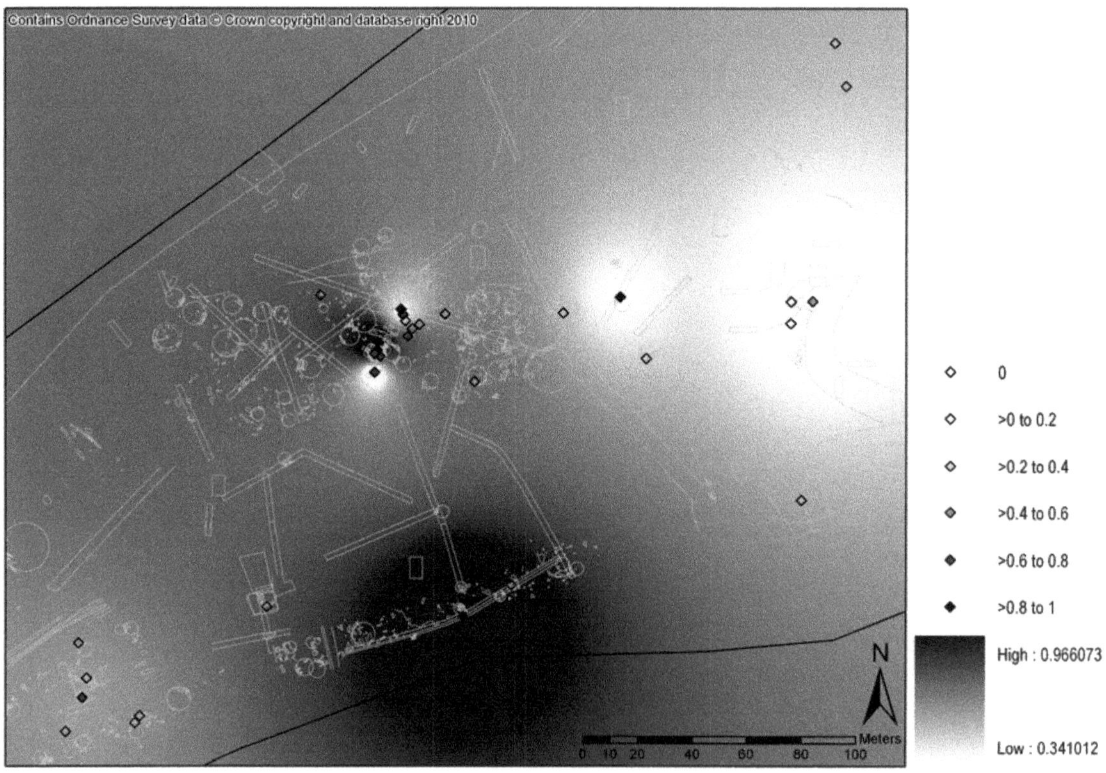

FIG 6.34 - INVERSE DISTANCE WEIGHTED INTERPOLATION OF BAYESIAN PROBABILITIES: 2250-2000 CAL BC.

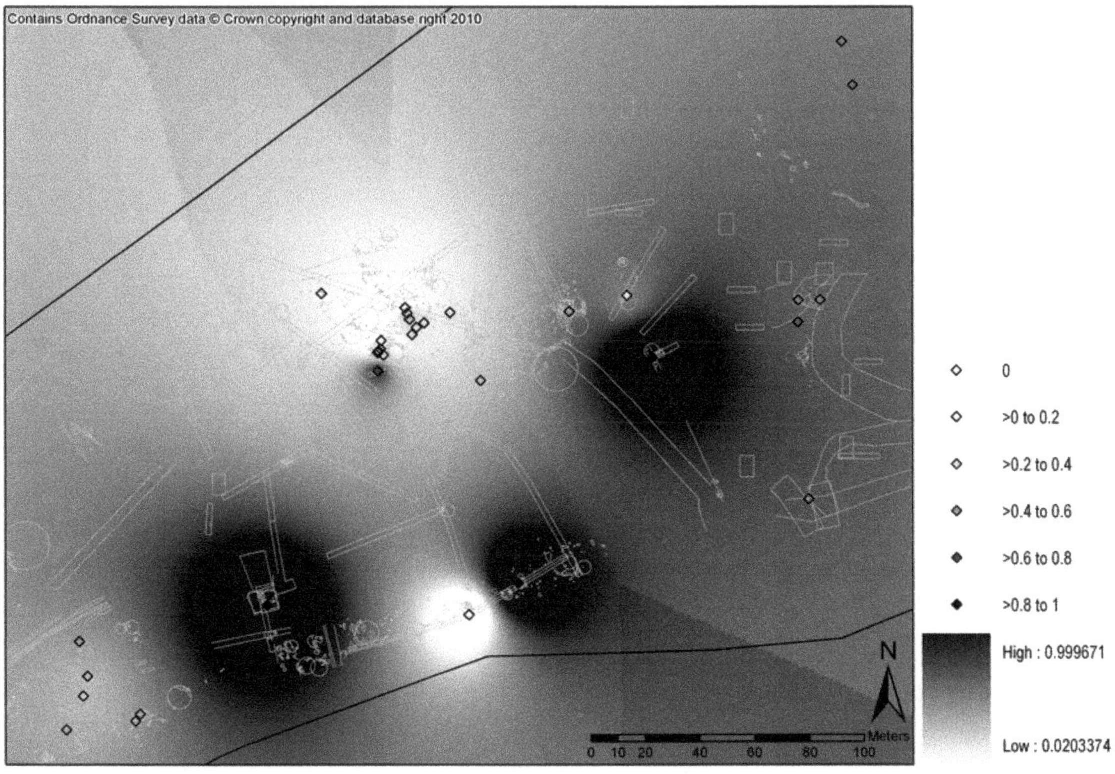

FIG 6.35 - INVERSE DISTANCE WEIGHTED INTERPOLATION OF BAYESIAN PROBABILITIES: 2000-1614 CAL BC.

and spatially-distributed sampling regimes may be realised; a possible sampling strategy in this regard will be outlined in Chapter 8. This is perhaps particularly important when dealing with the type of archaeological intervention prior to commercial quarrying of which Willington is a relatively small scale representative. Where whole sectors of past landscapes are to be excavated, unique opportunities arise to reconstruct the past on a greater spatial scale than is usually the case. If the full potential of this reconstructive exercise is to be achieved, then sampling strategies for dating evidence need to reflect both geographic and chronological concerns. In the alternative, spatially-aware archaeologists will remain trapped in the rut of dealing primarily with agglomerated period data of low temporal resolution, and true spatio-temporal analysis will remain an unfrequented arena.

Moving on to the archaeological interpretative results of the TGIS analysis, several points arise. Firstly, it would appear that tree-throw usage activity began earlier in the south-western area of the site, at some time between 3750 and 3500 cal BC. The majority of activity in this thematic phase does, however, fall within the 3500 to 3000 cal BC period as identified by the excavator. With the fire clearances, three or four distinct peaks are seen in the radiocarbon evidence. The first of these is maybe caused by intrusive material, but the other three are more convincing. It is possible, however, that these three peaks simply reflect a pattern caused by the radiocarbon sampling process: more data points could have subsumed this cyclical picture of clearance. Without any evidence to the contrary, however, we can suggest that the result is likely to be a genuine one. Also, support is provided for the excavator's conclusion that the Burnt Mound I thematic phase had a high probability of being contemporary to the fire clearances. Although undoubtedly more speculative, my analysis suggests that the tree-throw activity may either have taken place in one large clearing that spread from the south-west, or that it took place in two separate clearings, with the south-western being earlier. Furthermore, the subsequent fire clearance appears to be more opportunistic and piecemeal than directed and systematic. Finally, the ability of the TGIS to combine the spatial with the temporal has proven useful in considering questions about residuality / intrusion. Understanding this issue is crucial to archaeological interpretation and, as such, any tool or methodology that can add rigour to the assessment of residuality is of considerable benefit.

In conclusion, the application of the TGIS to the Willington material has provided evidence for the interpretive potential inherent in the TGIS for the study of intra-site archaeology and radiocarbon dating evidence. If a new site's sampling strategy was constructed around an assumption of integration into the TGIS, with more samples taken and / or over a wider spatial spread, then stronger results would probably be produced. Despite this, the successful integration of a dataset that was not based upon such a strategy has proven the wider applicability of the TGIS to the analysis of intra-site material. It has provided evidence for spatial patterns in the data and helped to consider questions concerning residuality / intrusion. These results could probably have been achieved without the use of TGIS, but the key point is that the software made the process easier to undertake and much easier to visualise spatially. Leaving intra-site GIS behind, the next question to ask is how the TGIS performs with inter-site data.

CHAPTER SEVEN: POTTERY IN ROMAN NORTHAMPTONSHIRE

In this chapter, I shall demonstrate how the TGIS can be used to facilitate and enhance the analysis of a regional archaeological dataset. The specific case study selected consists of an examination of the dates of pottery collected during field survey conducted over a large region, taking a probabilistic approach to the elucidation of any spatial patterning. A key aim of this chapter is to explore pathways for extending formal spatial analysis into the temporal dimension, an important objective of TGIS (see Chapter 4).

Background

The dataset used in this chapter comprises a corpus of Roman pottery evidence collected via field survey and collated by Dr. Jeremy Taylor for his PhD thesis (1996). The study area encompassed the Nene and lower Welland valleys, comprising the pre-1974 county of Northamptonshire. The original project aimed to create a landscape-based perspective on occupation around the Nene valley (which falls as a boundary area between archaeological cultural groupings) from the 4th century BC to the 4th century AD, through the bringing together of the diverse evidence for the region and the incorporation of new data. Taylor summarised his methodology as follows (Taylor 1996: 2, 102-103):

1. To incorporate the diverse information available from published excavations into a study of the spatial development of activity foci in relation to their immediate landscape context;
2. To integrate this with survey data to better understand the spatial organisation of several blocks of landscape across the region, in order to study changing patterns of land use and social interaction;
3. To examine additional field-walked and excavated data from the county to assess whether particular landscape forms extended across the region or if other landscapes might have existed in areas outside those of the detailed study.

It is the third of these stages that is of particular interest here. The database was constructed of field walked material collected by David Hall from across Northamptonshire since the 1970s, with pottery scatters taken as the material correlates of past activity foci in the landscape (Taylor 1996: 111, 185). This type of field survey data has material of a fundamentally different nature to that provided by excavation, generally being of greater spatio-temporal span but at lower resolution. As such, it provides us with the possibility of studying inter-site patterns, but rarely provides complete coverage. Therefore, interpolation is needed to fill in the gaps (Wheatley 1996: 275). Specifically, the survey in this case formed an extensive exercise in mapping the location and general extent of settlements and other foci, through a program of field-walking available arable fields at 30m line intervals and recovering concentrations of material that pre-dated AD 1400 (Fig 7.1). The wide intervals between field-walkers and the single visit to each parish (usually covering no more than 10% of each area) restricted the coverage of the survey. The material for the Iron Age and Roman periods was recorded according to a site-oriented methodology (i.e. material was recorded when it was deemed to form a site) but with a poorly defined density threshold for site definition, biasing the data towards relatively high density scatters and with no consideration of temporal variation in circulation. This probably caused an under-representation of particular periods and scales and created a bias towards locations where activity was long-lived, and / or ceramic deposition was high and focussed, and / or where robust fabrics were produced (Taylor 1996: 186-187).

Inter-site comparisons are only possible if we assume that the recovered data form a representative sample of the material deposited irrespective of the size of the place, an approach that Taylor argued could be criticised but which had undoubtedly provided interesting results in the past (Taylor 1996: 187-188). In order to consider spatio-temporal variation through the period of interest, a database comprising 232 groups of artefacts was analysed according to the fabric series and grouped into 33 wares on the basis of fabric, form, provenance and chronology (Table 7.1, page 151). The artefacts were then quantified by weight and sherd count, with the sherd count being favoured in the analysis that follows; this was the variable used in Taylor's analysis and it is more simple statistically, in the sense that it has not been subject to mathematical calculation (in contrast to density). The chronological framework was constructed from published excavation reports and the wares then divided into a series of chronological periods, on the grounds of the degree of overlap between them - with some wares overlapping much more than others (Table 7.2, page 152; Fig 7.2). The boundaries between phases were inserted where least overlap was apparent (Taylor 1996: 188-189).

Wares 28 and 33 were used as catch-all groups for fabrics that could only be dated Early to Late Roman at best and were not used in Taylor's interpolations. With regard to the periods used, the greatest overlap was between the Mid / Late and Late periods. Various (unspecified) problems suggested that perhaps these two periods should have been grouped together, but there was no time for Taylor to experiment with this. The specific question which needed to be answered through interpolation was the extraction of the underlying large scale patterns usually associated with the overall exchange area for pottery from a particular source (i.e. patterns of supply and distribution). Achieving this would allow localised variation (i.e. sites with a greater or lower than expected proportions of a given ware) to be discovered that could then be linked to differences in the status or function of individual locales (Taylor 1996: 189, 196, 200).

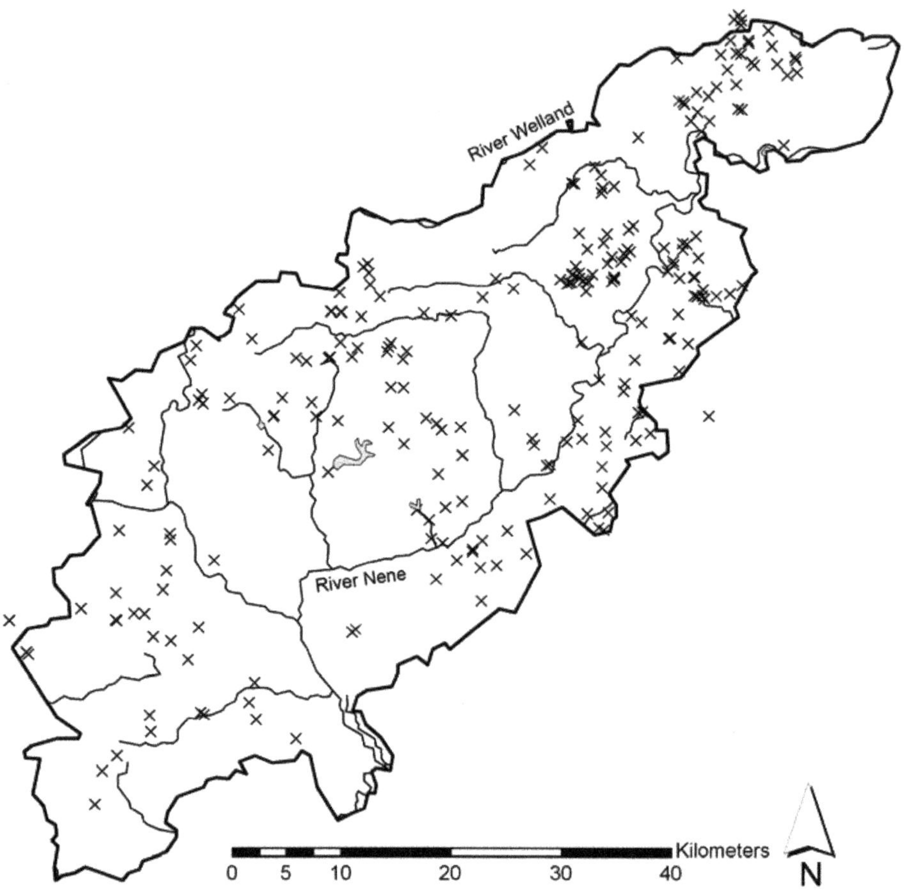

Fig 7.1 - Field survey site locations for Iron Age and Roman periods.

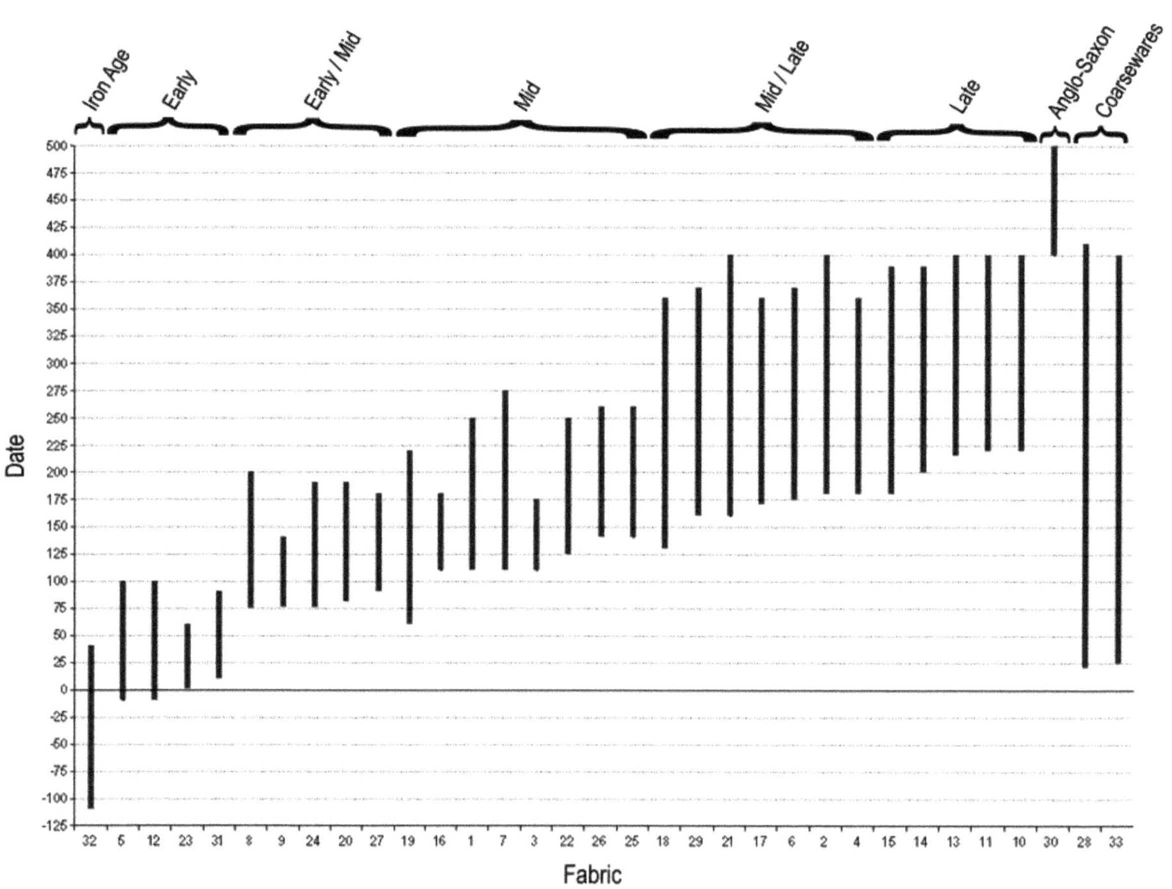

Fig 7.2 - Graph of ceramic types grouped into phases.

Trend surface analysis was the interpolative method settled on to achieve this, as it attempts to split each data point across an area into two components: the regional trend and purely local effects. It does this by fitting a mathematical best-fit surface (a polynomial function) using ordinary least squares regression analysis, with the trend surface values being interpreted as the regional pattern, and with residuals above or below the surface showing local effects. As such, trend surfaces can be used as crude models of overall pottery supply patterns through time across a region, and they can then be employed to remove regional biases and allow better comparison of what constituted high or low levels of ceramic deposition at a site (Taylor 1996: 200-201; Bailey and Gatrell 1995: 168).[1] Trend surface analysis deliberately generalises the surface to discover the underlying trend. As such, it will almost never exactly honour the observed data for two reasons: measurement error and the fact that most phenomena are subject to more than one trend-producing process. The main difficulty when undertaking this form of interpolation is deciding upon the form of the trend to be modelled, as the data are not used to help select the model. Trend surface analysis should be treated with care, as there is no strong reason to assume that the phenomenon of interest varies spatially in such a simple fashion. In essence, it has most merit as an exploratory technique when used to indicate broad trends and relationships, without too much emphasis on formal hypothesis tests for the models produced (O'Sullivan and Unwin 2003: 256, 265; Bailey and Gatrell 1995: 171).

Trend surface analysis has been widely, if not always appropriately, used by many different interest groups (Bailey and Gatrell 1995: 168) and certainly has archaeological potential. Archaeological distributions are often difficult to interpret due to being the result of many interlocking factors and the technique has the potential to extract broad regional trends from these complex pictures. Analytically, the peaks in an archaeological trend surface may be due to patterns of fieldwork or may carry further significance. Further, the polynomial generalisations produced form an approximation of past spatial processes and should not be confused with the past processes themselves. Hodder and Orton stated that the main difficulty with the use of trend surface analysis in archaeology comes when assessing the quality of the interpolation produced: formal tests such as the analysis of variance are not appropriate, since the necessary assumptions cannot be justifiably applied to most archaeological data. They suggested comparison with randomly generated data as a possible method for performing this assessment, but I find it difficult to see what this would achieve. Other problems associated with the archaeological usage of trend surface analysis are the need for sufficient data points, edge effects caused when data points do not extend to the edges of the study area, and possible effects caused by the clustering of data points (Hodder and Orton 1976: 155, 158, 162-164). As such, although useful, trend surface analysis in archaeology should be undertaken with appropriate caution.

With insufficient data to create trend surfaces for his Iron Age and Anglo-Saxon phases, Taylor created a series of surfaces for the other five phases. He used computer software to fit quadratic polynomial surfaces to the data, as these were shown to produce a significantly better fit than simple surfaces. As a general rule, the surfaces were more reliable towards the centre of each plot and any results for outside the area of the county should be treated sceptically due to the lack of data points (Taylor 1996: 201-202). In order to make the interpolations directly comparable with those produced using the TGIS, I recreated Taylor's surfaces in ArcGIS using the same tool and parameters as for the TGIS results. The specific parameters used were a polynomial function with a maximum power of two (i.e. quadratic) and using linear regression (the logistic alternative being applicable to binary data only). The broad trends are the same as Taylor's results, but carry a little less local detail. I can only think that this is because Taylor's surfaces were not in fact quadratic, but cubic or even quartic, as his surfaces did not possess the smoothed form of a quadratic surface. This is not a problem, as the key consideration is how they compare to the TGIS results (which will follow later) and the broad trends remain present. The five following maps show the resulting trend surfaces replicating Taylor's original analysis.

Taylor interpreted the Early phase (Fig 7.3) as suggesting a considerable divide along the northern side of the Nene valley. In the Early / Middle phase (Fig 7.4) he saw a marked re-orientation of production and the supply of pottery prior to the Roman conquest, a pattern that coincides well with known pottery production sites during this period. The Middle period (Fig 7.5) showed a similar pattern and could have been affected by the chronological overlap with the preceding phase. Taylor noted that the Mid / Late phase (Fig 7.6) was dominated by Nene valley colour coats, but that this pattern was reversed in the Late phase (Fig 7.7). He opined that the mirror effect may have been due to the contemporaneity of fourth century Nene Valley (in Mid / Late) and Oxfordshire (in Late) colour coats, but that the lower values for the latter would have been subsumed under the trend of the former in the Mid / Late interpolation. Taylor concluded that the trend surfaces showed a series of changing patterns over time and space in the quantity of pottery deposited across the region, as the foci of production and redistribution changed (Taylor 1996: 201-205). He then went on to study the residuals at each site above and below each trend surface, and interpreted what those results meant in terms of supply and distribution of pottery (Taylor 1996: 205-215). It is not my intention in the remainder of this chapter to challenge or refine Taylor's final interpretations. Instead, the hypothesis of this thesis is to test how the TGIS produces different (and more robust) results than conventional GIS analysis. If the TGIS trend surfaces are different from those produced

[1] http://webhelp.esri.com/arcgisdesktop/9.2/index.cfm?TopicName =Performing_trend_analysis

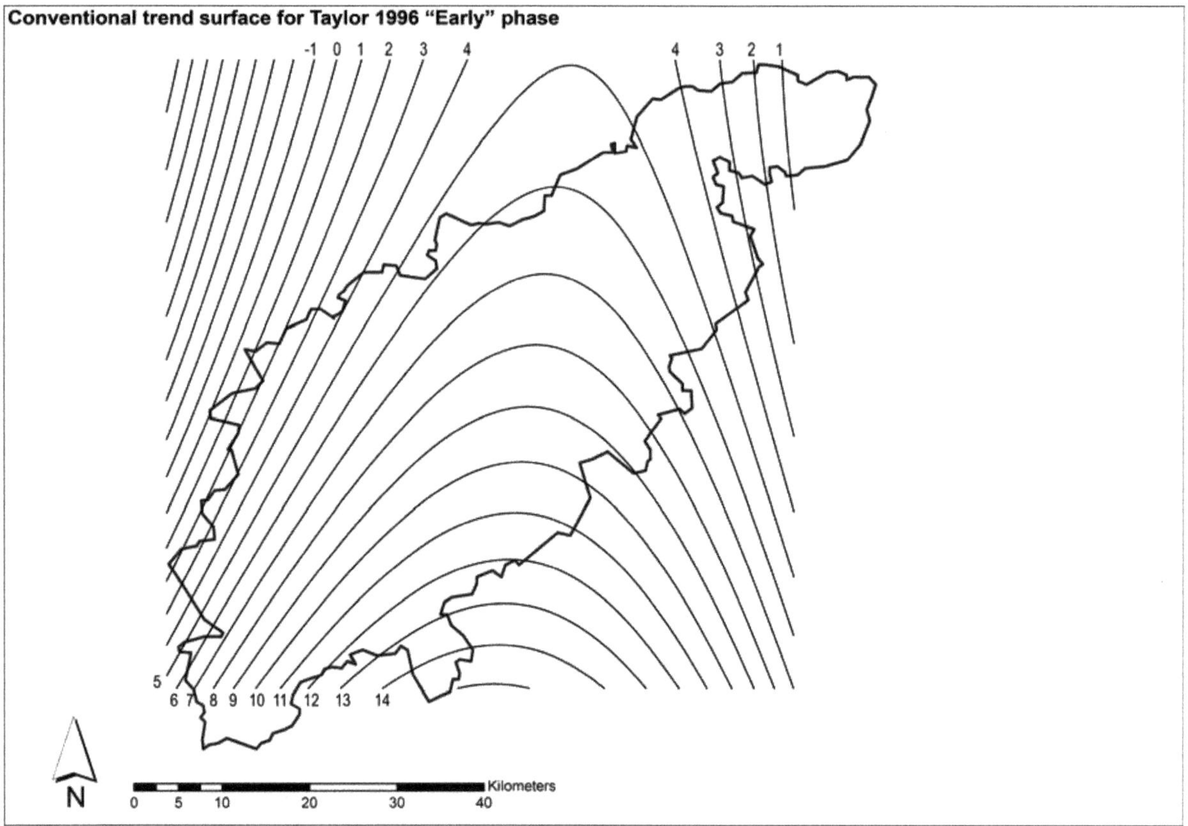

FIG 7.3 - TREND SURFACE FOR EARLY PHASE BASED UPON FABRIC CLASSIFICATION.

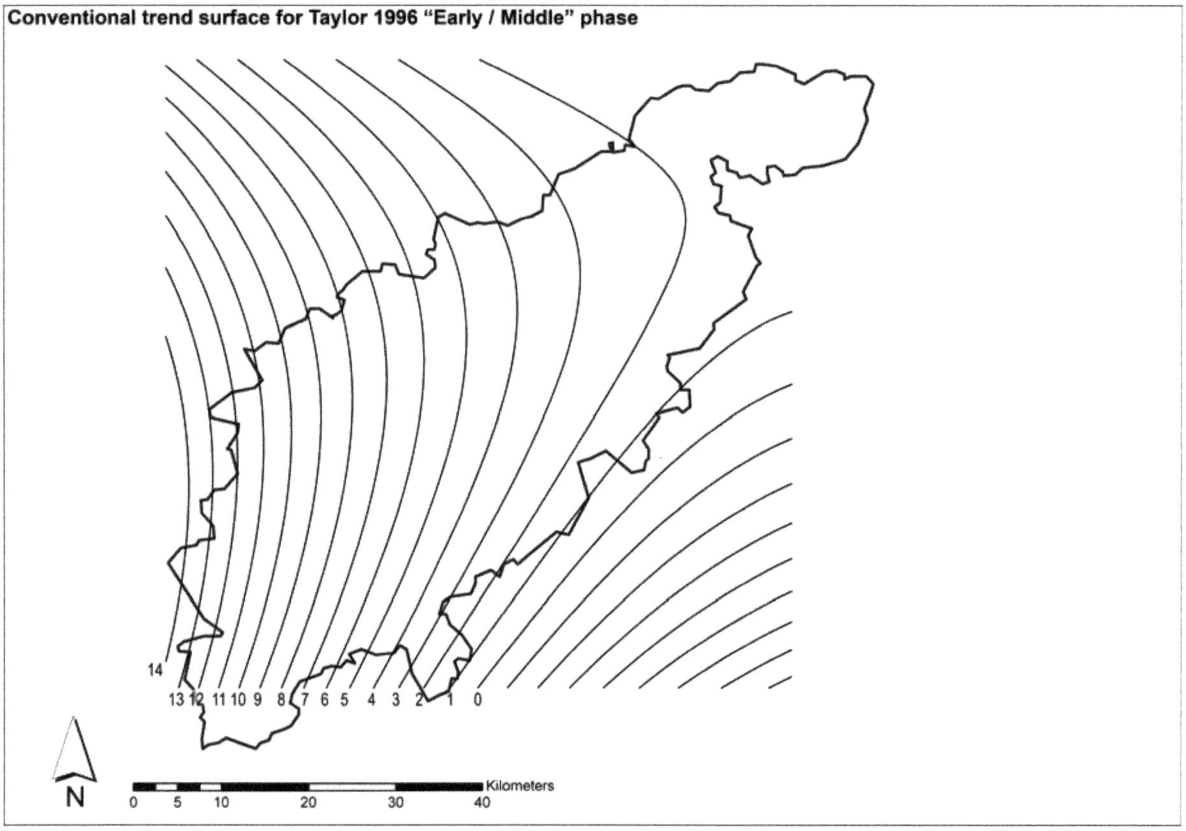

FIG 7.4 - TREND SURFACE FOR EARLY / MIDDLE PHASE BASED UPON FABRIC CLASSIFICATION.

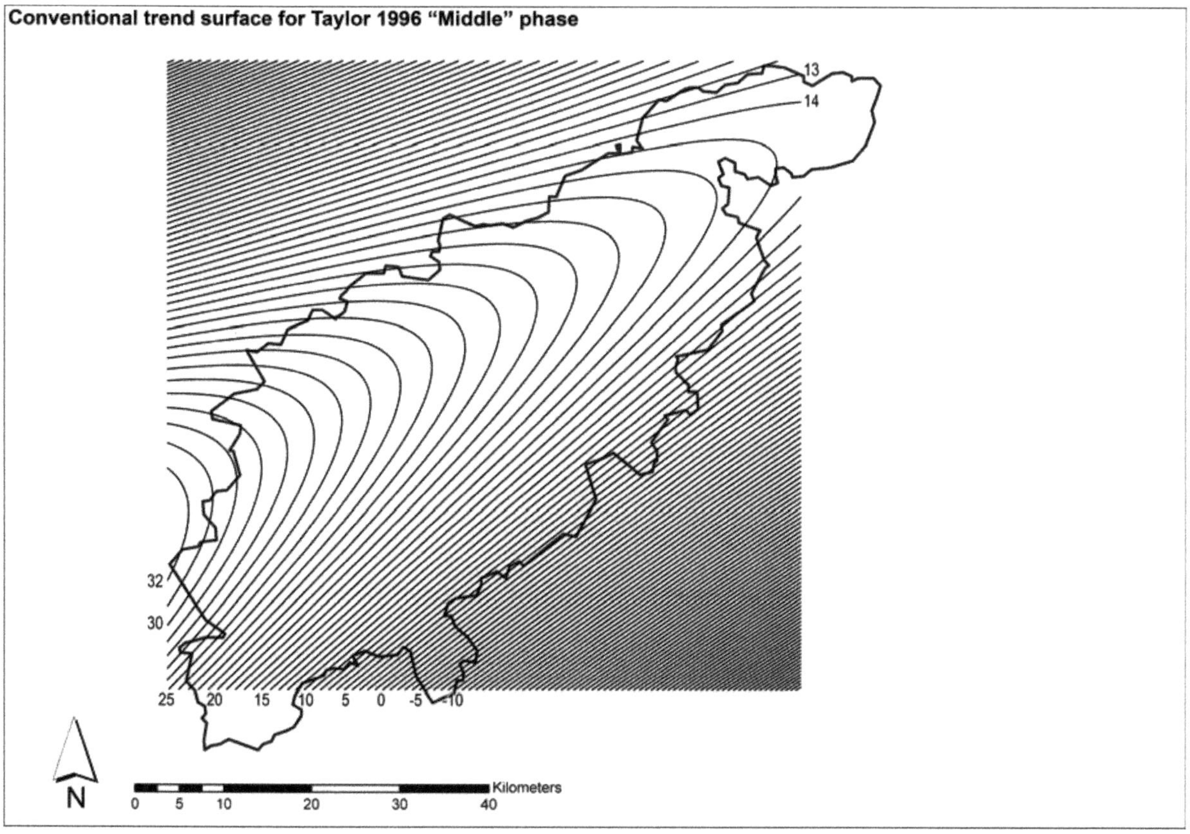

FIG 7.5 - TREND SURFACE FOR MIDDLE PHASE BASED UPON FABRIC CLASSIFICATION.

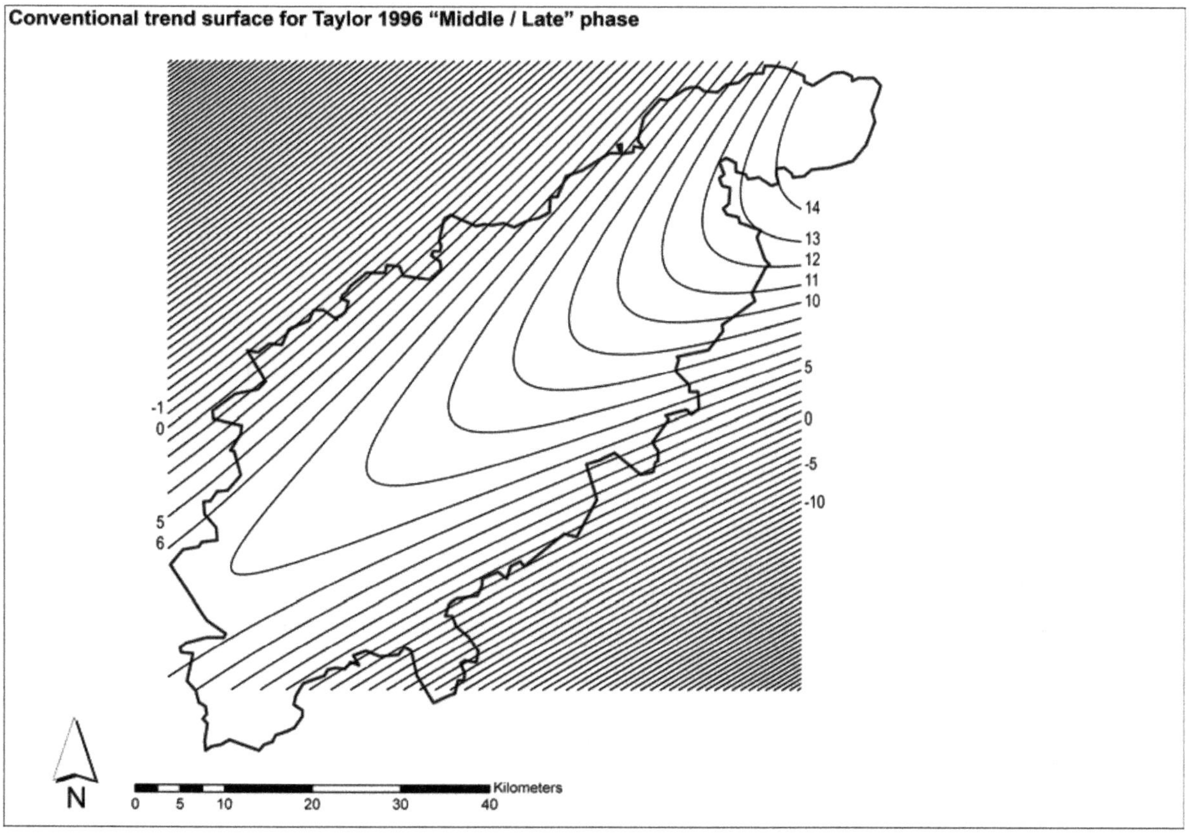

FIG 7.6 - TREND SURFACE FOR MID / LATE PHASE BASED UPON FABRIC CLASSIFICATION.

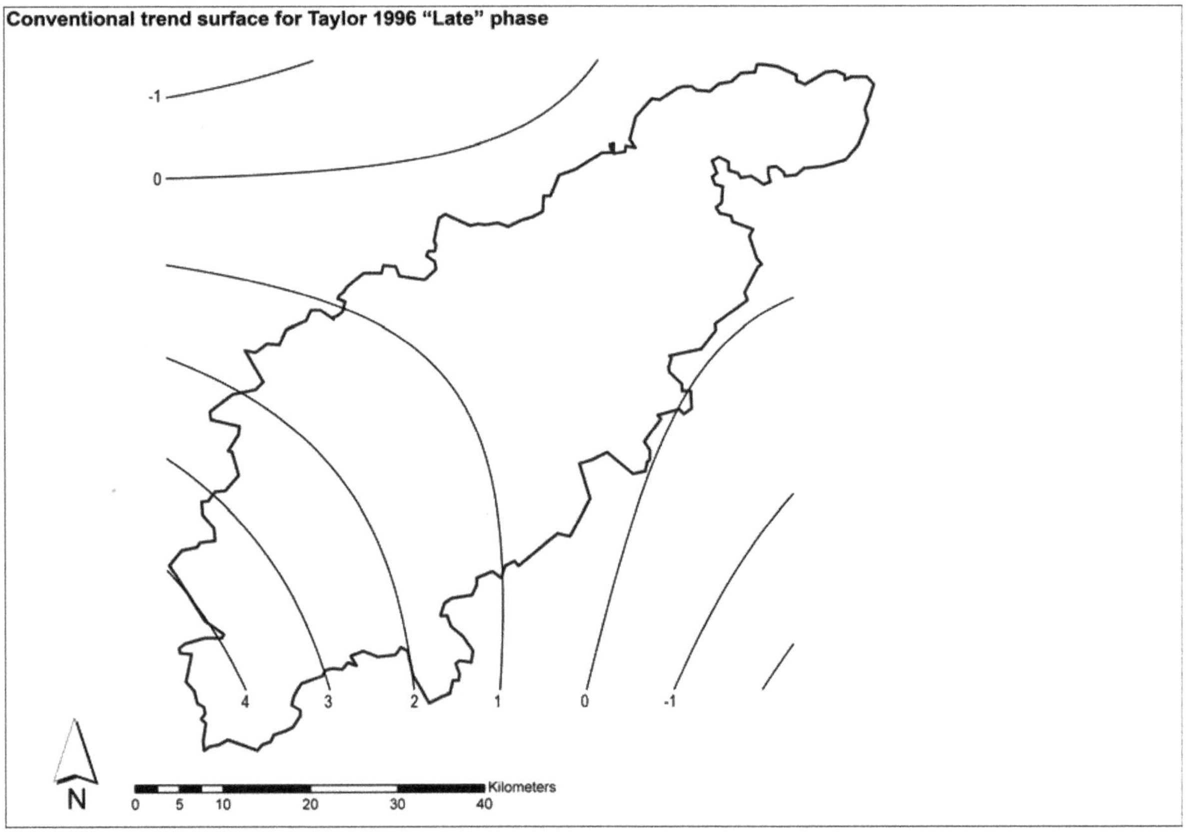

FIG 7.7 - TREND SURFACE FOR LATE PHASE BASED UPON FABRIC CLASSIFICATION.

by Taylor, then it is fair to say that the residuals will also be different: however, these lie beyond the scope of the current work.

The TGIS approach

There are two major issues with the work undertaken by Taylor that the application of the TGIS can help to resolve. The first issue is caused by the overlapping chronological phases and the separation out of different, partly contemporary, fabric types into different phases. The second issue stems from the fact that Taylor's phasing was not able to take into account the large numbers of sherds (1,070) belonging to fabrics 28 and 33 that could not be dated to better than a four century time-span. Coarsewares are generally much less variable, and thus diagnostic, in terms of form than finewares and, as a result, may only be dated at a much lower temporal resolution. However, it has been demonstrated by Trément that only taking account of the more closely datable finewares can produce a seriously flawed image of past activity. The graphs below (Fig 7.8) show very clearly that the overall ceramic profile (including coarsewares) for the site of the villa at Sivier paints a very different picture of significant activity from the first to sixth centuries AD when compared against the picture painted by the finewares of activity petering out and ending around AD 450 (Trément 2000). Therefore, it is important to be able to include less closely-dated pottery types within the spatio-temporal analysis of ceramic field survey evidence. The use of the TGIS enables archaeologists to deal with these two issues, through the application of a probabilistic model of ceramic temporality.

The implementation of field survey evidence in the TGIS

Certain issues arose in integrating the data into the TGIS. Firstly, as the TGIS is only able to cope with objects possessing a single date range, it was necessary to extract the individual sherd counts for each site and each fabric into a new, much larger data table. With up to 33 fabrics spread across 245 sites, this resulted in a long list of 2,097 individual pottery dates, each with its own location, sherd count, and date range (taken from Table 7.1, page 151). This data table was then added to the map in the TGIS for analysis. The large size of the table and the resultant large numbers of spatially concurrent dating objects had two significant implications. Firstly, it meant that processing times in the TGIS were much longer than those associated with the Willington dataset (see Chapter 6). This made it more difficult to quickly test out new ideas, but the time taken still remained within a reasonable level for patient analytical work - taking no longer than five minutes to process a calculation.

The second result was that it became necessary to sum the results of the TGIS output in order to analyse the data further and to return it to the situation where a single point represented a single site. With 33 fabrics, there could be up to 33 spatially concurrent dates on each site location. These stacks of dates with differing values made the visual examination of the raw TGIS data impossible

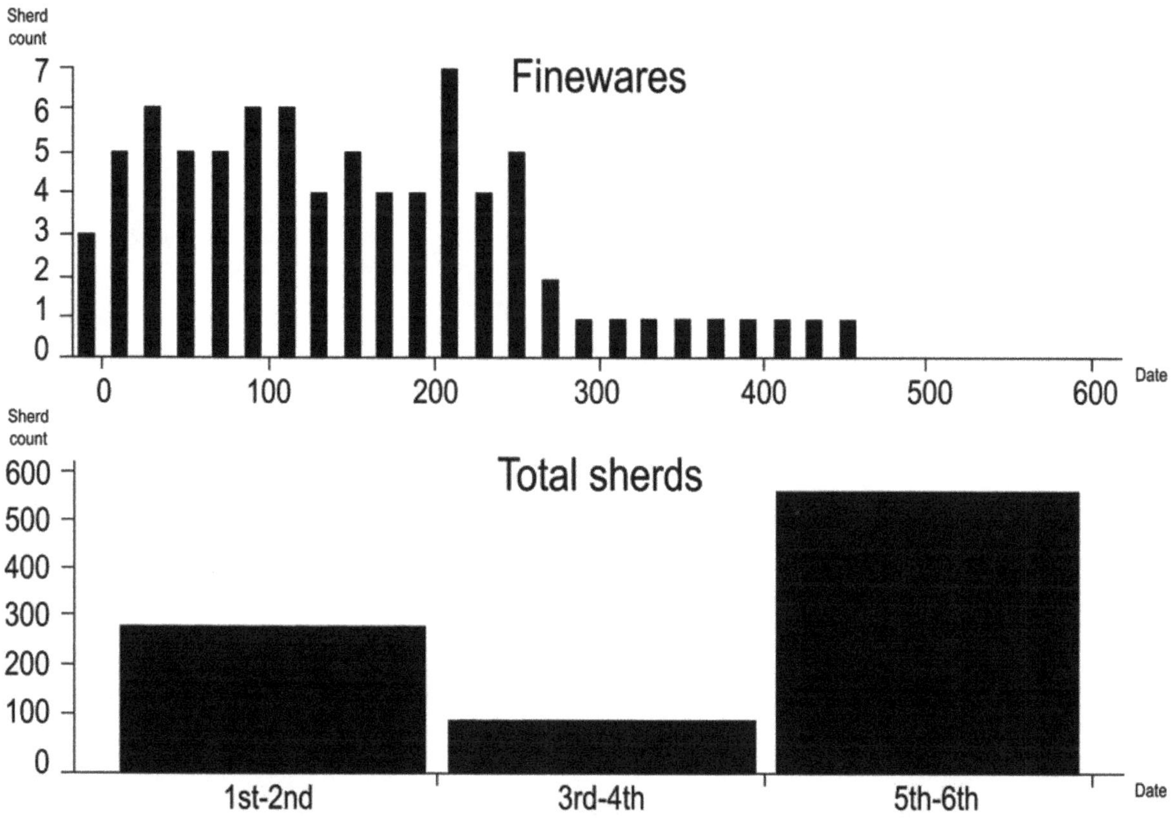

FIG 7.8 - COMPARISON OF THE TEMPORAL PROFILES FOR FINEWARES AND THE ENTIRE CERAMIC ASSEMBLAGE FOR THE VILLA AT SIVIER, FRANCE; THE HORIZONTAL AXIS SHOWS DATE, THE VERTICAL SHOWS SHERD COUNT. (ADAPTED FROM TRÉMENT 2000: FIG 9.16).

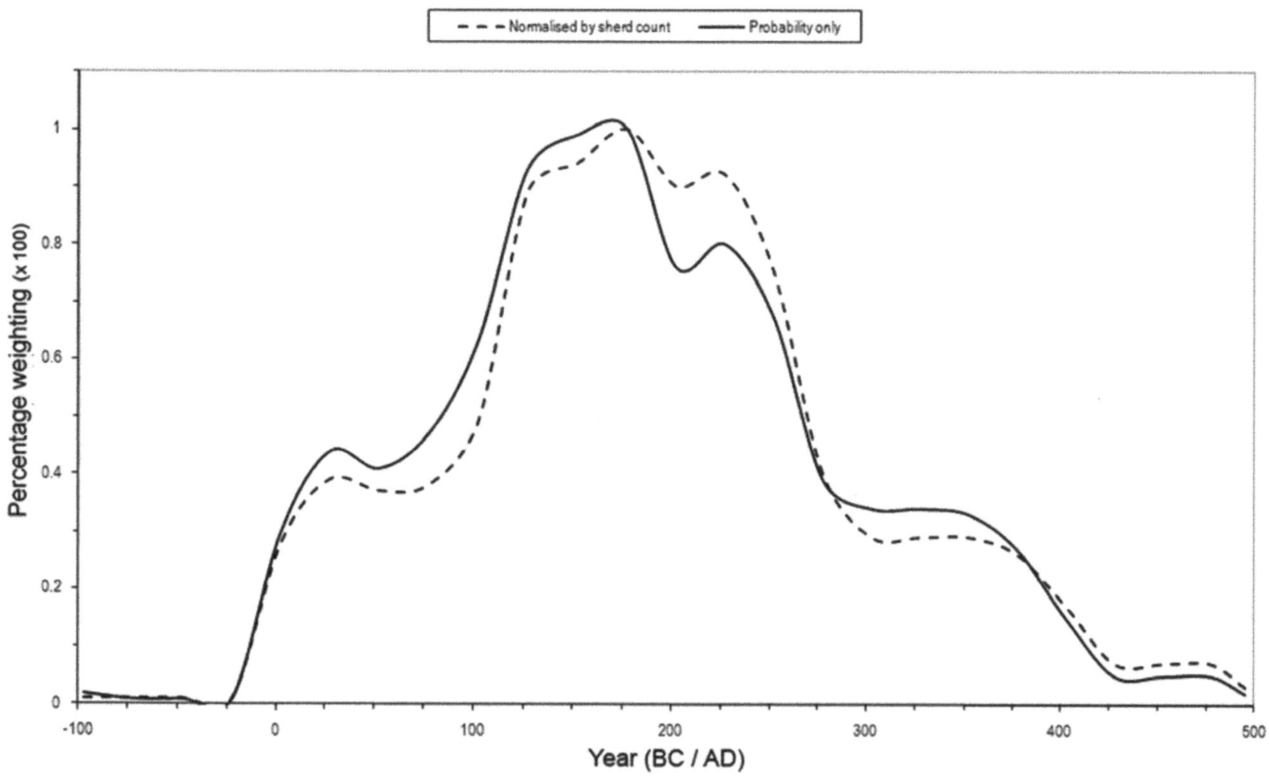

FIG 7.9 - COMPARISON OF PERCENTAGE WEIGHTING (1.0 = 100%) OF DATASET PROBABILITY PROFILES FOR RAW PROBABILITY DATA AGAINST THE SAME DATA NORMALISED BY SHERD COUNT.

and also prevented the creation of interpolated trend surfaces. As such, the sum result process within the TGIS was used to sum the output probabilities of each data point where they were in the same spatial location. These summed probabilities were multiplied by the sherd count for each date, as a weighting factor and to ensure that the trend surfaces took into account the increased likelihood of a fabric's presence being representative of past activity levels according to increase in sherd count. This does not take into account any possible concerns over increased fracture rates of sherds on different sites due to past agricultural conditions, but raw sherd count data was preferred in order to produce comparable results to Taylor's analysis. However, the flexibility of the approach means that it is straightforward for the user to use other weighting factors if desired. Comparison of the profiles for the percentage weighting of the summed raw probability data for the dataset as a whole against the same data normalised by sherd count, illustrates that including sherd counts results in a relatively small but still significant effect and, as such, its inclusion is essential to the analysis (Fig 7.9).[1]

Analysis and results

The TGIS analysis began with a re-examination of Taylor's chronological phasing in the TGIS to enable a direct comparison against his results. After that, an alternative phasing was tested in order to try and eliminate the overlaps between the different phases and to see if a different chronological basis might produce alternative results. Finally, the TGIS Analyst tool was used to compare the internal probability profiles of different sub-regions of the overall dataset to consider regional temporal variation in activity levels.

Taylor's phasing in the TGIS

In order to enter Taylor's phasing into the TGIS, it was necessary to define a minimum and maximum year for each of the five periods, as follows: Early - 100 BC to AD 50; Early / Middle - AD 50 to AD 175; Middle - AD 130 to AD 260; Mid / Late - AD 175 to AD 400; Late - AD 275 to AD 400. These five sets of dates were entered into the TGIS and output layers produced, weighted through multiplication by sherd count and summed where spatially concurrent. Trend surfaces (quadratic linear) were created using ArcGIS's Trend tool[2] for these new layers, and contour maps then created and overlaid. The following five maps show the results for each period.

Comparing (Fig 7.15) the TGIS trend surface for the Early phase (Fig 7.10) against Taylor's equivalent (Fig 7.3), we can see that the overall direction of the trend is very similar, but that the width of the higher probability area is somewhat narrower and the peak value shown is a little lower. Moving onto the Early / Middle phase (Fig 7.11) and its conventional comparator (Fig 7.4), we see a much more substantial difference. Although the general direction of the trend is again similar, this time the width of the trend peak is much narrower and of significantly higher value. For the Middle phase (Fig 7.12) and its equivalent (Fig 7.5), the peak values and direction are similar, but the TGIS trend is much more elongated. With the Mid / Late phase (Fig 7.13) and its analogue (Fig 7.6), the picture painted is quite distinct. Whereas the Taylor surface peaks in the north east, the TGIS surface shows a higher value peak that covers the length of the county from south west to north east. Finally, considering the Later phase (Fig 7.14) against its comparator (Fig 7.7) shows a similar degree of difference. This time, the Taylor peak is in the south west, whereas the TGIS surface again covers the county, with a peak in the north east.

The TGIS output is based upon a probabilistic model. Whereas the conventional output takes each pottery style as a whole, but split into different phases, the TGIS output essentially assigns each pottery style a probability based upon the likelihood of it falling within each phase. In effect, the output forms a fuzzy set (see Chapter 5), with each pottery date having a probabilistic membership degree of belonging to each phase. This membership degree is then multiplied by the sherd count for each item on each site to give weight to the level of deposition. This gives a stronger weighting to large groups of pottery over smaller groups, and helps to balance out the difficult problem of the dichotomy between smaller amounts of closely dated ceramics and larger amounts of pottery dated to a lower temporal resolution. For example, consider a site with 50 sherds of a coarseware dated to AD 100-500 and 10 sherds of a fineware dated to AD 150-200. If we were to consider that period of AD 150-200, the fineware would have a probability of 1.0, which would give a membership degree of 10.0 when multiplied by sherd count. The coarseware would have a probability of 0.125, giving a membership degree of 6.25. When summed together to reflect the occurrence of both in the same spatial location, this theoretical site would receive a weighting of 16.25. Looking then at the period AD 250-300, the coarseware would again receive a final weighting of 6.25 (and the site in turn). However, this would have been overlooked in the conventional fineware based model and zero probability of activity would have been assigned to that site at that time. It could be argued on common sense grounds that the fineware should take interpretative precedence due to its narrower dating (as per Johnson 2004a: see Chapter 5 above), but from a purely probabilistic point of view, that would be incorrect.

I believe that the TGIS model thus gives a truer picture of the probability of past activity. It is apparent from the comparison of the Taylor surfaces and the TGIS surfaces (Fig 7.15), that the individual influence of particular groups of pottery (specifically the Nene Valley and Oxfordshire colour coats) is lessened and integrated into

[1] Percentage weighting is preferred here because comparison of the shape of the graph is the important factor - we are interested in the internal probability profiles of the two factors rather than the raw figures.

[2] http://webhelp.esri.com/arcgisdesktop/9.2/index.cfm?TopicName=Trend

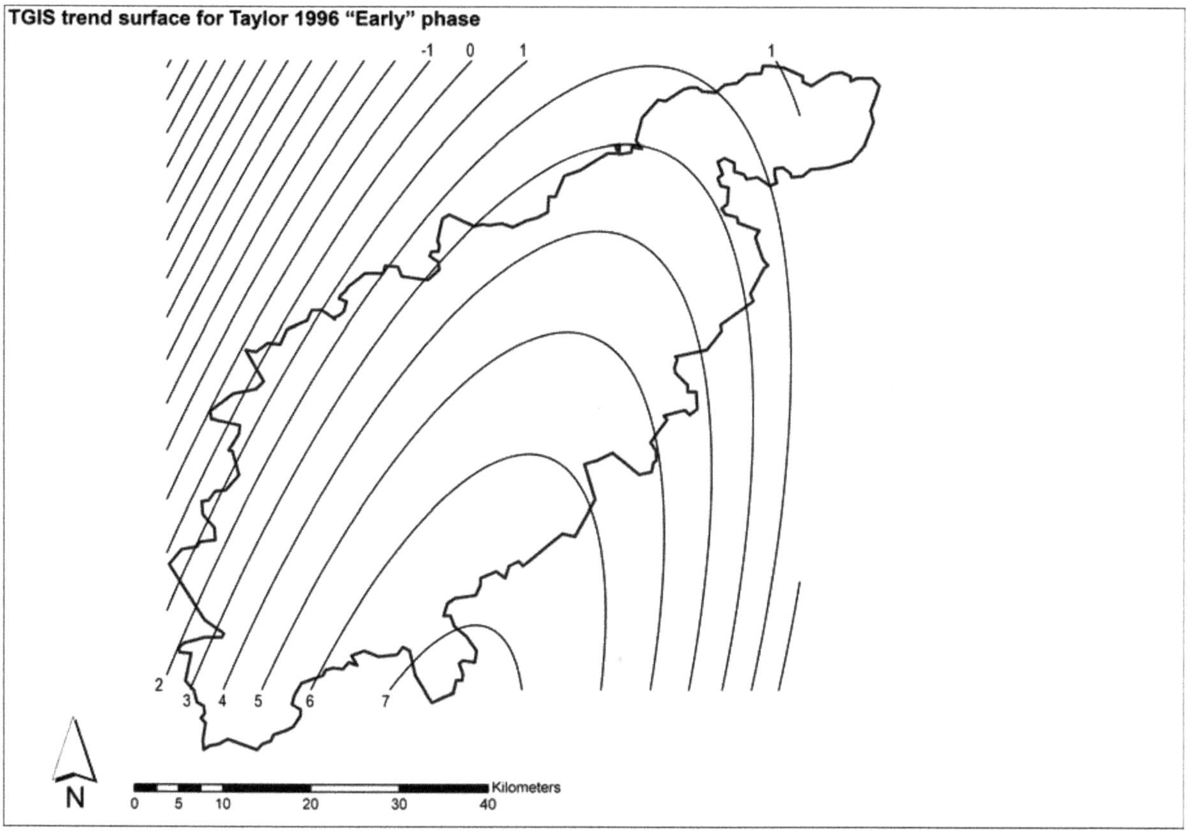

FIG 7.10 - TREND SURFACE FOR EARLY PHASE (100 BC TO AD 50) BASED UPON TGIS PROBABILITY OUTPUT NORMALISED BY SHERD COUNT. THE POINT DATA LAYER SHOWS THE SUMMED TGIS OUTPUT, SHADED ACCORDING TO THE PARAMETER OF SUMMED PROBABILITY MULTIPLIED BY SHERD COUNT. THE TREND SURFACE IS SHOWN AS ISOLINES.

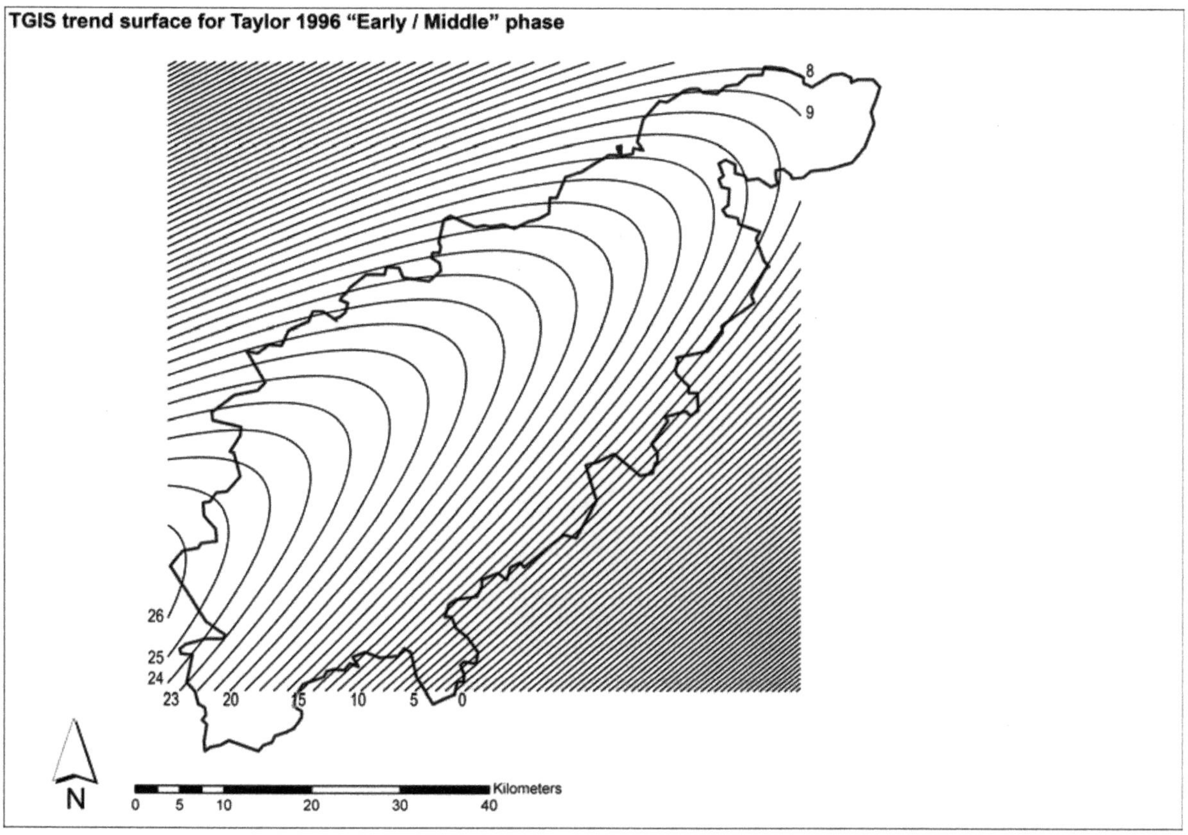

FIG 7.11 - TREND SURFACE FOR EARLY / MIDDLE PHASE (AD 50 TO AD 175) BASED UPON TGIS PROBABILITY OUTPUT NORMALISED BY SHERD COUNT. LEGEND EXPLANATION AS PER FIG 7.10.

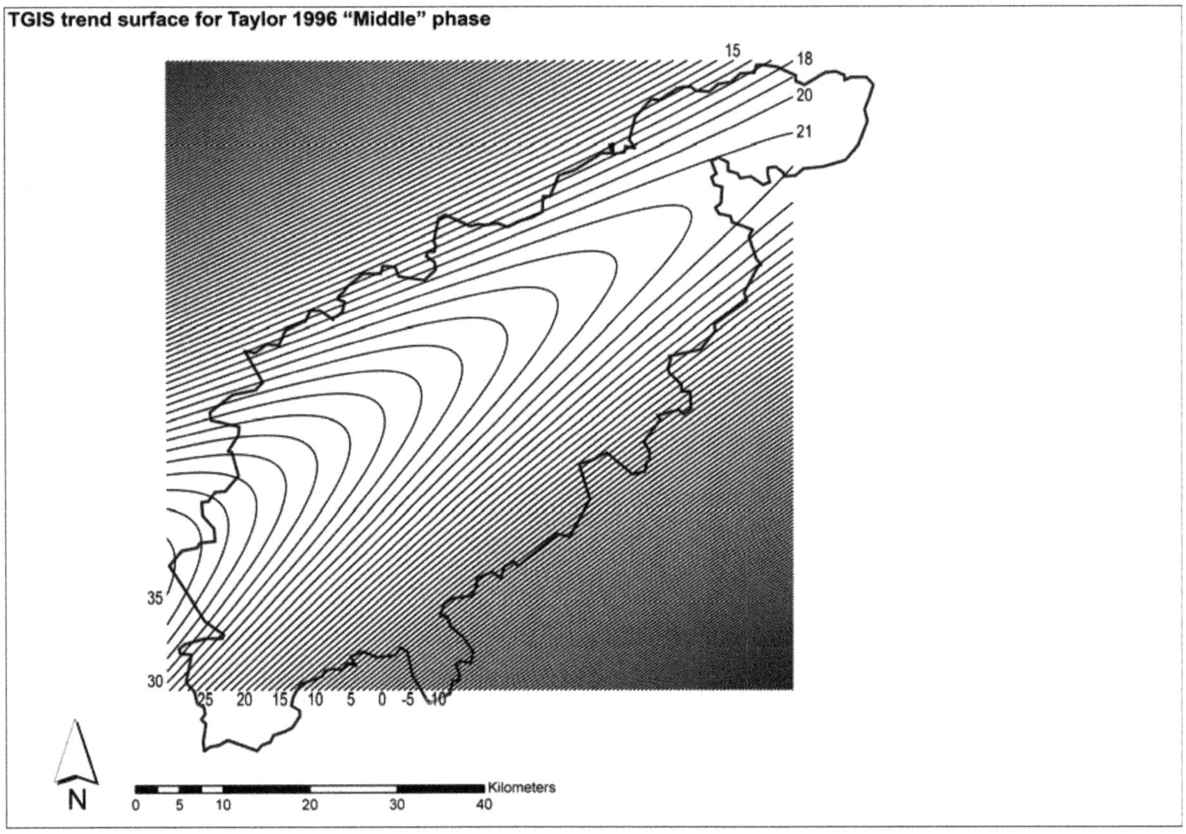

FIG 7.12 - TREND SURFACE FOR MIDDLE PHASE (AD 130 TO AD 260) BASED UPON TGIS PROBABILITY OUTPUT NORMALISED BY SHERD COUNT. LEGEND EXPLANATION AS PER FIG 7.10.

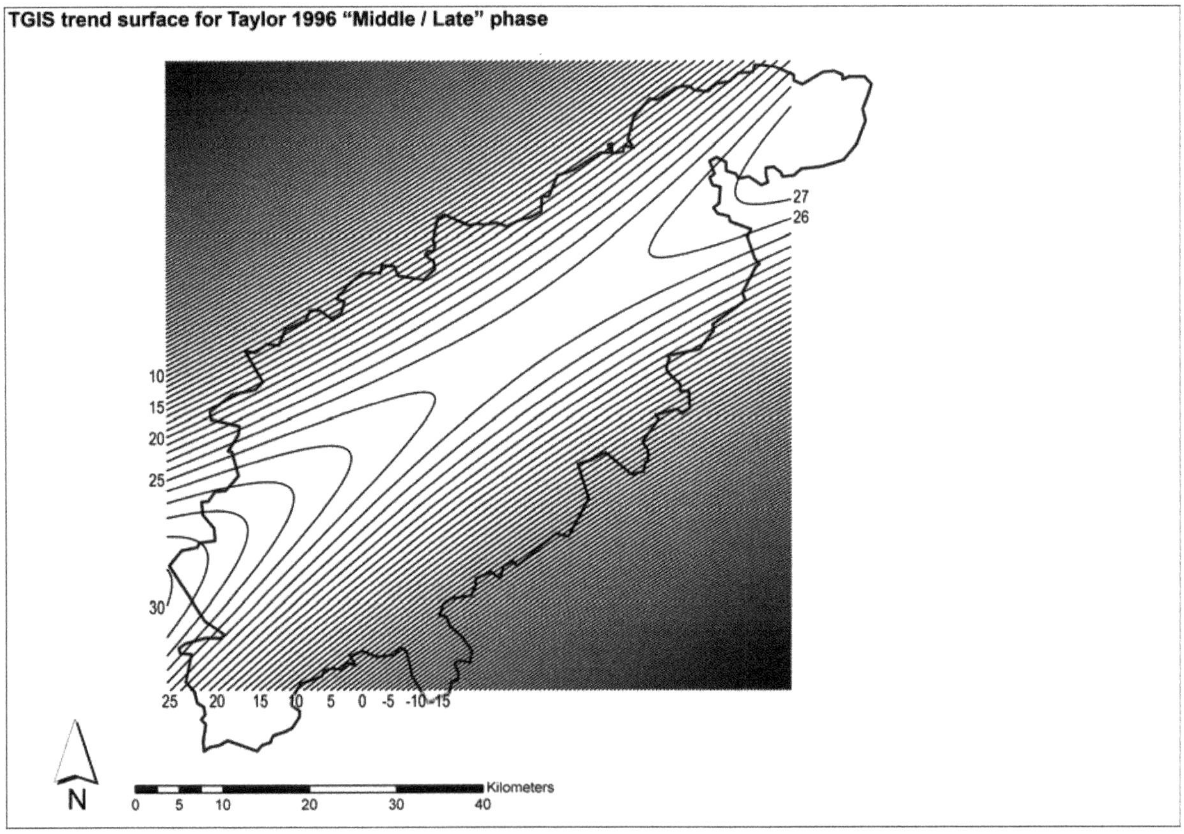

FIG 7.13 - TREND SURFACE FOR MID / LATE PHASE (AD 175 TO AD 400) BASED UPON TGIS PROBABILITY OUTPUT NORMALISED BY SHERD COUNT. LEGEND EXPLANATION AS PER FIG 7.10.

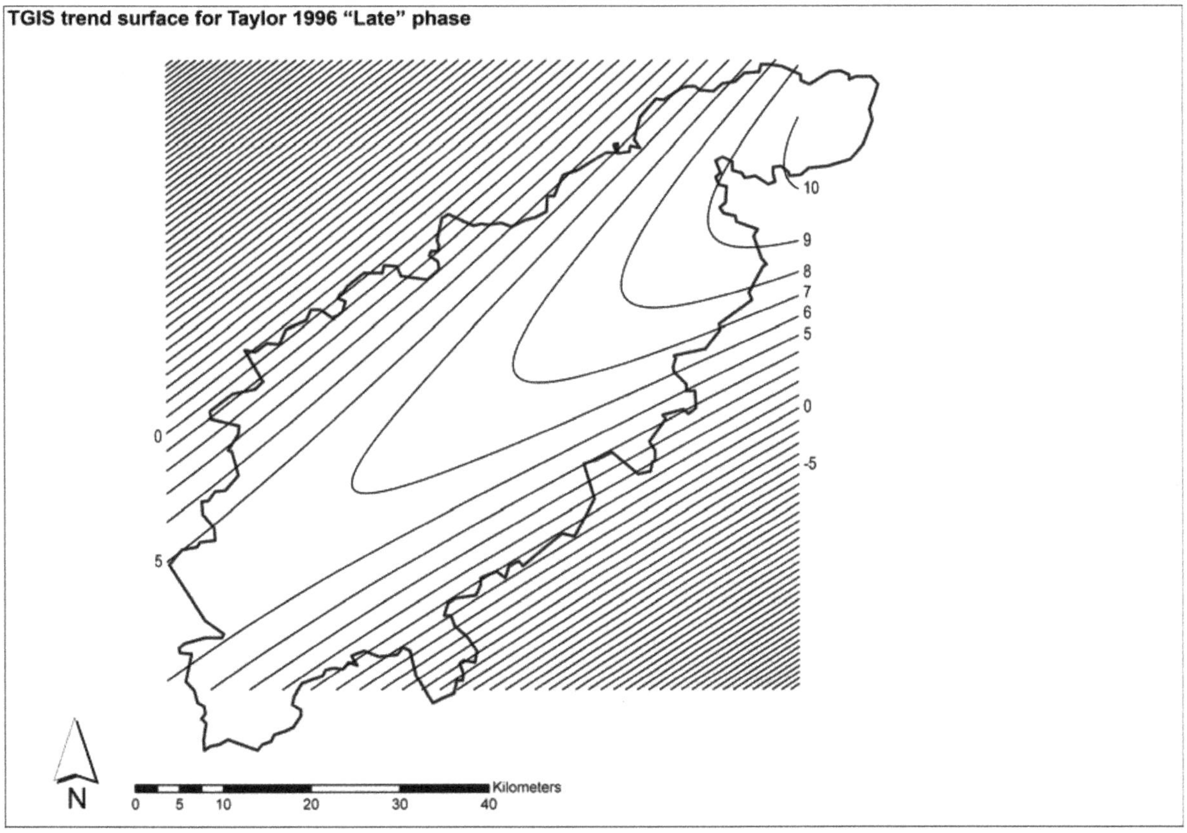

FIG 7.14 - TREND SURFACE FOR LATE PHASE (AD 275 TO AD 400) BASED UPON TGIS PROBABILITY OUTPUT NORMALISED BY SHERD COUNT. LEGEND EXPLANATION AS PER FIG 7.10.

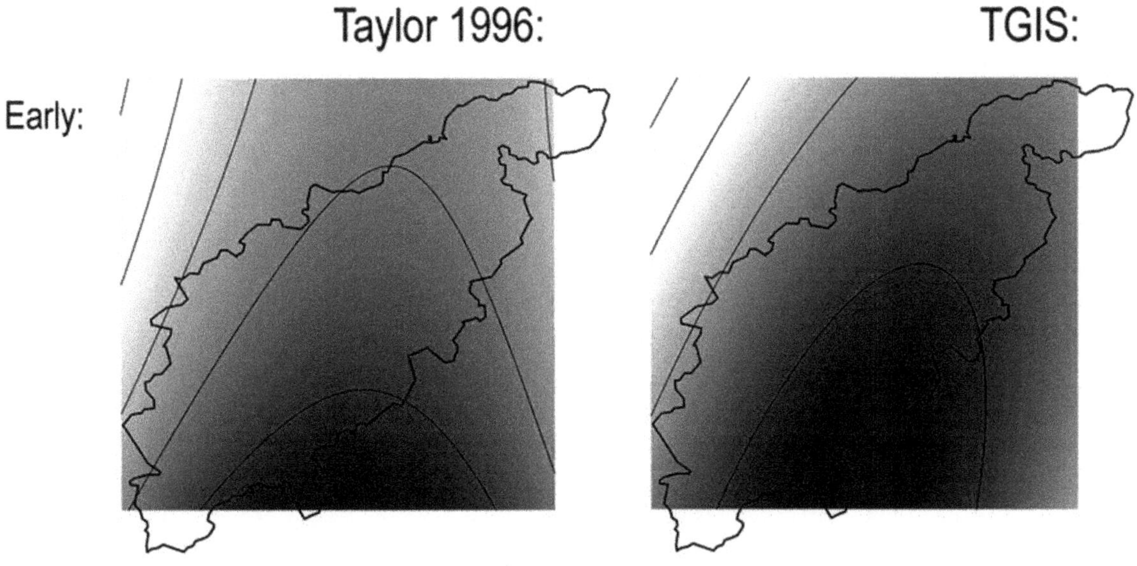

FIG 7.15 (A) - COMPARISON OF TAYLOR 1996 AND TGIS TREND SURFACES FOR EACH TAYLOR 1996 PHASE. CONTOUR LINES AT 5 UNIT INTERVALS. FOR MORE VALUES AND MORE DETAIL, SEE ORIGINAL FIGURES.

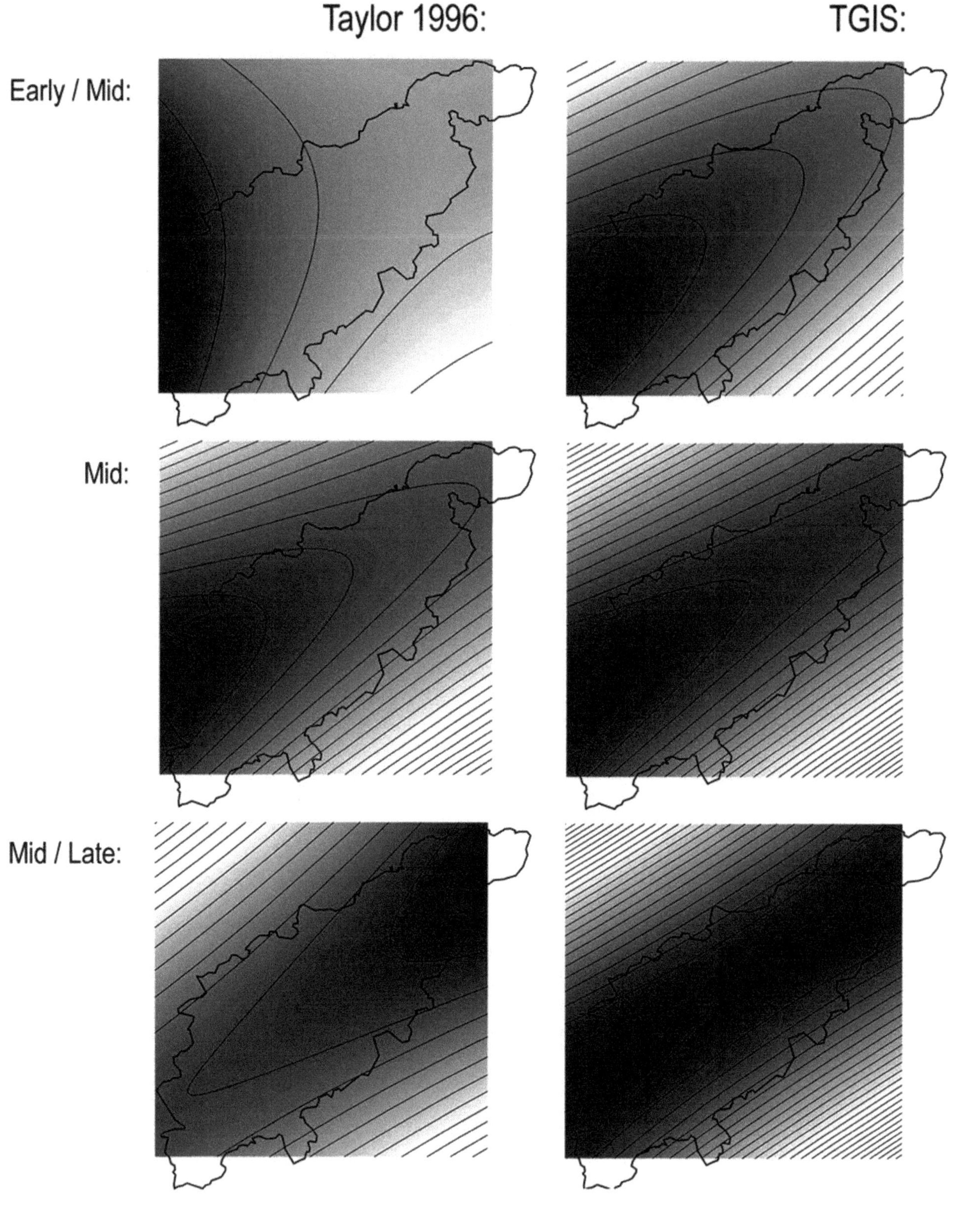

FIG 7.15 (B) TO (D)- COMPARISON OF TAYLOR 1996 AND TGIS TREND SURFACES FOR EACH TAYLOR 1996 PHASE. CONTOUR LINES AT 5 UNIT INTERVALS. FOR MORE VALUES AND MORE DETAIL, SEE ORIGINAL FIGURES.

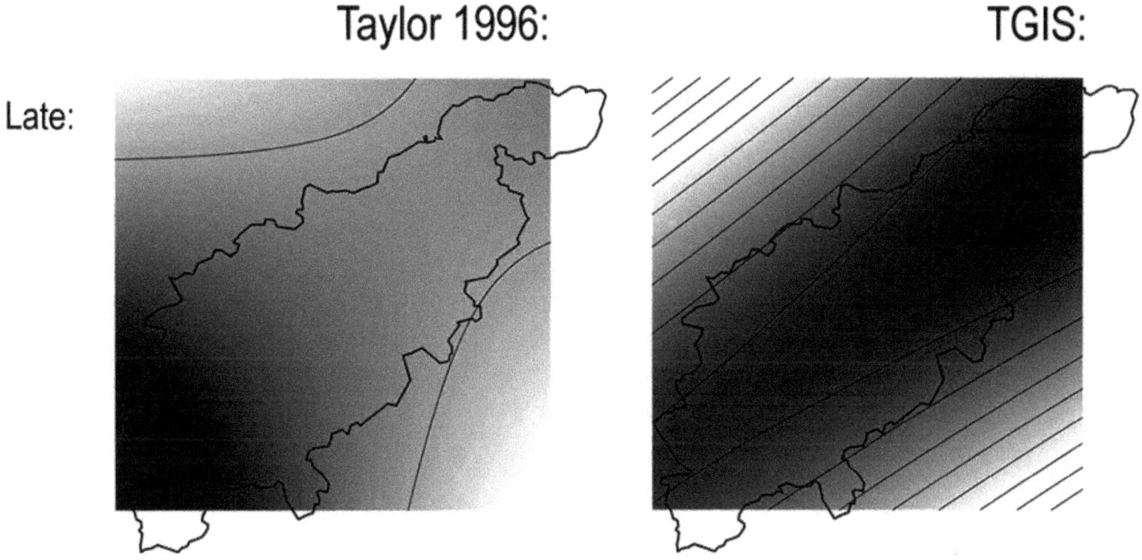

FIG 7.15 (E) - COMPARISON OF TAYLOR 1996 AND TGIS TREND SURFACES FOR EACH TAYLOR 1996 PHASE. CONTOUR LINES AT 5 UNIT INTERVALS. FOR MORE VALUES AND MORE DETAIL, SEE ORIGINAL FIGURES.

the larger trend through the use of the fuzzy membership model. The inclusion of the less tightly-dated coarsewares (fabrics 28 and 33) in the TGIS results helps to strengthen the argument for the TGIS methodology (and output) being more representative of the probability of past ceramic deposition as all fabrics are allowed an influence on all phases within which they fall temporally, whether partially or as a whole. The individual effect of any particular fabric on the regional trend is still present, but proportionate to its levels of deposition. The results produced are both more robust and more useful than those produced in Johnson's experiment with aoristic analysis (2004a) and in Lock and Harris's p(use) model (2000), as they are not swamped by undiagnostic material and there is less capacity for fudging the evidence to fit preconceived ideas.

Alternative phasing produced using Analyst tool

If we accept that the probabilistic model presents a stronger case for being representative, the question that arises is why then persist with overlapping phases? The answer is that we do not have to: the Analyst tool has a function built into it that allows users to construct a series of phases based on the rate of change of the overall probability profile (or a subset of that profile) for the data layer (or layers) in question. This is best explained and illustrated through the following example. The Analyst tool was run with a 25 year resolution to produce a graph of the summed probability of all of the data points (partly) contemporary to each time-slice. The Form Periods system was then used to group the individual time-slices together into larger phases based upon a rate of change of 25% of the maximum summed probability; the 25% figure being chosen so as to produce a series of phases that reflected relatively large changes in activity levels across the county.

The tool works as follows: the TGIS moves through each time-slice whilst keeping account of the current summed probability. When that summed probability reaches 25% of the maximum summed probability (or crosses zero probability), a new phase is begun and the current summed probability reset. This iterative process produces a phasing based directly upon the rate of change in the overall probability profile, with each phase involving a change in the overall weighted probability of 25%. For the current dataset, the following phases were defined by the TGIS: 100 BC to 10 BC; 10 BC to AD 90; AD 90 to AD 115; AD 115 to AD 240; AD 240 to AD 265; AD 265 to AD 415; AD 415 to AD 500 (Fig 7.16). These phases were then run in the TGIS and trend surfaces created as before.

For the period 110 BC to 10 BC (Fig 7.17), the values of the trend surface are too low for any analytical weight to be placed upon the results, other than to say pottery recovery was itself low. The values of the trend surface represent the average sherd count that might be expected on a typical site at any particular location: clearly, a maximum value of less than one sherd is so close to zero that in essence no spatial trend is present. For the period 10 BC to AD 90 (Fig 7.18), a clear trend is shown with its peak in the south. For the period AD 90 to AD 115 (Fig 7.19), the trend peak moves around somewhat to the west. It is notable that this particular period is short in duration and seems to represent a distinct, rapid increase in the level of pottery deposition across the region: this appears to coincide with the start of production of the upper Nene greywares. For the period AD 115 to AD 240 (Fig 7.20), the trend peak is in the same position, but lengthens into the centre of the county. For the period AD 240 to AD 265 (Fig 7.21), a very similar pattern is apparent. This phase is again distinctly short in duration, but this time seems to represent a significant fall in the level of pottery deposition. For the period AD 265 to

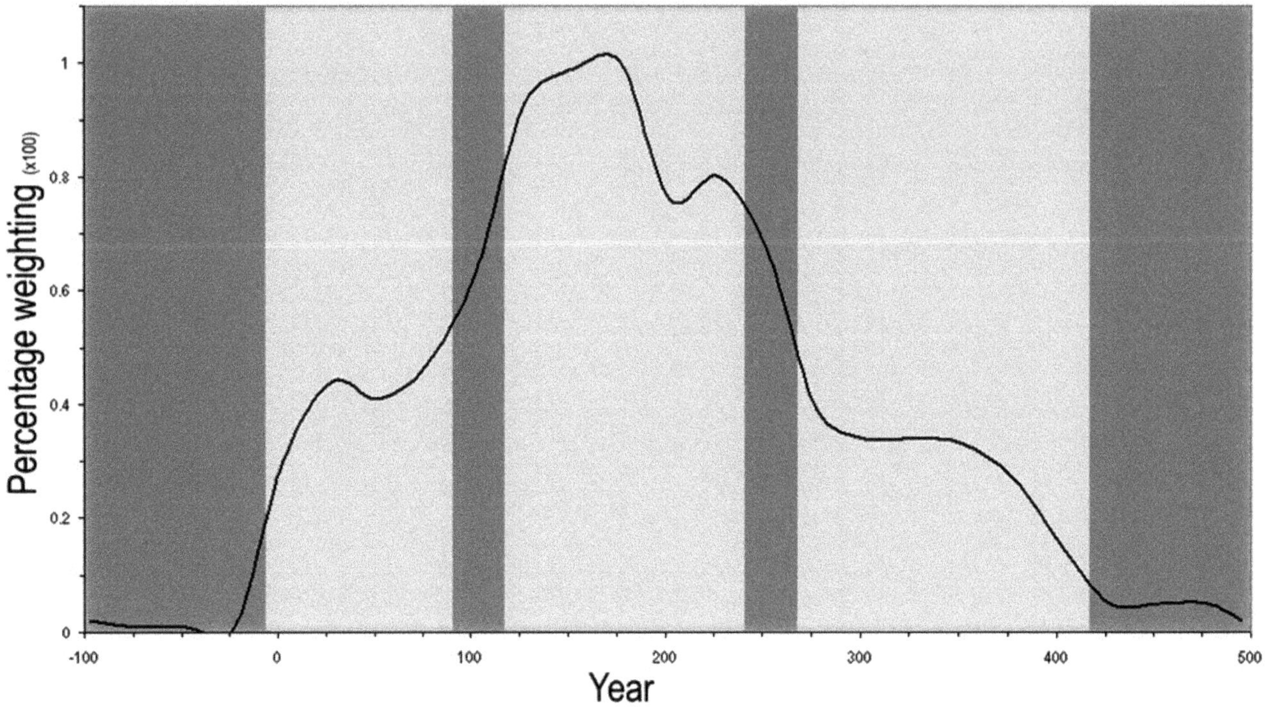

FIG 7.16 - PHASES DEFINED BY TGIS ANALYST TOOL BASED UPON 25 YEAR TIME-SLICE SUMMED PROBABILITY PROFILE (THE BLACK CURVE) AT 25% RATE OF CHANGE. ALTERNATE PHASES DEPICTED BY DARK AND PALE GREY BLOCKS. NOTE THAT SOME OF THE DETAIL OF THE GRAPH IS DETERMINED BY THE SMOOTHING ALGORITHM USED BY EXCEL.

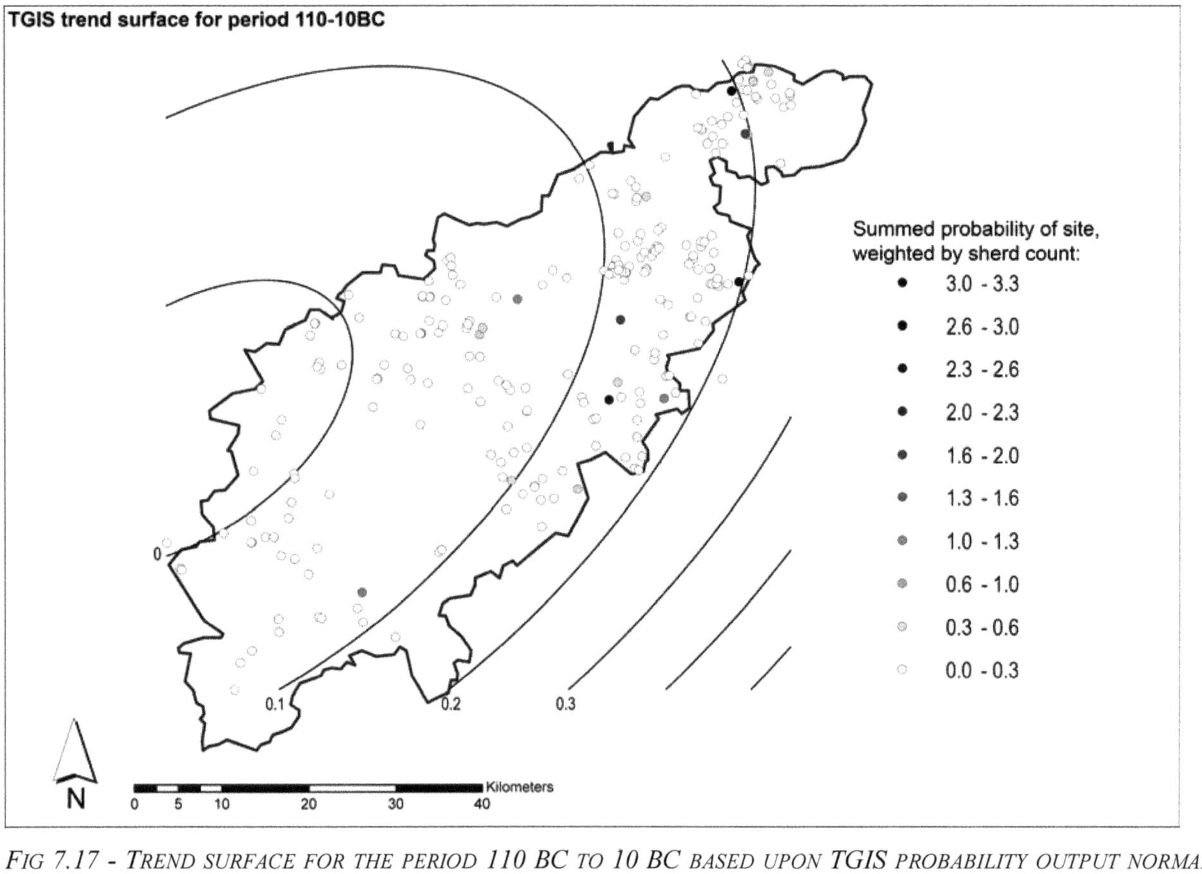

FIG 7.17 - TREND SURFACE FOR THE PERIOD 110 BC TO 10 BC BASED UPON TGIS PROBABILITY OUTPUT NORMALISED BY SHERD COUNT. THE POINT DATA LAYER SHOWS THE SUMMED TGIS OUTPUT LAYER, SHADED ACCORDING TO THE PARAMETER OF SUMMED PROBABILITY MULTIPLIED BY SHERD COUNT. THE TREND SURFACE IS SHOWN AS ISOLINES.

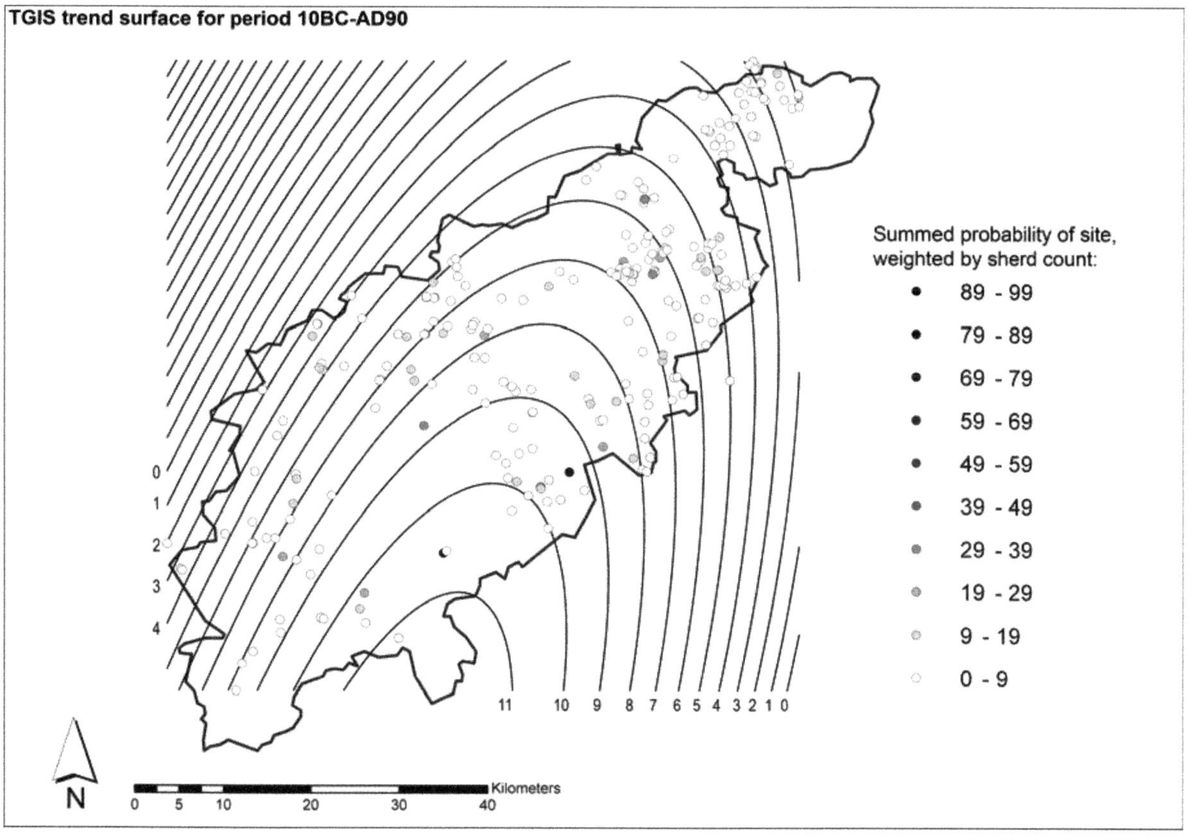

FIG 7.18 - TREND SURFACE FOR THE PERIOD 10 BC TO AD 90 BASED UPON TGIS PROBABILITY OUTPUT NORMALISED BY SHERD COUNT. LEGEND EXPLANATION AS PER FIG 7.17.

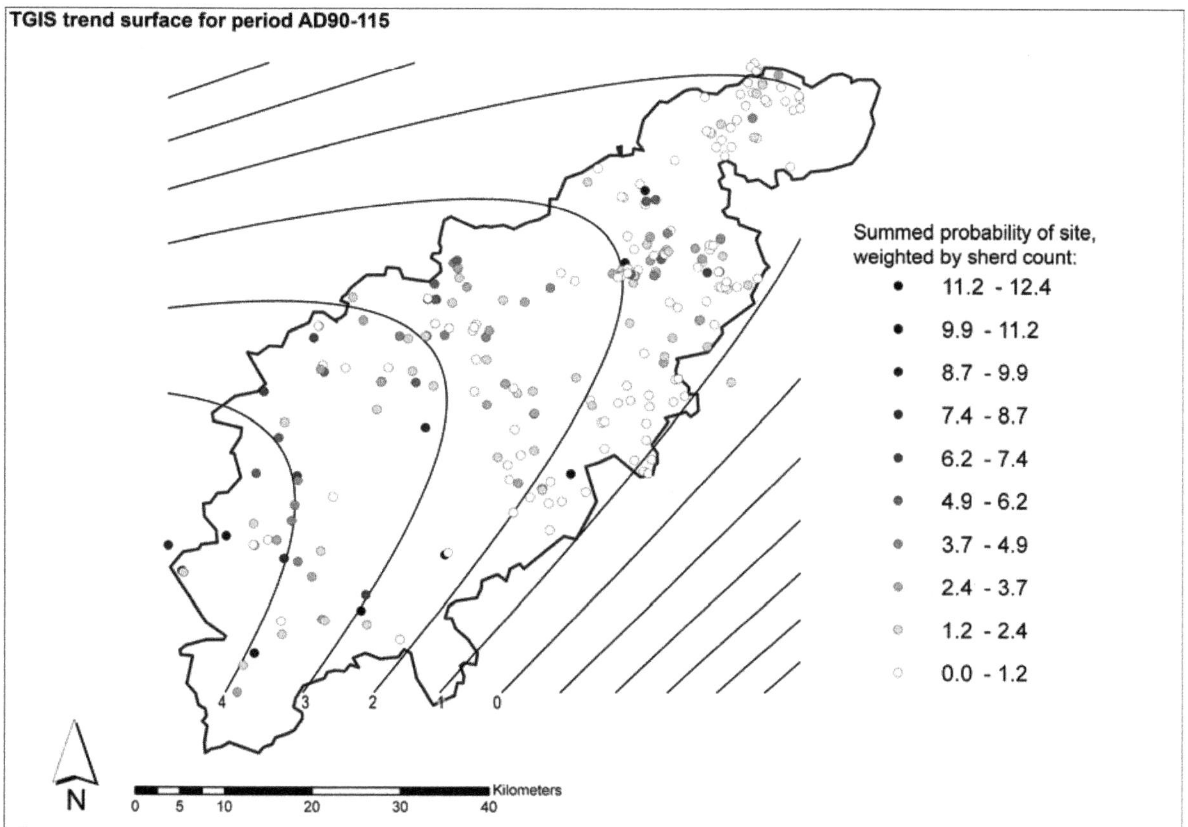

FIG 7.19 - TREND SURFACE FOR THE PERIOD AD 90 TO AD 115 BASED UPON TGIS PROBABILITY OUTPUT NORMALISED BY SHERD COUNT. LEGEND EXPLANATION AS PER FIG 7.17.

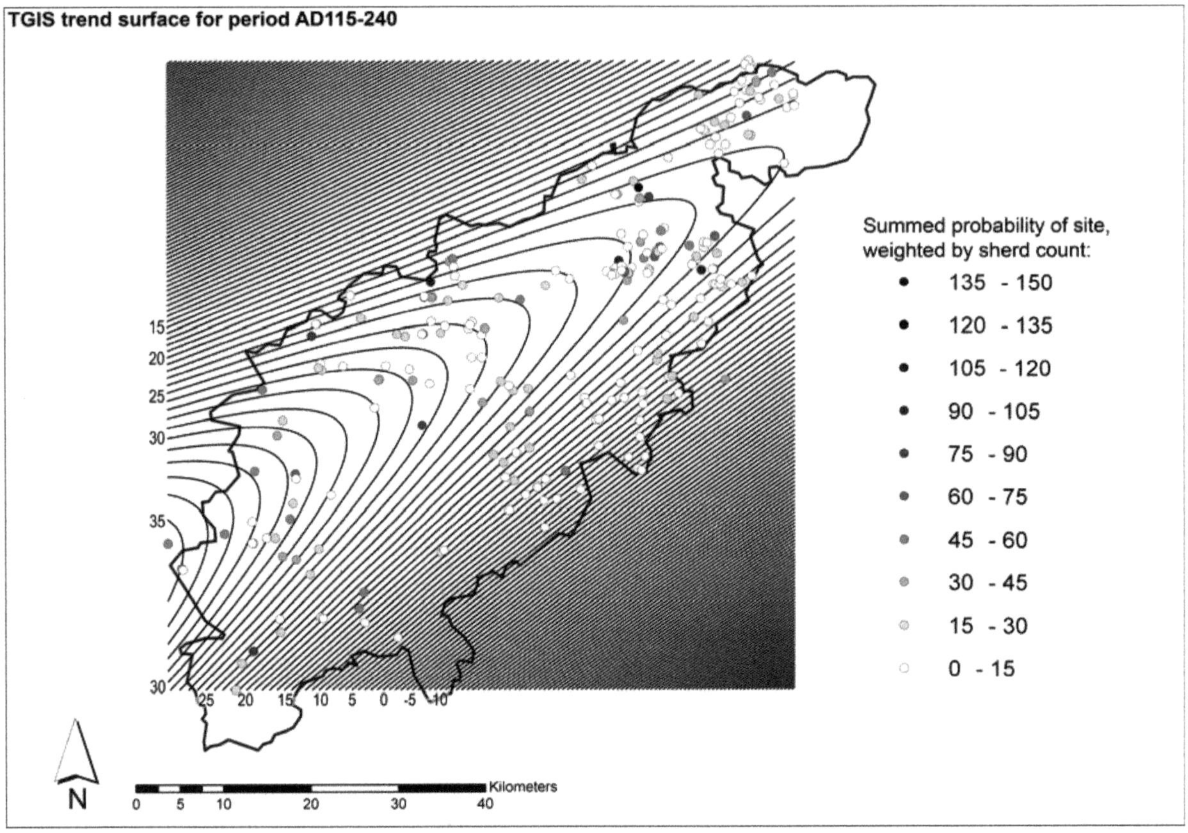

FIG 7.20 - TREND SURFACE FOR THE PERIOD AD 115 TO AD 240 BASED UPON TGIS PROBABILITY OUTPUT NORMALISED BY SHERD COUNT. LEGEND EXPLANATION AS PER FIG 7.17.

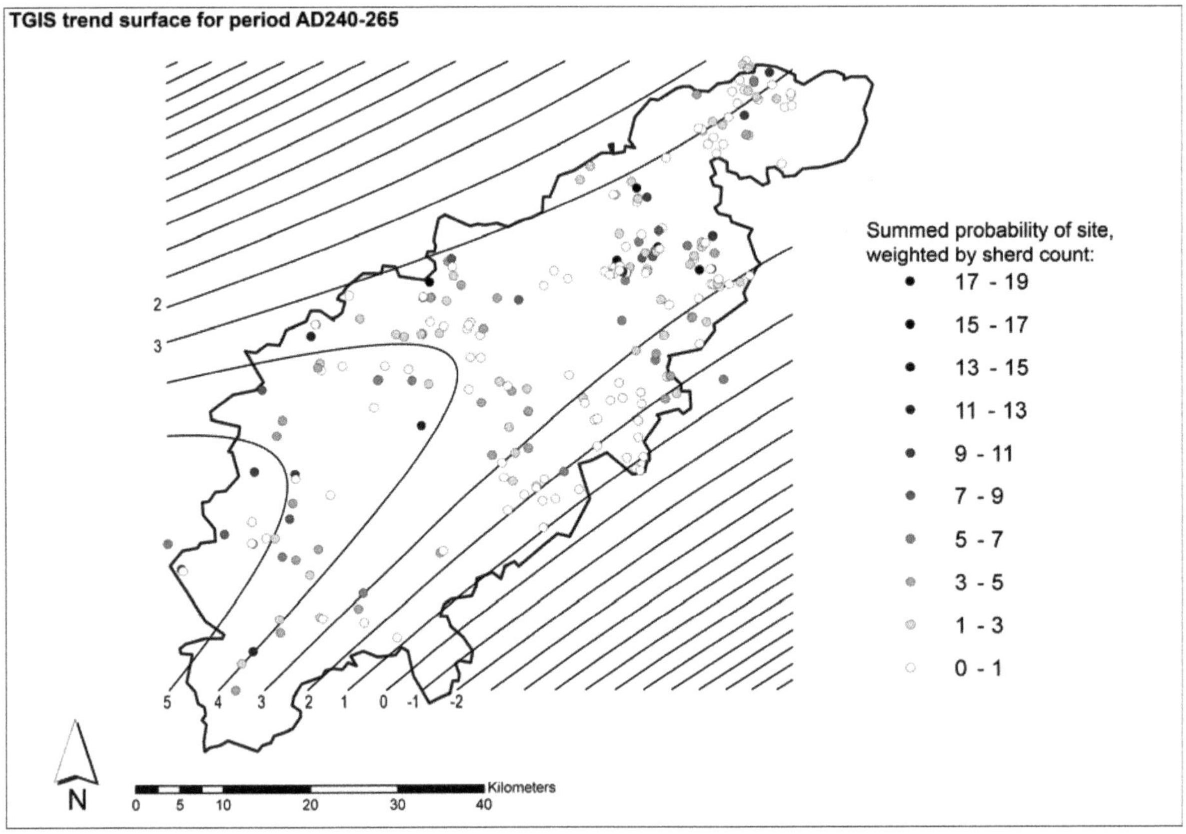

FIG 7.21 - TREND SURFACE FOR THE PERIOD AD 240 TO AD 265 BASED UPON TGIS PROBABILITY OUTPUT NORMALISED BY SHERD COUNT. LEGEND EXPLANATION AS PER FIG 7.17.

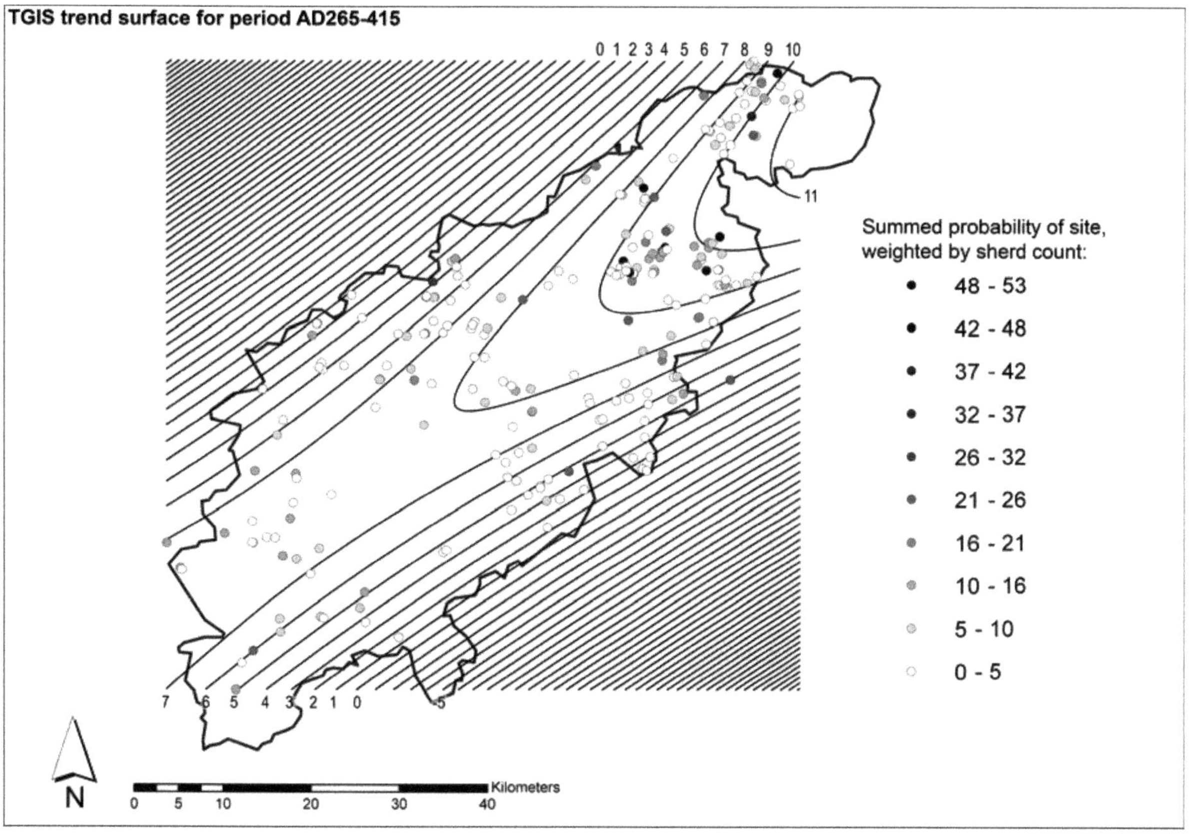

FIG 7.22 - TREND SURFACE FOR THE PERIOD AD 265 TO AD 415 BASED UPON TGIS PROBABILITY OUTPUT NORMALISED BY SHERD COUNT. LEGEND EXPLANATION AS PER FIG 7.17.

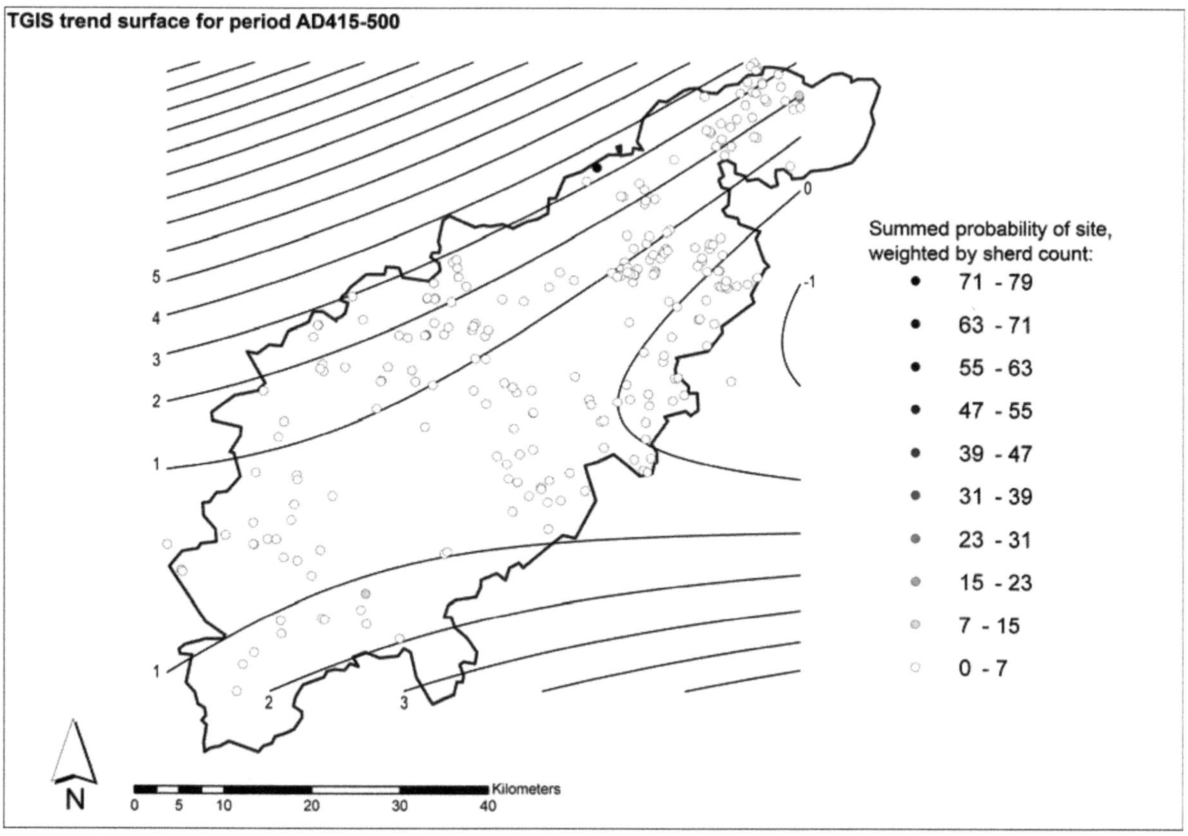

FIG 7.23- TREND SURFACE FOR THE PERIOD AD 415 TO AD 500 BASED UPON TGIS PROBABILITY OUTPUT NORMALISED BY SHERD COUNT. LEGEND EXPLANATION AS PER FIG 7.17.

AD 415 (Fig 7.22), the high values of the trend now cover most of the county along its longer axis, but with something of a peak towards the north east. Finally, for the period AD 415 to AD 500 (Fig 7.23), the values of the trend surface are again too low for any great analytical weight to be placed upon the results, other than to again say pottery recovery was low.

Comparing these results against those obtained for the Taylor phasing, two particular conclusions stand out. First, it appears that the phasing produced by the Analyst tool paints a smoother picture of change in the trend surfaces for pottery deposition across the county insofar as the pattern for each period seems to fall part way between its predecessor and successor, and an evolving pattern of change over time is produced. There is an element of this in the trend surfaces produced according to the Taylor phasing, but this is submerged beneath the issues arising from the overlapping time-slices. Secondly, the appearance of two very short phases in the Analyst tool phasing emphasises the very sharp increases and decreases in pottery deposition during those two periods of history. Thus, this arguably paints a stronger picture of the varying rate of change of this process (i.e. it sheds light upon the tempo of change as well as its spatial character). However, it is also possible that these apparent sharp increases are also partly due to the particular dating given to the various ceramic fabrics involved.

The choice between the two different phasings considered comes down to a single issue. That is, whether primacy is given to the ceramic chronology or to the ceramic distribution. The Taylor phasing is determined according to the dates of the various fabrics and is, as such, independent of the changing level of deposition occurring over time. The TGIS Analyst tool phasing is, by contrast, based strongly upon the changing rates of deposition for the specific data being studied (and also recovery rates during survey), but with an element of detail in the pattern that remains influenced to some degree by the ceramic chronology. As such, it is less clear cut in its influencing factors than the Taylor phasing, but paints a truer picture of change in the past. It certainly removes the problem of overlapping time-slices. The choice for users essentially comes down to which methodology they are conceptually more comfortable with. I would argue that the TGIS Analyst tool methodology presented here offers a powerful alternative to existing approaches, with its basis in ceramic deposition rates specific to the area of study making it the more robust of the two phasing approaches presented.

Analysing regional variation

As a final example of the potential value of the TGIS when studying regional survey data, the Analyst tool was used to consider regional variation in the overall probability of ceramic deposition within the county. To do so, the region was divided into three blocks along its longer axis of roughly equal size and according to where apparent divisions could easily be inserted between aggregations of data points (i.e. the three main site clusters in the study area) (Fig 7.24). These sub-regions were then used to make a spatial selection of the objects falling within each sub-region and the TGIS Analyst tool again used to output the summed probabilities based upon 25 year time-slices and normalised through multiplication by sherd count. The results were exported to Excel and then considered graphically.

The north-east sub-region contained 108 sites, the central 102 sites, and the south-west 35 sites. When comparing the total summed probabilities for the three sub-regions (Fig 7.25), the lower number of sites in the south-west is apparent. However, the south-west also appears to contain sites with a higher level of deposition than the other two sub-regions, as it clearly has a higher weighting than might be expected due to its relatively low site count. The overall trends over the centuries are broadly comparable, painting a picture of large increases in activity (by which I mean deposition levels) around c.25 BC and AD 100, and decreases in activity around AD 250 and AD 400. However, there are some significant differences in specific detail. First, it is apparent that activity levels increased around AD 100 more greatly in the north-east than in the other two sub-regions, with the north-east overtaking the central area as the focus of activity; a prime position that it maintained until the end of our period of interest. It is also apparent that the slight slump in activity around AD 200 in the central and south-west sub-regions is not present in the north-east, where activity continued to increase. After AD 300, the north-east sub-region saw disproportionately high levels of activity, although still at a much reduced level than during the preceding centuries.

The percentage weightings for each region (Fig 7.26) enable the consideration of how this variation in activity compared between sub-regions in terms of their own individual internal probability distributions. This graph reinforces certain of the previous conclusions. In particular, the primacy of the north-east sub-region after AD 200 remains especially apparent, as does the relative importance of the other areas prior to that date. This graph also makes clear the broad comparability of the shape of the probability pattern for the south-west and central sub-regions, with two exceptions being the first century AD primacy of the central area and the more gradual rate of fourth / fifth century AD decline in the south-west.

It is, of course, also possible to construct graphs like the latter for each and every site in the dataset. As an example of this, such an analysis was performed on the site of Southwick (Fig 7.27), as this seemed to show some variation from the trend in the previous analyses (consider Figs 7.11-14 and 7.19-21). The graph produced (Fig 7.28) compares the percentage probability weighting profile for the site against the regional trend (both multiplied by sherd count). The picture painted is one of a below average profile prior to about AD 100, with the site

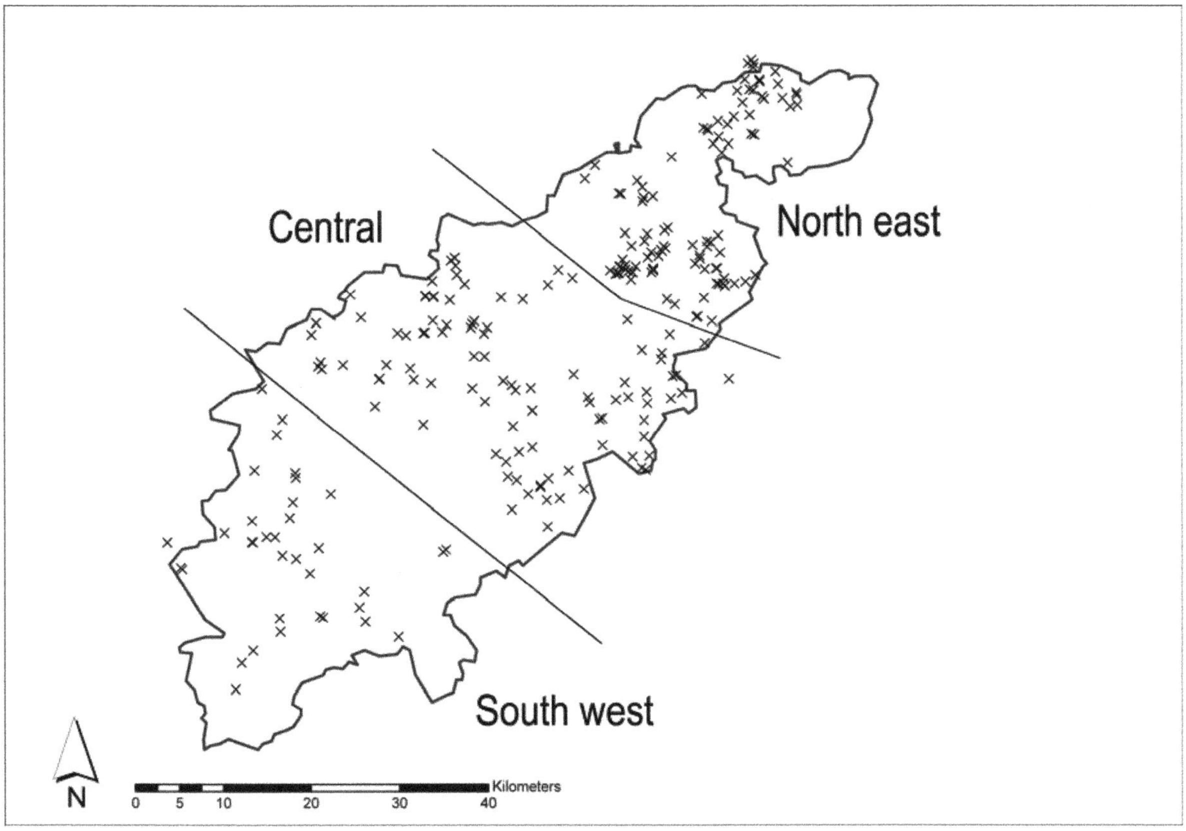

FIG 7.24 - SUB-REGIONS OF NORTHAMPTONSHIRE USED FOR FURTHER ANALYSIS.

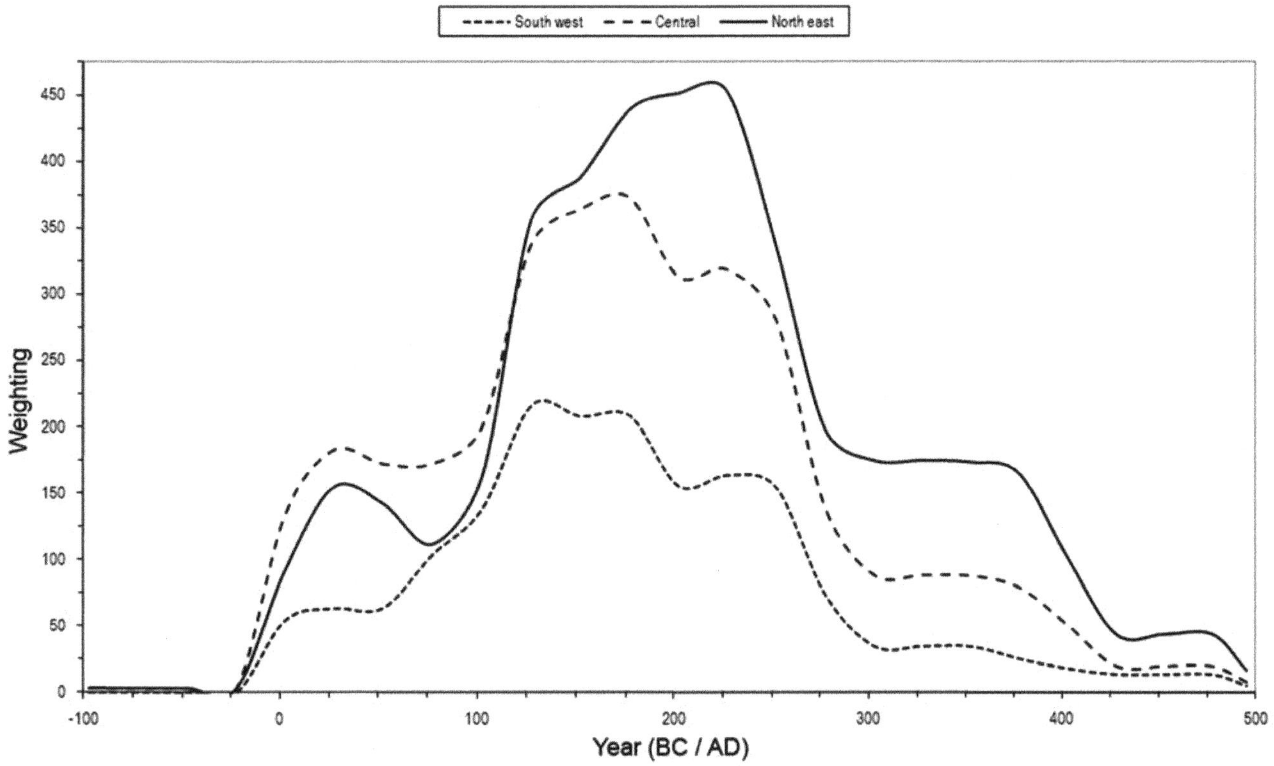

FIG 7.25 - COMPARISON OF TOTAL SUMMED PROBABILITIES FOR THE THREE REGIONS OVER TIME.

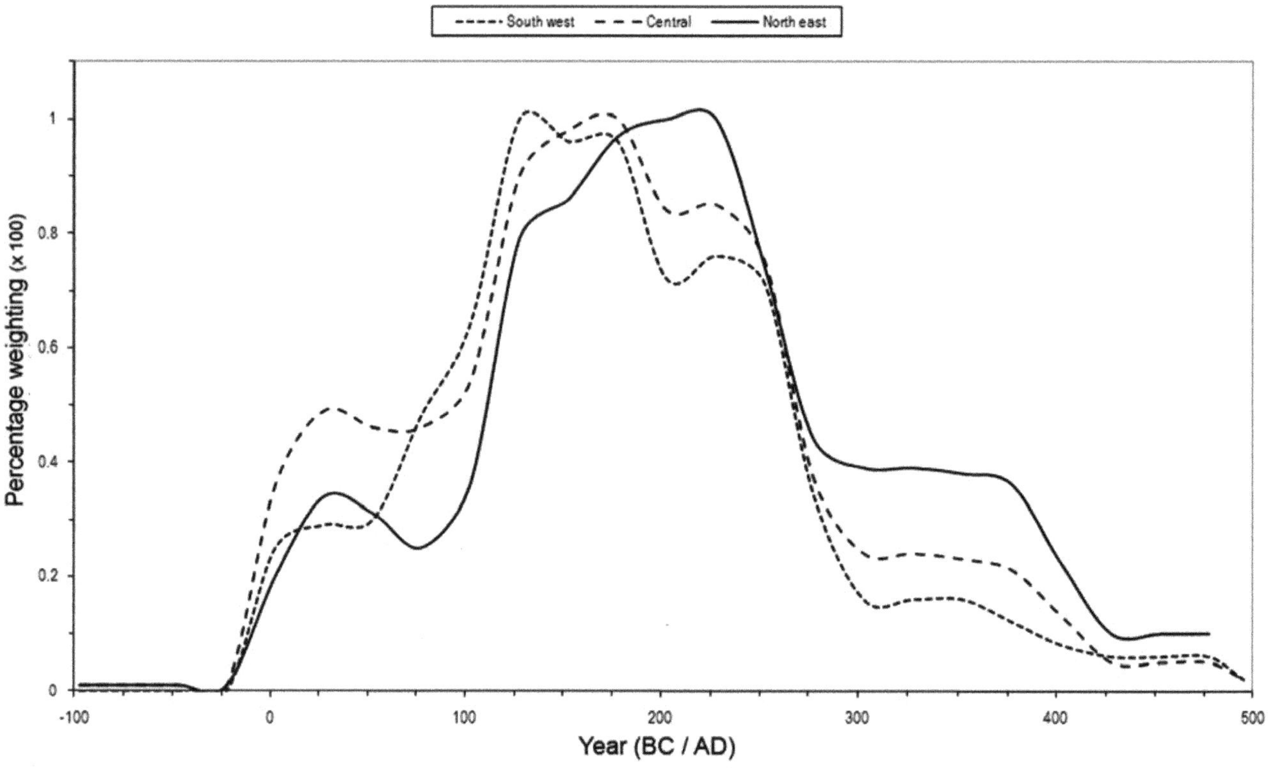

FIG 7.26 - COMPARISON OF PERCENTAGE (1.0 = 100%) SUMMED PROBABILITIES FOR THE THREE REGIONS OVER TIME.

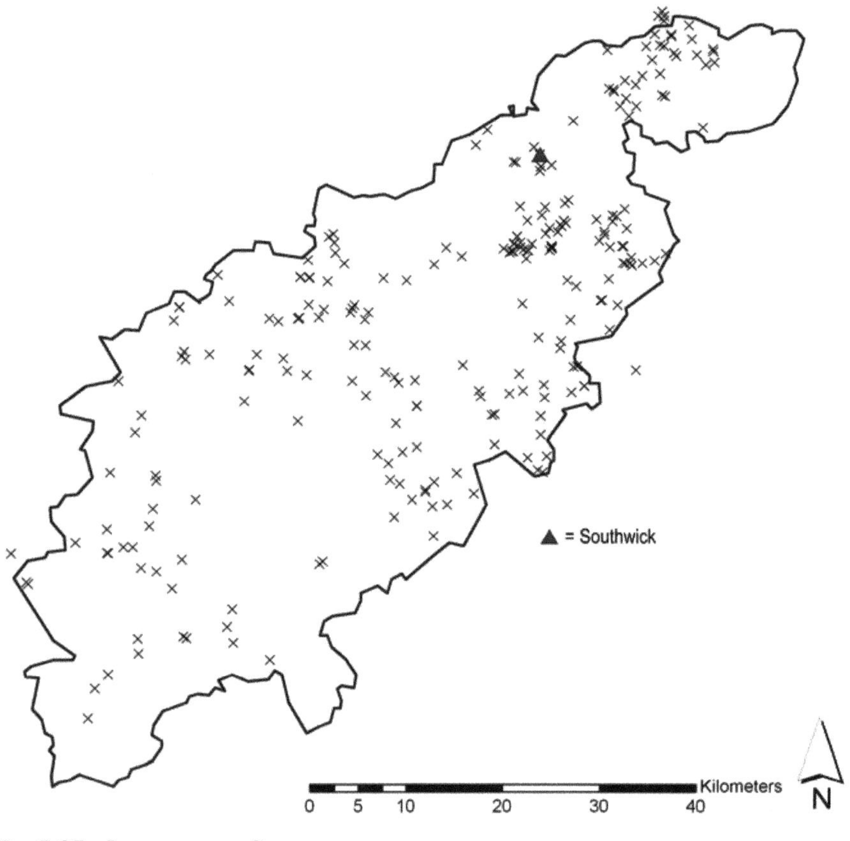

FIG 7.27 - LOCATION OF SOUTHWICK, SHOWN AS A TRIANGLE.

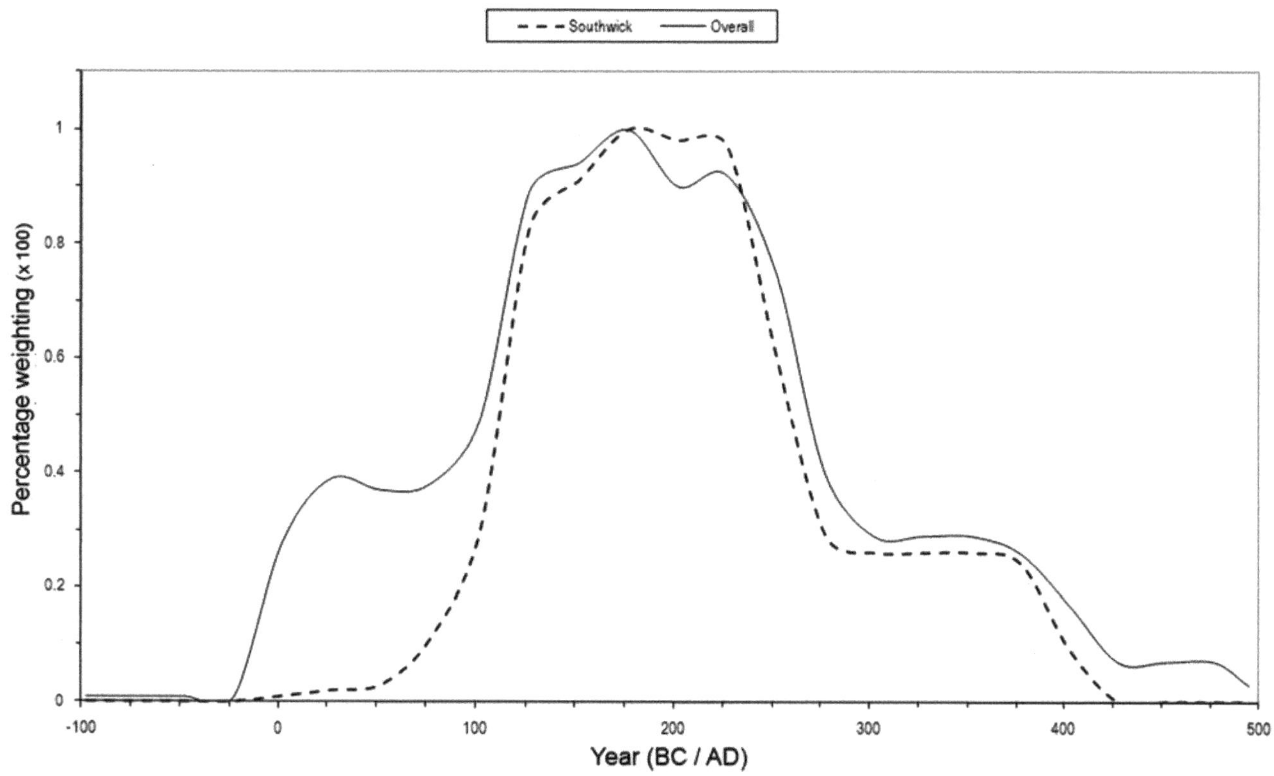

FIG 7.28 - COMPARISON OF PERCENTAGE (1.0 = 100%) SUMMED PROBABILITIES FOR SOUTHWICK AGAINST THE REGIONAL TREND OVER TIME.

close to matching the trend after that time until around AD 400. It is thus apparent that activity on the site was largely limited to the Roman period (at least so far as ceramic deposition was concerned), but with an internal pattern close to average for activity during the Roman occupation. However, the previous maps show that the actual levels of activity (as opposed to the pattern) were greater than the trend.

These analyses of temporal probability produced using the TGIS Analyst tool draw inspiration from and build upon the graphs produced by Trément (2000) as mentioned above (see Fig 7.8). They make it possible to consider in more detail the temporal probability profiles for ceramic deposition for a region, a subset of a region, or a site, and their comparison against analogue and broader trends. It is also possible that working with these tools might end up increasing the accuracy of our ceramic chronologies and could form a step in the road towards calibration of pottery dates (alongside work such as Going 1992 as discussed in Chapter 3). Increased resolution in our typological dating can only increase the value and accuracy of this type of analysis.

Conclusions

Whereas the previous chapter demonstrated the potential of the TGIS (a potential that could be fully met by intra-site datasets collected based upon a sampling strategy defined with TGIS implementation in mind), this chapter has demonstrated the power of the TGIS when analysing a regional dataset such as is more commonly studied using GIS methods. The TGIS provides a tool which enables the resolution of some of the chronological problems associated with the GIS analysis of data collected through field survey, and also provides a tool that increases the potential for the study of the temporality of such data.

Specifically, the probabilistic nature of the TGIS makes it possible to, in essence, create fuzzy sets for each phase in the user's model, based upon the percentage probability of a ceramic fabric falling within said period. Through this, the shackles which bind our phasings to the tyranny of typological periods are broken. As such, overlapping time-slices are no longer necessary and, furthermore, it becomes possible to integrate all of our dating evidence within our models, including ceramic types that might traditionally be split between more than one conventional phase and types that might previously have been discounted due to their low dating precision. Some researchers might be uncomfortable with this escape from conventional boundaries, but I do believe that this offers a productive new methodology.

Naturally, although the periods defined for each ceramic type in the typology no longer form the basis of our phasing in this new system, the typology still holds a strong influence over the results: the temporality of the analysis stems from that source (and ultimately from the dating evidence used in its definition). Improvements and adjustments to the typology would still manifest themselves in the TGIS results. However, weight can also be given to rates of deposition in the determination of phases, producing a phasing that more closely reflects changing socio-economic conditions in the past, rather than being based upon interpretative decisions made in the present day. This is not an argument for any perceived greater objectivity in the TGIS analysis (the ceramic dates themselves still being determined in the conventional manner), simply an argument for it being more representative. However, it would perhaps be possible to feed this evidence for rate of change back into the original typology and use it to improve the precision of the original dates. It may even prove possible to produce a typological dating system that varies in its associated dates according to geographical location within a region, based upon local prevailing economic conditions. This would build upon the work previously undertaken on seriation of assemblages by Marsh (1981), in which he made an analytical comparison of Samian ware deposition against supply.

In addition, the Analyst tool provided within the TGIS creates the possibility of studying the probability time-lines of a region, sub-region or site, either on their own terms or against their contextual background. This has in turn provided an important step forward in the study of the temporality of field survey data, especially if potentially combined with the trend surface residuals in a detailed analysis. Together, these advantages have helped in the essential task of introducing temporality to spatial analysis.

Moving on to the archaeological results of this case study, when compared against Taylor's trend surfaces, the TGIS trend surfaces show some significant differences. In particular, the Mid / Late and Late phases both show the influence of two pottery sources in the TGIS trend surfaces, whereas each map showed the effect of a different pottery source in the Taylor trend surfaces. Taylor had suggested that, if combined, the trend in his Late phase would have been subsumed beneath the stronger trend in his Mid / Late phase (Taylor 1996: 204). However, as noted, the shape of the TGIS trend surfaces for these two phases shows that this was not the case when approached probabilistically. With the alternative phasing produced using the TGIS Analyst tool, a more representative picture of the rate and nature of change over time was produced. Finally, the exploration of sub-regional temporality using the TGIS Analyst tool showed a clear shift in the focus of influence from the central area towards the north-east around AD 100, a change which persisted until the end of the Roman period.

In conclusion, the aim of the two case studies explored in this and the preceding chapter was to prove both the potential and power of the TGIS. With appropriate sampling regimes and with accurate data, I am certain that the TGIS can greatly enhance the analytical process. As the two case studies have shown, even with less than ideal data, very interesting results have been achieved and with considerably greater ease than would previously have been the case. The next, and final, chapter will thus explore the achievements and potentials of the TGIS in greater detail, and discuss possible future avenues of development.

CHAPTER EIGHT: ANALYSIS AND CONCLUSIONS

Analysis

The central aim of this work was to construct a fuzzy temporal GIS for explicitly archaeological usage that took into account recent developments in the understanding of the nature of archaeological time. The preceding chapters have discussed the theoretical and practical backgrounds to such formulations, previous work on introducing time into GIS, the construction of the new archaeological TGIS, and the outcomes of using that TGIS in the analysis of two archaeological case studies. It is now, finally, the time and the place to consider whether the software produced fulfils the central desires enshrined at the outset of this book (Chapter 1).

In good archaeological fashion, we shall begin with the evidence on (or in) the ground and work our way upwards; that means beginning with the archaeological results of the two case studies. The Willington dataset consisted of a series of radiocarbon dates taken from a site that principally involved Neolithic activity around tree-throws, Neolithic / Bronze Age fire clearance, and Bronze Age burnt mound activity (Chapter 6). The implementation and analysis of this data in the TGIS produced several interesting results. Firstly, the spatial pattern of the tree-throw activity was shown to begin earlier in the south-western area of the site, consisting of either one enlarging clearing or two separate clearings with the south-western being the earliest to form. Secondly, episodes of fire clearance were shown to have three or four distinct peaks, and the interpolations produced for those periods could be interpreted as an approximation of the probability of clearance having taken place in any particular area over time. This clearance appeared likely to be opportunistic and piecemeal, rather than concerted and systematic. Further, the TGIS results provided support for the excavator's conclusion that Burnt Mound I was contemporaneous to the fire clearance activity. Finally, a useful conclusion that arose through the explicit linking of the spatial to the temporal was the issue of residuality. Essentially, spatial proximity can be used as a test of the likely influence of residuality / intrusion. This will be discussed further below and was a result that could only easily have been achieved using TGIS.

The Roman Northamptonshire dataset consisted of a large corpus of field-walked pottery dating from the later Iron Age to the early post-Roman periods (Chapter 7). A series of trend surfaces for this material had been previously created by Dr Jeremy Taylor for his PhD thesis, based upon phasings taken from the ceramic typology and, thus, the various periods contained different pottery fabrics and overlapped temporally to varying degrees (Taylor 1996). Implementation and analysis of this data in the TGIS meant that the different fabrics no longer needed to be separated out into different overlapping phases, as their membership of each phase would be treated probabilistically. It also meant that less precisely dated coarsewares could be included in the analysis, to avoid misleading results (see Trément 2000).

The end result showed the relative influence of each different pottery source present during any particular period, whereas before they had been variously separated out into different phases. Thus a truer picture was painted of pottery supply in any particular period. Essentially, the various phases could be conceived of as fuzzy sets, with each pottery type possessing a membership degree determined probabilistically: this helps escape from the totalising nature of conventional chronology (see below). Furthermore, an alternative and arguably more representative phasing was produced based upon the rate of change over time in the ceramic evidence (specifically, a compound variable based upon the typology and rates of deposition). Finally, the TGIS Analyst tool was used to compare overall supply patterns for three sub-zones of the overall region, demonstrating a shift in the centre of influence from the central area to the north-east around AD 100, a conclusion that broadly agreed with Taylor's original interpretation.

Having briefly summarised the archaeological results of the two case studies, it is next necessary to discuss in more detail the software produced and the particular strengths (and weaknesses) discovered through its practical application. At the heart of the TGIS lies its approach to handling temporal probabilities. Four methods for calculating the probability of an archaeological date falling within a particular time period were included: a straightforward percentage; probability tracing a normal distribution; a simple triangular rendering of decreasing *terminus post quem* probability; and probabilities output from the calibration software, OxCal v.4.0. Of these, the first and the last were thoroughly tested through the case studies. The various features of the TGIS that were not extensively tested using case study material will be discussed further below. These four calculation methods provide the coverage to enable the TGIS to handle the probabilities of the vast majority of archaeological dating methods (discussed in detail in Chapter 3).

Probabilities are recorded as attributes of the layer in question, from 0.0 (0%) to 1.0 (100%), with a precision of two decimal places (i.e. 1%). This is to maintain simplicity, as probabilities of a better precision than 1% are unlikely to be interpretatively relevant for archaeological data. The recording of probabilities as attributes of this nature is straightforward and should prove conceptually simple for users used to working with GIS and scientific dates. Further, probabilities can be displayed using any of the appropriate cartographic tools provided within ArcGIS. The simplest way to do this is through colour and this is the methodology programmed into the TGIS. Alternative methods, such as sound (with harmony and disharmony perhaps conveying probability), might be more engaging to the user but might also be considered gimmicks. Colour is the most straightforward, understandable, and quantifiable solution. Furthermore, through treating time as an attribute in this way, the 2D GIS effectively becomes 2.5D, in the same way as when elevation is introduced as an attribute. This is an issue that will be considered more fully below.

The percentage probability served as a catch-all for types of date where the internal probability profiles are unknown. In the context of this work, it was used for the probabilities of the ceramic dates in the second case study. In the future, it may become possible to apply a type of calibration to pottery data, perhaps based upon economic cycles (Going 1992), or upon supply patterns for individual sites or regions (Marsh 1981): this would provide ceramic dates with a more complex and accurate internal probability profile. However, as it stands today, this is not the case and the percentage methodology is best used. That methodology was calculated as the percentage of the time-frame associated with any particular date that overlapped the time-frame currently selected in the TGIS. This is a mathematically simple and robust operation. Working with these percentages is more representative of the nature of the data being studied than is the case when working with dates that cannot be split and appropriately weighted within their internal time-frame. No errors or specific difficulties were discovered during the use of this element of the TGIS.

Moving on to the radiocarbon probability calculation method, the internal probabilities of these dates were taken directly from the calibration software. In the context of this work, this was used for the radiocarbon dates in the first case study. The specific internal probabilities for each date are output from OxCal and copied into a table. This table is then added to the map document in ArcMap and linked to the radiocarbon geographic data layer using a one-to-many relate (based upon the laboratory code). As such, when the TGIS wishes to calculate a percentage for a specific date, it simply has to look up all of the probability slices that fall within the currently selected time-frame and sum them, correcting for the temporal width of the OxCal slices. Exactly the same system is used for both conventionally calibrated radiocarbon dates and the output of Bayesian models created in the calibration software. It should be noted that chronological Bayesian models are based upon the stratigraphic relationships between the dates that they contain. As such, this allows some consideration of stratigraphic temporality to enter into the TGIS, albeit indirectly through the calibration software. Again, no errors or specific difficulties were discovered during the use of this element of the TGIS.

The TGIS was designed so that different types of dating evidence could be separated out into different layers and analysed separately. This system was instituted so that different probability calculation methods could be used for different types of date, and would also allow users to investigate any intrinsic temporal bias within a particular layer when contrasted against other layers. Although, the two case studies explored did not contain multiple types of dating evidence, this feature was utilised in the first case study to separate out different thematic layers for individual analysis, and to enable comparisons between the conventional radiocarbon dates and the Bayesian model. The system was found to work well, but some difficulties were uncovered. Most importantly, it would have been useful to be able to analyse several layers at the same time without having to switch between them. This is one important way in which future versions of the TGIS could be improved, perhaps including the ability to offset certain layers due to discovered biases.

Secondly, a bug was discovered in the software that meant that the OxCal probabilities were not being selected properly when switching between layers. This proved intractable. It had been decided in any event that it was too complex to allow the TGIS to save OxCal settings when switching between layers, and therefore a reset procedure was integrated to prevent layer switching without reloading when working with OxCal probabilities. This is not a particular problem, as it merely introduces some additional stages when dealing with such data, but it is a limitation that would ideally need correction in future iterations of the TGIS. However, the fundamental issue arising from the testing of the multilinear nature of the TGIS was the difficulty encountered in locating an appropriate dataset with which to fully test this particular element. This, in part, illustrates the need for archaeologists to consider using TGIS-oriented data capture strategies, an issue that will be discussed at the end of this chapter.

Perhaps the most important feature of the TGIS is the way in which it enables temporality to be introduced into spatial analysis, in effect spatialising temporal uncertainty. It does this through the writing of the probability outputs to the attribute table for the layer being studied and, if desired, to a new layer summed according to spatial concurrence. The second of these also allows the inclusion of a weighting (called normalisation in the TGIS) factor. As a result, this probability field (or the normalised field) can be used as a z-factor in many of the spatial analysis tools built into ArcGIS. For example, interpolation algorithms require a value field to be interpolated. In the two case studies in this book, two types of interpolations were undertaken: ArcGIS's Inverse Distance Weighted tool was used in Chapter 6 and trend surfaces were created in Chapter 7. Both of these methodologies functioned well with the data provided, as the results of those two chapters demonstrate. I am confident that there would be many other forms of spatial analysis to which this temporal influence could be applied. The complete absence of time from conventional spatial analysis is a very significant shortfall in its ultimate validity and, as a result, this element of the TGIS offers a very important step forward for archaeologists working with GIS.

The final feature of the TGIS that was tested as part of the case studies was its ability to create cumulative probability profiles. This feature is intentionally limited in scope (e.g. it cannot take account of OxCal probabilities as summing those is considered to be unproductive - see Chapter 5), but has considerable potential. It is most securely used as a rapid method for testing the influence of different factors or regions on the overall temporal probabilities of a dataset (or subsets of the same). This is clearly shown by the graphs in Chapter 7 that demonstrate both the importance of including sherd counts in my analysis (Fig 7.8) and the varying influence of different regions within Northamptonshire over time (Figs 7.24 and 7.25). However, the possibilities for this tool are considerable. For example, it would be possible to study the relative cumulative probabilities for every site and sub-region in a dataset, across multiple temporal scales (by varying the width of the summed time-slices). It would then be possible to test different phasings for a dataset, through the use of the "form periods" tool, based upon varied scales and rates of change. The flexibility of this approach offers something of a fresh take on archaeological chronology and could prove distinctly fruitful, particularly in areas such as calibration of ceramic dating evidence (see above).

It did not prove possible to utilise some features of the TGIS using the case study data. These features have been tested mathematically and all function correctly, but were either not applicable to the material in question or did not prove relevant to the particular questions being asked. The TGIS is able to calculate probabilities according to a *terminus post quem* model, a normal distribution model, or for single year dates (which are assigned a probability of 1.0 if they fall within the selected period, no matter which probability model is used). As previously argued, any precision of greater than one year in an archaeological date (e.g. the season of a dendrochronological date) ought to feed into different interpretive threads than the year portion of such a date; for example, some of the charred remains of Burnt Mound II at Willington suggested that it had had an early- to mid-summer usage period (Chapter 6). Essentially, questions in regard to seasonality are in respect of a different type of temporality than questions in regard to chronology. Seasonality could be analysed in GIS using conventional methodologies (i.e. by the recording of relevant attributes) and is not an area with which the TGIS can offer any particular advantages. As such and in alliance with the models previously discussed, these three probability models cover the full gamut of current commonly-used archaeological dating techniques[1] (Chapter 3). Finally, the TGIS also records the temporal topological relationship between each date and the selected period (see Chapter 4 for more on temporal topology). This is a useful piece of information, but it was not considered relevant to the particular questions asked of the case study data. Nevertheless, its inclusion in the TGIS only serves to extend its useful functionality. Through the recording of this topological information alongside the probability information, users are able to use the TGIS to visualise and analyse change across their data layers.

Conventional TGIS as created by geographers for their specific purposes were discussed in Chapter 4. They were dismissed for archaeological usage for several reasons:

1. Their reliance on data that is very temporally precise, often down to the nearest minute or second;
2. Their inability to deal with data expressed as a temporal range;
3. Their concentration on issues such as animation and temporal query languages;
4. Their frequent obsession with real-time updating and resulting data quality issues.

Geographers have acknowledged the first two of these issues, but done little to tackle them. It is also those two issues in particular that have made their TGIS totally inadequate for archaeologists. This is because our temporal data only very rarely fit into this rigid model. Archaeological time is usually expressed as a range of possible dates, often with a varying probability within that range: inherently precise clock time simply cannot express this fuzzy complexity. The third issue has created some useful seeds for future development, but in areas that are only tangentially related to archaeological concerns, especially when it is remembered that the problems caused by the first two issues have restricted our access to these developments. The final issue simply illustrates one of the ways in which archaeological interest in TGIS is often incompatible with geographers' particular areas of interest.

Undertaking the analysis presented in the two case studies in this book would not have been possible using a geographers' TGIS. Their systems provide a useful forum for the study of precise temporal data, but would not cope with our much more nebulous connection to modern clock time: the archaeological Dalí's clock of the title. It might prove possible to use conventional TGIS to study some archaeological material (e.g. a cemetery dataset, where the date of every grave was known to within a single year), but the vast majority is excluded by the specific problems listed above. In this regard, the TGIS produced for this work is clearly of much greater value to archaeologists, insofar as it has been specifically designed to deal with the first two critical issues with the geographers' model. Hopefully, in its demonstration of the possibilities inherent in taking a less strict view of time as modern clock time, the archaeological TGIS might encourage commercial GIS developers to adopt a wider, less precise viewpoint when creating their own new software.

It is also necessary to compare the TGIS created for this work with the only other non-case specific archaeological TGIS, Ian Johnson's TimeMap (Chapter 4). TimeMap

[1] In Chapter 3, I promised to include the ability to deal with typological, numismatic, radiocarbon, thermoluminescence and dendrochronological dates.

has three main elements: its ability to interpolate the extent of a spatial phenomenon between known time-slices (i.e. periods where the phenomenon has a known extent); its ability to deal with the animation of temporal data; and its ability to integrate with the internet for the display of maps and to obtain remotely hosted data. As such, TimeMap can do certain things that the TGIS created for this work cannot: in particular, animation and its web-integration. However, although TimeMap is able to deal with data that might fall within a temporal range, it does not make any calculations of the probability of a date actually falling within any particular part of its range. Thus, its ignorance of probability and its lack of the tools provided by a larger GIS package means that it cannot provide the important step forward in respect to introducing temporality to spatial analysis that the work presented here can.

TimeMap is, in essence, restricted by its mischaracterisation of archaeological temporality as snapshots. This snapshot-based model is borrowed from geographers' TGIS, particularly in the work of Langran (1992). It relies upon the assumption that archaeological data are easily divided up according to a series of sequential slices of time, but this is an assumption that only works comfortably with agglomerated period data. The ability to deal with that type of data (e.g. a sequence such as Iron Age farmsteads, Roman villas, Saxon farms) is adequately provided by conventional GIS, as the temporality of such data is of too large a scale to be fruitfully analysed within TGIS. Beyond that, the snapshot model only works in some limited circumstances where the full state of a variable is known at various specific stages: most examples of this are historical or geographical rather than archaeological (e.g. the changing borders of the Roman empire or changing sea levels). In its raw form, archaeological data consists of dates and stratigraphy, a data model that is incompatible with TimeMap's snapshot temporality (Chapter 4).

Essentially, the TGIS and TimeMap answer different needs. TimeMap deals well with many of the ways in which a TGIS can be used to communicate with the general public, through the internet and museum displays. It is able to produce attractive animations of an empire's shifting borders or of the changing influence of different cultural groups. It also has a strong emphasis on issues that are undoubtedly important, but only tangentially related to core TGIS concerns: in particular, its integration with the ECAI (Johnson and Wilson 2003). By contrast, the TGIS produced for this work has a far greater emphasis on the concerns of archaeologists using GIS for their everyday work. It has a stronger focus on the nature of archaeological temporality and its associated dating methodologies, but is less expert when it comes to the communication of results and ideas to the wider public. TimeMap and the TGIS are in many ways different and yet complimentary tools.

As discussed in Chapter 5, the standard probability model integrated into the TGIS has some similarities to the aoristic method proposed by Johnson (and others) as a route into dealing with temporal probability (2004a). Aoristic analysis works by splitting the time-line for a dataset into a series of time-slices of equal width. Objects have a data range and are given a probability of 1.0 divided by the number of time-slices which they overlap. The methodology has potential, but is restricted by its fixed time-slice width and the imprecision inherent in its calculation method (Chapter 5). The TGIS escapes these two specific difficulties by providing full flexibility over period selection and more accurate probability calculation, including the suite of complex probability calculation methods. As such, it provides a clear development from and improvement over the aoristic method.

It may be recalled that I outlined two areas that I saw as essential for the development of specifically archaeological TGIS: the ability to deal with the inherent uncertainties of archaeological dates; and the ability to approach integration of different dating techniques (Chapter 4). The first of these is the easier to approach and, as outlined above, the TGIS produced has achieved considerable ability and flexibility in the handling of a wide range of uncertainties. The second issue is trickier. Essentially, the solution implemented in the TGIS is two-fold. Firstly, this is in the conception of phase as fuzzy set. Any particular period of interest can be entered into the TGIS for analysis and a membership degree calculated probabilistically for every date in the layer in question. Secondly, that layer could contain many types of date for direct comparison through a straightforward percentage analysis, or different techniques can be separated out onto different layers and then analysed using more accurate probability calculation methods. With intelligent human input and analysis, the TGIS thus provides the tools to help consider this complex question. Furthermore, as mentioned above, this conception of phase as fuzzy set is one way of moving beyond the totalising approach encapsulated in conventional chronology building. The recording of temporal topology as implemented in the TGIS and the provision of a simple temporal buffering methodology in the Set button only help increase the versatility of this solution when considering questions about contemporaneity.

At the start of Chapter 4, we discussed some of the issues arising out of the continuing growth of information technology in archaeology. From this, I concluded that, first and foremost, the TGIS should be useful to archaeologists, help them answer their temporal questions (whilst not trying to do their job for them), and be as simple and transparent as possible in its methods and function. This is where the importance of treating time as an attribute in the GIS becomes most apparent. The resultant 2.5D approach means that it was possible to implement the TGIS in ArcGIS, a conventional and very widely used GIS package. However, this is not a 2.5D implementation that prohibits the user from using elevation data in conjunction with the temporal data:

ArcGIS's functionality in this regard remains available and, further, the TGIS could also be adapted to function with a 3D GIS to produce, essentially, a 3.5D solution. Moving to a fully 3D (2D space plus 1D time) or 4D solution would have required extensive adaptation and, as a result, the software would have begun to move away from the familiar bounds of everyday GIS. This would result in only more expert GIS users being willing or able to use the TGIS, thus doing a great disservice to the democratic aims of the project. We shall return again to this issue below.

Further, the two case studies prove that the TGIS is certainly not a toy and can produce genuine archaeological results that were not obvious when conventional methods were employed. The data required to feed the beast is somewhat complex, but no more so than is the case with most other modern archaeology. It is true that, through its use, the TGIS has fostered a desire for more data: however, it is important to stress that this desire is for data gathered *in a different way* rather than inherently more complex data (see below). Digital archaeology should be approached with theoretical consideration and self-criticism as to whether it is appropriate. The idea is to answer questions in new ways where our data are suitable, not to become beguiled by technology and use digital tools just because they are there: when badly used, tools can unduly colour the nature of our interpretation (Chapter 4). The case studies in this book have demonstrated how the TGIS can aid in answering typical interpretive questions, whilst also demonstrating that these questions still need intelligent input from the user to formulate and to decipher (see Chapters 5-7): the TGIS **helps** archaeologists to answer their questions - it does not answer their questions **for** them.

Continuing our ascent of the interpretive ladder, we finally return to the theoretical issues with which this book began. Several of the theoretical strands discussed in Chapter 2 have had a strong influence on the construction of the TGIS and/or are explorable through usage of the TGIS. It should be recalled that my aim was not to produce a new theory of archaeological time, but to create a TGIS that could handle as many existing theoretical models as possible. We shall begin with one of the simplest temporal models: evolutionism. This model has strong links with conventional chronology and, as such, is heavily embedded in modern clock time. However, the TGIS contains no inherent uni-directional picture of temporal change, as an evolutionary model might suggest.

This type of temporality is far more obvious in other TGIS that place a greater emphasis on animating change over time, whereas the TGIS created for this work provides flexible tools to its users for the selection of different time-frames and time-slices. Thus, it would be possible to pursue an evolutionary or conventionally chronological model in the usage of the TGIS, but it is not restricted to such. Time and chronology have become blurred in archaeological thought to the detriment of both, and we must embrace new forms of temporality if we are to progress as a discipline. New models do not abandon chronology wholesale, but take into account different conceptions of time and, as a result, suggest that chronology should not be the only conception of time used by archaeologists. As such, a TGIS that only dealt with conventional chronology (a C-GIS) would remain mired in the problems caused by following modern clock time and, as archaeologists move forward with their temporal models, might become obsolete soon after its birth.

Secondly, we come to the A- and B-series models of time. The A-series is time as perception, with a present, a past and a future - or reproduction, retention and protention. The B-series is time as recorded, moving ever forward at a constant rate (at least in every day experience, assuming no trips are taken towards the speed of light). Within the TGIS, both of these models of time can be seen. The tools for selection of periods are based on the B-series model. The A-series is present in the recording of temporal topology, where objects are displayed only according to their topological relationship (as opposed to their scalar relationship) to the current period of interest. This is only as is appropriate, as neither A- nor B-series can exist without the other in human experience. However, the base model of time in the TGIS is B-series in nature, and forms a foundation for further A-series models of temporal perception that might be built on top of this in the future.

As previously discussed, the TGIS contains a strong element of multilinearity in its inclusion of the ability to study different layers according to different temporal attributes and probability models. Although not strictly an example of the application of non-linear dynamics[1] to archaeology, this specific aspect of the TGIS developed has certainly taken some inspiration from this field of temporal theory. This is, in particular, its emphasis on mathematical calculation of temporal probability, and its entirely flexible approach to temporal scale. The user can study his or her data over any temporal scale that they wish and, through the use of the summed probability and period forming tools, create any number of period phasings based upon different rates of change. However, ultimately all of that functionality remains within the control of the user and any accusations of determinism or slavery to the model that might be levelled at non-linear dynamics[2] are much less applicable to the TGIS. This temporal scalar variability also makes it open and straightforward for users to pursue the three-scaled analysis typical of the *Annales* approach to their data using the TGIS, if so desired, especially the type of directed analysis seen in Bintliff's problem history (1991b).

[1] A highly structured form of disorder that redefines evolution and provides the ability to adjust scale to data.

[2] However, non-linear dynamical models are usually seen as tools rather than reality and tend to have speculative conclusions.

Finally (for now), we come to Gavin Lucas's desire to take archaeological temporality beyond the boundaries of conventional chronology (2005). The biographical approach advocated by Lucas is not explicitly present in the TGIS, although it might prove possible to use the TGIS to aid in such an analysis: in most cases, however, the biographical approach is too complex to be applied to all archaeological material and is problematic due to its avoidance of the overwhelming issue of contemporaneity (Bailey 2006). However, the TGIS does provide an alternative route into the construction of chronology. In particular, in the introduction of a spatial aspect to the consideration of questions about residuality (more on which below) and to the interpretation of dating evidence. The conventional construction of chronology has an understandable and entirely correct emphasis on the vertical spatial dimension (i.e. stratigraphy), but does not take enough account of horizontal spatial factors. The TGIS can introduce this horizontal dimension. Chronology remains integral to archaeological questions, but is itself sterile without those questions. We must improve the ways in which we engage with it, to the greater good (Whittle *et al.* 2007: 124):

> *Chronology is not an optional extra nor should it be seen as the preserve of myopic specialists in either scientific or material culture fields. It lies at the heart of a better understanding of agency, of locating what people in the past did at particular times and places.*

We shall return to chronology and other models of archaeological temporality below.

Conclusions

To bring the book to a conclusion, I shall now look in some detail at the ways in which the TGIS could be developed and improved in the future, before finally drawing to a close with a discussion of the most important conclusions that have been drawn from issues that arose through the process of creating this software. Several of these issues arise from the tension created when it was decided to implement the TGIS in ArcGIS. The 2.5D approach enables the dissemination of the TGIS via ArcScripts,[1] and straightforward installation and usage (and, if desired, adaptation in ArcGIS's VBA editor) by the majority of GIS-using archaeologists. This democratic ideal should help to foster a desire for TGIS in the archaeological community, a central concern of this project (Chapter 5): TGIS will over time become the rule and, as a result, archaeologists need to be able to access them and adapt them without too much trouble. However, the 2.5D approach has also resulted in various compromises that would require more bespoke solutions to solve, as we shall see below.

First, is the ability to animate changing patterns over time, functionality that has usually been considered a key component of any TGIS (Chapter 4). However, animation proved impossible within the confines of ArcGIS: due to both the software not rendering its maps smoothly and the need to save data to the layer table, resulting in any animation being discontinuous at best. In any event, animation is not necessarily an essential concern for archaeological TGIS. Animation is the simplest method of representing time and, as such, has perhaps acquired an undue orthodoxy as the way to "do TGIS". By contrast, the key archaeological TGIS issues are those approached by the software produced for this work: the ability to deal with probability, multilinearity, and chronology. Animation is useful for public dissemination of results (as per TimeMap), but not helpful for spatial analysis and other core GIS-utilising tasks. As such, whilst it would be desirable (and no doubt expected) to integrate an animation procedure into future versions of the TGIS, it should not be a key concern.

However, the ignorance of the TGIS with regards to stratigraphy (other than in the use of Bayesian modelled data) is a more pressing issue. The study of stratigraphy lies at the heart of chronology building (Chapter 3) and a system to deal with this vertical space in additional to horizontal space would be a key adaptation for a future version of the TGIS. Again, I do not believe that this would be easily achievable within the confines of ArcGIS. An ideal solution might involve split-screen rendering, with the map on one side and the section diagrams or Harris matrix on the other. Objects relating to dates would appear in both displays, linked to the same database object. As such, when a period of interest was selected in the TGIS, the probabilities would be calculated and painted to both the map and the stratigraphic diagram. Clicking on or querying objects would cause selection of the same objects to occur in both displays. Such an approach is beyond the constraints of the current work. Unlike the current TGIS, this split-screen concept could not work in ArcMap. It might prove possible to create such software using either ArcGIS Engine or (alternatively) GRASS / QGIS, both of which are more adaptable (by their very nature). Also, doing so would increase the complexity of the software and take it further away from the democratic aims of the current project, in which a key concern has been to encourage archaeologists to experiment with TGIS within a familiar environment (Chapter 5). However sophisticated a tool is, it is only useful if used routinely.

The future implementation of a more complex TGIS in the open source software package GRASS would be in keeping with a growing trend in both archaeology and the GIS community generally, with many open source projects now attracting government funding. For example, Oxford Archaeology, one the largest field units in the UK, has been migrating over to open source software to reduce costs and encourage the growth of

[1] http://arcscripts.esri.com/

such solutions.[1] QGIS and GRASS provides a vector-based GIS that is highly customisable in C++ and Python and with a large user community (Steiniger and Bocher Forthcoming). In essence, commercial products tend towards being easier to use and more readily understood, but open source products are easier to obtain on cost grounds and easier to adapt. If the TGIS were to be taken down the route towards greater complexity, then this balance would no longer be weighted towards its current emphasis on ease of comprehension and the advantages of an open source basis would become the more telling. Sections of the code produced could be imported into alternative software, particularly the most important sections: the probability calculation algorithms.

Another key area for future development of the TGIS would be through the introduction of a system to deal with the duration of objects. Currently, date objects in the TGIS are treated as single events of uncertain date. However, by their very discovery in the present, it is apparent that all recovered archaeological evidence does not cease to be once its moment of creation has passed (unlike the Higgs boson[2]). This duration element is a key component of Lucas's biographical use-life approach (Chapter 2) and is especially important for larger archaeological features, such as earthworks. A fully functional TGIS would, therefore, allow users to set a duration for objects. In this vision, dating objects would gain colour or appear once their minimum possible date was reached, would be coloured according to probability (as currently), but as they pass beyond their maximum date they would then receive another colouring whilst the selected period of interest remained earlier than their ultimate duration date. Once the ultimate duration was reached, the objects would become uncoloured or disappear. Perhaps, when recording temporal topology, future versions of the TGIS could also record the degree to which an object fell before or after the selected period, to help consider questions about continuity / discontinuity.

This system would be straightforward to implement in the TGIS, as it would involve no programming more complex than in the current TGIS. It has not been implemented in the TGIS as of yet for two reasons. First, the data studied in the case studies presented had no such known durations. This would likely be a common occurrence: durations might often need to be estimated (Chapter 3), and perhaps some function to deal with probability of presence / visibility would also be necessary. Second, the types of object likely to possess useful duration evidence would be earthworks, buildings and the like. These are not types of archaeological object which the TGIS is currently designed to work with. The TGIS is intended for use with dating evidence that consists of specific dated objects: this would include items like radiocarbon dates, other scientific dates, ceramic dates, and numismatic dates. These types of object, although possessing durations, are not usually employed in a duration-oriented manner in archaeological analysis (whether they should be or not is a separate issue). Within current archaeological interpretive methodologies, duration is more important to larger archaeological groupings: contexts rather than finds. As such, if the TGIS was to be expanded to usefully deal with duration, it would also need to be expanded to deal with the dating of such groupings, rather than just objects found within them. This is, of course, entirely desirable and achievable. As with stratigraphy, however, it would again increase the complexity of the software solution. As a final note, a useful addition to future versions of the TGIS would be the ability to handle dates expressed as a mean plus error band.

These future potential adaptations to the TGIS would finally make the software useful for the consideration of our final theoretical model of archaeological temporality. Here, we return again to social time (Bayliss *et al.* 2007: 2):

> *...we need to move from the measurement of elapsed time, to a sense of successive events, and then to how people experienced the flow of time and saw themselves in time...*

Recent theoretical and practical approaches to social time were discussed in Chapters 2 and 3. The current TGIS has been structured around core datasets and questions seen to lie at the base of subsequent interpretations and theoretical foundations. However, imagine a future TGIS that was able to deal with probability, multilinearity, chronology, stratigraphy, and duration across a whole suite of archaeological dating objects and wider groupings. Such a TGIS should prove applicable to the challenge laid down by Bayliss *et al.* above, and might discover Ingold's rhythm and resonance in the archaeological record (1993). It would certainly prove useful to the types of specific case-study analysis undertaken in this regard by Richard Bradley (2002) and hopefully to less relatively clear-cut analyses of social time, perhaps employing Gosden's intriguing System of Reference (1994). This is an important avenue for future archaeological exploration and could perhaps one day prove to be the ultimate crowning glory of archaeological TGIS.

To conclude, we return again to the specific archaeological conclusions of this work and a further consideration of their implications. To begin, the Willington case study involved an assessment of residuality based upon spatial location. As mentioned earlier, this is an important area in which the TGIS has provided a useful and unexpected new methodology for the consideration of a common chronological problem. Spatial proximity can be taken as a clear indication of increased likelihood of different groups of material being residual or intrusive. This is based upon an assumption

[1] http://www.ubuntu.com/products/casestudies/oxford-archaeology; http://thehumanjourney.net/index.php?option=com_content&task=view&id=128&Itemid=141

[2] http://news.bbc.co.uk/1/hi/sci/tech/3546973.stm

that objects that are closer in space are more likely to be related or similar, as per Tobler's First Law of Geography (Tobler 1970). Of course, archaeological objects being widely spread across space does not mean that no residuality exists between groups, but it does reduce the probability. This important element of the functionality of the TGIS was only discovered through its application to real-world material, and it can be hoped that further such methodologies and potentials might be discovered through its continuing use.

Secondly, the Willington case study also involved the application of the TGIS to the study of Bayesian modelled material. This system for improving radiocarbon chronologies is another important continuing development in archaeological thinking and practice (e.g. Bayliss *et al.* 2007). Introducing a horizontal spatial element to this type of study is an important consequence of the application of the TGIS. Archaeological chronology relies upon vertical stratigraphy, but the horizontal dimension is also important to interpretation and analysis. Use of the TGIS begins to allow us to take this more three-dimensional viewpoint. Implementation of a stratigraphic component to the TGIS could only aid in this. New chronologies will become an increasing force in archaeology, particularly for prehistoric periods, and the TGIS can aid in the interpretation and analysis of these fresh results.

Next, we return to the theme of introducing temporality to spatial analysis. This element has been discussed in detail above, but the conclusions need re-emphasis. Conventional GIS is short-served by its ignorance of time. Time is integral to all that humanity does and has done and GIS that takes no account of this important dimension is incomplete. Previous attempts to bring together temporality and GIS have fallen short of properly extending spatial analysis into the fourth dimension, except in very case-specific situations. With its emphasis on mathematical probability, the TGIS has begun to make it possible to approach this integration, albeit through a 2.5D solution. Therefore, this is perhaps the most important aspect and greatest success of this new software.

Finally, if we are to achieve the best possible results for this new spatio-temporal analysis, we as archaeologists need to think carefully about our data capture strategies. At present, sampling strategies take great account of stratigraphic positioning, as they should. However, if TGIS are to achieve their potential, sampling strategies also need to take into account the horizontal. This is not to say that the vertical should be discounted, but it would be prudent to apply both considerations, in much the same way as the emergence of GIS caused archaeologists to think more critically about the quality and quantity of spatial data collected. Specifically, with both of the case studies undertaken for this book, the dating material recovered tended to focus around particular areas. This would partly (and possibly largely) be a result of where the material was to be found and the gaps in coverage may be genuine, but it would be prudent to attempt to achieve greater horizontal coverage. This would allow the construction of far more precise interpolations and other spatio-temporal analyses. At the very least, an indication of absence of evidence would allow TGIS users to determine whether areas of no data are genuinely such, or simply areas of no sampling. As stated above, if the TGIS is to do all that it might, data capture strategies need re-formulating to bear in mind future TGIS implementation.

An example of this could be the application of a grid to the choice of dating material. The ideal would then be to take at least one or two dating samples (or more where appropriate) from each cell of a grid laid out across the site (chosen within the grid square according to stratigraphic considerations). The resolution of the grid could be adjusted to reflect the number of dating samples for which processing could be funded. Any cells that had no appropriate samples could be recorded as archaeologically barren, and areas that had material that could have been sampled but was not (due to cost or time constraints) could also be recorded. As a result, researchers studying the material in a TGIS would then know which areas of low probability or no data were likely to be the result of true trends, and which might in fact be gaps in the available dating evidence. Such a system would make for far more robust spatio-temporal analysis (and could help inform where costs limitations are pressing).

Thus we come to the final conclusion. The TGIS constructed for this work has achieved many of its aims and proven its worth through its application to real archaeological material and problems. Furthermore, a clear path for future development has been outlined. There is nothing in that agenda that is not ultimately achievable and that would not prove beneficial to both archaeological GIS and archaeology as a whole. Too much work on TGIS in archaeology has been theoretical rather than practical: this project restores some of that balance. The use of TGIS has the potential to revolutionise many areas of archaeological practice, in particular where spatial analysis is applied whilst now taking into account temporality. Our discipline is integrally temporal; time and chronology are our bread and butter. As such, the novel and creative use of TGIS is an area in which we ought to have much to say and discuss. The challenge is to grasp the nettle and immerse ourselves in this new technology: it is hoped that the research discussed here offers the first steps on that pathway.

TABLES

Technique:	*Major datable materials*:[1]	*Age range (1,000 years BP)*:[2]	*Uncertainty*:
Potassium-argon	Volcanic	1 to 1,000+	Error band (±1,000 to ±10,000 years)[3]
Uranium-series	Stalagmitic calcite; volcanic; shells; tooth enamel; bone etc.	Present to 90	Error band
Fission-track	Volcanic; glass; obsidian; slag	0.5 to 1,000+	Error band (±10%)[4]
Electron spin resonance	Stalagmitic calcite; tooth enamel; shell	c.1 to 100+	Error band (< ±7% for teeth with low uranium content)[5]
Amino acid racemization	Tooth enamel; shell; bone etc.	1 to 1,000+	Error band (±5 to 10%) or calibrated[6]
Obsidian hydration	Obsidian	Present to 1,000+	Error band (minimal if relative dating of samples from same site)[7]
Quartz hydration[8]	Quartz	0.1 to 100	Error band (25-30%)
Archaeomagnetism	Spatially fixed baked clays; burnt rocks[9]	0 to 8[10]	Error band (±10 to ±100) years[11]
Magnetostratigraphy	Volcanic	90 to 1,000+	Error band (±10,000 years)[12]

TABLE 3.1 - SUMMARY OF DATING TECHNIQUES NOT DIRECTLY IMPLEMENTED IN THE TGIS.

[1] Aitken 1990: 3 - Fig 1.1
[2] Aitken 1990: 4 - Fig 1.2
[3] Aitken 1990: 123
[4] Aitken 1990: 135
[5] Taylor and Aitken 1997: 229
[6] Aitken 1990: 209-11
[7] Aitken 1990: 218; Taylor and Aitken 1997: 314
[8] Ericson *et al.* 2004
[9] Aitken 1990: 225; Gose 2000: 409, 420
[10] Aitken 1990: 4 - Fig 1.2
[11] Aitken 1990: 251; Sternberg 1997: 323
[12] Sternberg 1997: 323

Year band (Cal BC):	**3750-3500**		**3500-3250**		**3250-3000**		**3000-2750**	
LabCode:	*TGIS_Prob*	*TGIS_Topol*	*TGIS_Prob*	*TGIS_Topol*	*TGIS_Prob*	*TGIS_Topol*	*TGIS_Prob*	*TGIS_Topol*
Comb_[1328]<64>	0.00	After	0.00	After	0.00	After	0.00	After
Comb_[291]<25>A	0.05	Overlap Max	0.95	Overlap Min	0.00	Before	0.00	Before
GrA-31770	0.00	After	0.35	Overlap Max	0.63	Overlap All	0.01	Before
GrA-31785	0.00	After	0.00	After	0.00	After	0.00	After
GrA-31786	0.00	After	0.00	After	0.00	After	0.00	After
GrA-31799	0.76	Overlap Max	0.24	Overlap Min	0.00	Before	0.00	Before
GrA-31801	0.00	After	0.39	Overlap All	0.61	Overlap All	0.00	Before
GrA-31803	0.00	After	0.00	After	0.00	After	0.00	After
OxA-14483	0.00	After	0.37	Overlap All	0.63	Overlap Min	0.00	Before
OxA-14484	0.00	After	0.22	Overlap All	0.77	Overlap All	0.01	Before
OxA-15047	0.03	Overlap Max	0.92	Overlap All	0.05	Overlap Min	0.00	Before
OxA-15084	0.00	After	0.30	Overlap Max	0.65	Overlap All	0.03	Overlap Min
OxA-15110	0.00	After	0.00	After	0.00	After	0.00	After
OxA-15116	0.47	Overlap Max	0.52	Overlap Min	0.00	Before	0.00	Before
OxA-15899	0.98	Overlap All	0.02	Overlap Min	0.00	Before	0.00	Before
OxA-15900	0.00	After	0.34	Overlap Max	0.64	Overlap All	0.02	Overlap Min
SUERC-7596	0.00	After	0.30	Overlap Max	0.65	Overlap All	0.04	Overlap Min
SUERC-7597	0.00	After	0.36	Overlap All	0.64	Overlap All	0.00	Before
SUERC-7607	0.99	Overlap Min	0.00	Before	0.00	Before	0.00	Before
SUERC-8156	0.00	After	0.38	Overlap Max	0.61	Overlap All	0.00	Before
Year band (Cal BC):	**2750-2500**		**2500-2250**		**2250-2000**		**2000-1915**	
LabCode:	*TGIS_Prob*	*TGIS_Topol*	*TGIS_Prob*	*TGIS_Topol*	*TGIS_Prob*	*TGIS_Topol*	*TGIS_Prob*	*TGIS_Topol*
Comb_[1328]<64>	0.00	After	0.00	After	0.54	Overlap Max	0.44	Overlap All
Comb_[291]<25>A	0.00	Before	0.00	Before	0.00	Before	0.00	Before
GrA-31770	0.00	Before	0.00	Before	0.00	Before	0.00	Before
GrA-31785	0.00	After	0.43	Overlap Max	0.56	Overlap All	0.00	Before
GrA-31786	0.00	After	0.00	After	0.73	Overlap All	0.26	Overlap All
GrA-31799	0.00	Before	0.00	Before	0.00	Before	0.00	Before
GrA-31801	0.00	Before	0.00	Before	0.00	Before	0.00	Before
GrA-31803	0.00	After	0.00	After	0.61	Overlap All	0.36	Overlap All
OxA-14483	0.00	Before	0.00	Before	0.00	Before	0.00	Before
OxA-14484	0.00	Before	0.00	Before	0.00	Before	0.00	Before
OxA-15047	0.00	Before	0.00	Before	0.00	Before	0.00	Before
OxA-15084	0.00	Before	0.00	Before	0.00	Before	0.00	Before
OxA-15110	0.00	After	0.01	After	0.98	Overlap All	0.01	Overlap Min
OxA-15116	0.00	Before	0.00	Before	0.00	Before	0.00	Before
OxA-15899	0.00	Before	0.00	Before	0.00	Before	0.00	Before
OxA-15900	0.00	Before	0.00	Before	0.00	Before	0.00	Before
SUERC-7596	0.00	Before	0.00	Before	0.00	Before	0.00	Before
SUERC-7597	0.00	Before	0.00	Before	0.00	Before	0.00	Before
SUERC-7607	0.00	Before	0.00	Before	0.00	Before	0.00	Before
SUERC-8156	0.00	Before	0.00	Before	0.00	Before	0.00	Before

TABLE 6.1 - PROBABILITIES OF TREE-THROW / USAGE DATES FROM 3750 TO 1915 CAL BC IN C.250 YEAR TIME-SLICES.

Year band (Cal BC):	3366-3250		3250-3000		3000-2750		2750-2500	
LabCode:	*TGIS_Prob*	*TGIS_Topol*	*TGIS_Prob*	*TGIS_Topol*	*TGIS_Prob*	*TGIS_Topol*	*TGIS_Prob*	*TGIS_Topol*
GrA-31787	0.00	After	0.00	After	0.00	After	0.00	After
GrA-31789	0.00	After	0.00	After	0.00	After	0.00	After
GrA-31796	0.18	Overlap All	0.60	Overlap All	0.22	Overlap Min	0.00	Before
GrA-31797	0.00	After	0.00	After	0.00	After	0.00	After
GrA-31800	0.00	After	0.00	After	0.00	After	0.00	After
OxA-15081	0.00	After	0.00	After	0.00	After	0.57	Overlap All
OxA-15082	0.00	After	0.00	After	0.00	After	0.00	After
OxA-15083	0.00	After	0.00	After	0.00	After	0.00	After
OxA-15898	0.36	Overlap All	0.63	Overlap All	0.00	Before	0.00	Before
SUERC-7592	0.00	After	0.00	After	0.00	After	0.65	Overlap All
SUERC-7593	0.00	After	0.00	After	0.00	After	0.00	After
SUERC-7594	0.00	After	0.00	After	0.00	After	0.00	After
Year band (Cal BC):	**2500-2250**		**2250-2000**		**2000-1750**		**1750-1614**	
LabCode:	*TGIS_Prob*	*TGIS_Topol*	*TGIS_Prob*	*TGIS_Topol*	*TGIS_Prob*	*TGIS_Topol*	*TGIS_Prob*	*TGIS_Topol*
GrA-31787	0.00	After	0.00	After	0.20	Overlap Max	0.77	Overlap All
GrA-31789	0.05	Overlap Max	0.93	Overlap All	0.02	Overlap Min	0.00	Before
GrA-31796	0.00	Before	0.00	Before	0.00	Before	0.00	Before
GrA-31797	0.00	After	0.00	After	0.20	Overlap Max	0.77	Overlap All
GrA-31800	0.00	After	0.65	Overlap All	0.34	Overlap Min	0.00	Before
OxA-15081	0.42	Overlap Min	0.00	Before	0.00	Before	0.00	Before
OxA-15082	0.00	After	0.55	Overlap Max	0.44	Overlap Min	0.00	Before
OxA-15083	0.00	After	0.00	After	0.98	Overlap All	0.02	Overlap Min
OxA-15898	0.00	Before	0.00	Before	0.00	Before	0.00	Before
SUERC-7592	0.33	Overlap Min	0.00	Before	0.00	Before	0.00	Before
SUERC-7593	0.01	After	0.93	Overlap All	0.06	Overlap Min	0.00	Before
SUERC-7594	0.00	After	0.00	After	0.48	Overlap Max	0.51	Overlap All

TABLE 6.2 - PROBABILITIES OF FIRE CLEARANCE DATES FROM 3366 TO 1614 CAL BC IN C.250 YEAR TIME-SLICES.

Year band (Cal BC):	3698-3500		3500-3250		3250-2500		2500-2250	
LabCode:	*TGIS_Prob*	*TGIS_Topol*	*TGIS_Prob*	*TGIS_Topol*	*TGIS_Prob*	*TGIS_Topol*	*TGIS_Prob*	*TGIS_Topol*
OxA-14481	0.99	Overlap All	0.00	Before	0.00	Before	0.00	Before
OxA-15046	0.04	Overlap Max	0.96	Overlap Min	0.00	Before	0.00	Before
OxA-15111	0.00	After	0.00	After	0.00	After	0.00	After
OxA-15112	0.00	After	0.00	After	0.00	After	0.01	Overlap Max
OxA-15113	0.00	After	0.00	After	0.00	After	0.09	Overlap Max
OxA-15114	0.00	After	0.00	After	0.00	After	0.00	After
OxA-15115	0.00	After	0.00	After	0.00	After	0.00	After
SUERC-7598	0.00	After	0.00	After	0.00	After	0.24	Overlap Max
SUERC-7602	0.00	After	0.00	After	0.00	After	0.00	After
SUERC-7604	0.00	After	0.00	After	0.00	After	0.06	Overlap Max
SUERC-7605	0.39	Overlap Max	0.59	Overlap Min	0.00	Before	0.00	Before
SUERC-7606	0.00	After	1.00	Within	0.00	Before	0.00	Before
SUERC-7909	0.00	After	0.00	After	0.00	After	0.32	Overlap Max

Year band (Cal BC):	2250-2000		2000-1897	
LabCode:	*TGIS_Prob*	*TGIS_Topol*	*TGIS_Prob*	*TGIS_Topol*
OxA-14481	0.00	Before	0.00	Before
OxA-15046	0.00	Before	0.00	Before
OxA-15111	0.33	Overlap Max	0.64	Overlap All
OxA-15112	0.98	Overlap All	0.01	Before
OxA-15113	0.90	Overlap All	0.00	Before
OxA-15114	0.95	Overlap All	0.04	Overlap Min
OxA-15115	0.64	Overlap Max	0.35	Overlap All
SUERC-7598	0.76	Overlap All	0.00	Before
SUERC-7602	0.91	Overlap All	0.09	Overlap All
SUERC-7604	0.93	Overlap All	0.00	Before
SUERC-7605	0.00	Before	0.00	Before
SUERC-7606	0.00	Before	0.00	Before
SUERC-7909	0.67	Overlap All	0.00	Overlap All

TABLE 6.3 - PROBABILITIES OF BURNT MOUND I DATES FROM 3698 TO 1897 CAL BC IN C.250 YEAR TIME-SLICES.

Year band (Cal BC):	2131-2000		2000-1750		1750-1500		1500-1250	
LabCode:	*TGIS_Prob*	*TGIS_Topol*	*TGIS_Prob*	*TGIS_Topol*	*TGIS_Prob*	*TGIS_Topol*	*TGIS_Prob*	*TGIS_Topol*
Comb_Trough	0.00	After	0.00	After	0.00	After	0.01	Overlap Max
GrN30408	0.00	After	0.00	After	0.00	After	0.00	After
GrN30409	0.00	After	0.00	After	0.00	After	0.00	After
GrN30410	0.00	After	0.00	After	0.00	After	0.00	After
GrN30411	0.00	After	0.00	After	0.00	After	0.00	After
GrN30412	0.00	After	0.00	After	0.00	After	0.26	Overlap Max
Poz18006	0.00	After	0.00	After	0.00	After	0.00	After
Poz18007	0.00	After	0.00	After	0.00	After	0.00	After
Poz18009	0.00	After	0.00	After	0.00	After	0.61	Overlap Max
Poz18010	0.00	After	0.00	After	0.00	After	0.02	Overlap Max
Poz18029	0.48	Overlap All	0.50	Overlap Min	0.00	Before	0.00	Before

Year band (Cal BC):	1250-1000		1000-921	
LabCode:	*TGIS_Prob*	*TGIS_Topol*	*TGIS_Prob*	*TGIS_Topol*
Comb_Trough	0.98	Overlap All	0.00	Before
GrN30408	1.00	Overlap All	0.00	Before
GrN30409	1.00	Overlap All	0.00	Before
GrN30410	0.80	Overlap All	0.18	Overlap All
GrN30411	0.97	Overlap All	0.03	Overlap All
GrN30412	0.73	Overlap Min	0.00	Before
Poz18006	0.52	Overlap Max	0.44	Overlap All
Poz18007	0.98	Overlap Max	0.02	Overlap All
Poz18009	0.37	Overlap Min	0.00	Before
Poz18010	0.98	Overlap All	0.00	Before
Poz18029	0.00	Before	0.00	Before

TABLE 6.4 - PROBABILITIES OF BURNT MOUND II DATES FROM 2131 TO 921 CAL BC IN C.250 YEAR TIME-SLICES.

Number:	*Description*:	*Date*:
1	Lower Nene valley greywares	AD 110-250
2	Lower Nene valley coloured and painted wares	AD 180-400
3	Lower Nene valley coloured and painted wares	AD 110-175
4	Lower Nene valley mortaria	AD 180-360
5	Early grog tempered wares	10 BC-AD 100
6	Soft pink grogged wares	AD 175-370
7	Upper and middle Nene greywares	AD 110-275
8	Early Roman oxidised fabrics	AD 75-200
9	Early upper Nene greywares	AD 75-140
10	Oxidised Oxfordshire wares	AD 220-400
11	Oxfordshire colour coated wares	AD 220-400
12	Early fine grog / shell tempered wares	10 BC-AD 100
13	Late Roman greywares	AD 215-400
14	Late Roman sandy oxidised wares	AD 200-390
15	Black Burnished Ware 1	AD 180-390
16	Second century grogged wares	AD 110-180
17	Mancetter Hartshill mortaria	AD 170-360
18	Lincolnshire mortaria	AD 130-360
19	South and central Gaulish Samian	AD 60-220
20	East Gaulish Samian	AD 80-190
21	Later Harrold shelly wares	AD 160-400
22	Cambridgeshire shelly wares	AD 125-250
23	Early shell tempered wares	AD 1-60
24	Second century sandy greywares	AD 75-190
25	Mid Roman greywares	AD 140-260
26	South / central Northamptonshire sandy wares	AD 140-260
27	White / cream sandy wares	AD 90-180
28	Other shell tempered wares	AD 20-410
29	Spanish amphorae	AD 160-370
30	Early - Middle Saxon wares	AD 400-500
31	Early Harrold wares	AD 10-90
32	Handmade Iron Age wares	110 BC-AD 40
33	Quartz tempered wares	AD 25-400

TABLE 7.1 - LIST OF POTTERY FABRICS AND THEIR RESPECTIVE DATES (AFTER TAYLOR 1996: 456-457, FIG 7.2; JEREMY TAYLOR, PERS. COMM.).

Phase:	***Fabrics***:	***Date***:
Iron Age	32	c.4th - 1st centuries BC
Early Roman	5, 12, 23, 31	c.1st century BC - mid 1st century AD
Early / Middle Roman	8, 9, 20, 24, 27	Later 1st century AD - c.AD 175
Middle Roman	1, 3, 7, 16, 19, 22, 25, 26	c.AD 130-260
Mid / Late Roman	2, 4, 6, 17, 18, 21, 29	c.AD 175 - mid 4th century AD and later
Late Roman	10, 11, 13, 14, 15	c.AD 275 - mid 4th century AD and later
Anglo-Saxon	30	5th - 8th centuries AD

TABLE 7.2 - LIST OF POTTERY PHASES AND THEIR RESPECTIVE DATES (AFTER TAYLOR 1996: 188).

BIBLIOGRAPHY

Adam, Barbara. 1994. "Perceptions of time." In Ingold, Tim (ed.) *Companion encyclopaedia of anthropology. Humanity, culture and social life.* London: Routledge, pp. 503-526.

Agrawal, D.P., Bhalakia, V. and Kusumgar, S. 1999. "Indian and other concepts of time: a holistic framework." In Murray 1999, pp. 28-37.

Aitken, Martin J. 1990. *Science-based dating in archaeology.* London: Longman.

Aitken, Martin J. 1997. "Luminescence dating." In Taylor and Aitken 1997, pp. 183-216.

Aldenderfer, Mark. 2005. "Statistics for archaeology." In Maschner, Herbert D.G. and Chippendale, Christopher (eds.) *Handbook of archaeological methods. Volume I.* Oxford: Altamira, pp. 501-553.

Allen, Peter. 1997. "Models of creativity: towards a new science of history." In Van der Leeuw and McGlade 1997b, pp. 39-56.

Alvey, Brian A.P. 1993. "Interpreting archaeology with Hindsight: the use of three dimensions in graphic recording and site analysis." In Harris *et al.* 1993, pp. 218-228.

Arroyo-Bishop, D. and Lantada Zarzosa, M.T. 1995. "To be or not to be: will an object-space-time GIS/AIS become a scientific reality or end up an archaeological entity?" In Lock and Stančič 1995, pp. 45-53.

Backhouse, Paul. 2006. "Drowning in data? Digital data in a British contracting unit." In Evans and Daly 2006, pp. 50-58.

Bagg, Janet and Ryan, Nick. 1997. "Modelling historical change in southern Corsica: temporal GIS development using an extensible database system." In Kemp, Zarine (ed.) *Innovations in GIS 4.* London: Taylor & Francis, pp. 42-55.

Bailey, Douglass. 1993. "Chronotypic tension in Bulgarian prehistory: 6500-3500 BC." *World Archaeology,* **25**(2), pp. 204-222.

Bailey, Geoff. 1981. "Concepts, time-scales and explanations in economic prehistory." In Sheridan, Alison and Bailey, Geoff (eds.) *Economic archaeology. Towards an integration of ecological and social approaches.* BAR International Series **96**, pp. 97-117.

Bailey, Geoff. 1983. "Concepts of time in quaternary prehistory." *Annual Review of Anthropology,* **12**, pp. 165-192.

Bailey, Geoff. 1987. "Breaking the time barrier." *Archaeological Review from Cambridge,* **6**(1), pp. 5-20.

Bailey, Geoff. 2006. "Time's arrow: the measurement and theory of archaeological time." *Antiquity,* **80**(309), pp. 717-720.

Bailey, Trevor C. and Gatrell, Anthony C. 1995. *Interactive spatial data analysis.* Harlow: Longman.

Baines, Andrew and Brophy, Kenneth. 2006. "What's another word for thesaurus? Data standards and clarifying the past." In Evans and Daly 2006, pp. 236-250.

Barceló, Juan A., Briz, Ivan and Via, Assumpció (eds.) 1999. *New techniques for old times: CAA 98.* BAR International Series **757**.

Barham, Anthony J. 1995. "Methodological approaches to archaeological context recording: X-radiography as an example of a supportive recording, assessment and interpretive technique." In Barham, Anthony J. and Macphail, Richard I. (eds.) *Archaeological sediments and soils: analysis, interpretation and management.* London: Institute of Archaeology, University College, pp. 145-182.

Barker, Graeme. 1991. "Two Italys, one valley: an *Annaliste* perspective." In Bintliff 1991a, pp. 34-56.

Barker, Philip. 1993. *Techniques of archaeological excavation.* 3rd edn. London: Batsford.

Barrett, John C. 1993. "Chronologies of remembrance: the interpretation of some Roman inscriptions." *World Archaeology,* **25**(2), pp. 236-247.

Bayliss, Alex, Bronk Ramsey, Christopher, Van der Plicht, J. and Whittle, Alastair. 2007. "Bradshaw and Bayes: towards a timetable for the Neolithic." *Cambridge Archaeological Journal,* **17**(1), pp. 1-28.

Beamish, Matthew G. 2009. *Neolithic and Bronze Age activity on the Trent valley floor. Excavations at Egginton and Willington, Derbyshire, 1998–1999.* Derbyshire Archaeological Journal, **129**, pp. 17-172.

Bender, Barbara, Hamilton, Sue and Tilley, Christopher. 1997. "Leskernick: Stone Worlds; Alternative Narratives; Nested Landscapes." *Proceedings of the Prehistoric Society,* **63**, pp. 147-178.

Binding, Ceri, May, Keith, and Tudhop, Douglas. 2008. "Semantic interoperability in archaeological datasets: data mapping and extraction via the CIDOC CRM." In Christensen-Dalsgaard, Birte, Castelli, Donatella, Ammitzbøll Jurik, Bolette, and Lippincott, Joan (eds.) *Research and Advanced Technology for Digital Libraries. 12th European Conference, ECDL 2008, Aarhus, Denmark, September 14-19, 2008. Proceedings.* Berlin: Springer, pp. 280-290. http://www.springerlink.com/content/h334h73737705708/

Bintliff, John (ed.) 1991a. *The Annales school and archaeology.* Leicester: Leicester University Press.

Bintliff, John. 1991b. "The contribution of an *Annaliste* / structural history approach to archaeology." In Bintliff 1991a, pp. 1-33.

Bradley, Matt. 2006. "Archaeological survey in a digital world." In Evans and Daly 2006, pp. 35-49.

Bradley, Richard. 1991. "Ritual, time and history." *World Archaeology,* **23**(2), pp. 209-219.

Bradley, Richard. 1998. "Ruined buildings, ruined stones: enclosures, tombs and natural places in the Neolithic of south-west England." *World Archaeology,* **30**(1), pp. 13-22.

Bradley, Richard. 2002. *The past in prehistoric societies.* London: Routledge.

Bradley, Richard. 2003. "The translation of time." In Van Dyke and Alcock 2003, pp. 221-227.

Braudel, Fernand. 1980. *On history.* London: Weidenfeld and Nicolson.

Brigham, Trevor, Watson, Bruce, Tyers, Ian and Bartkowiak, Ryszard. 1996. "Current archaeological work at Regis House in the City of London (part 1)." *London Archaeologist,* **8**(2), pp. 31-38.

Brown III, Marley R. and Harris, Edward C. 1993. "Interfaces in archaeological stratigraphy." In Harris *et al.* 1993, pp. 7-20.

Brown, Anthony G. 2000. "Floodplain vegetation history: clearings as potential ritual spaces." In Fairbairn, Andrew S. (ed.) *Plants in Neolithic Britain and beyond.* Oxford: Oxbow, pp. 49-62.

Bulliet, Richard. 1992a. "*Annales* and archaeology." In Knapp 1992a, pp. 131-134.

Bulliet, Richard W. 1992b. "Pottery styles and social status in medieval Khurasan." In Knapp 1992a, pp. 75-82.

Burch, Mark, Lees, Duncan, Hill, Julian, Rowsome, Peter, Jones, Sarah and Treveil, Paul. 1997. "Number 1 Poultry -- the main excavation: Roman sequence." *London Archaeologist,* **8**(5), pp. 127-136.

Burenhult, Göran (ed.) 2002. *Archaeological informatics: pushing the envelope. CAA 2001. Computer applications and quantitative methods in archaeology, proceedings of the 29th conference, Gotland, April 2001.* BAR International Series **1016**.

Burrough, Peter and McDonnell, Rachael A. 1998. *Principles of geographical information systems.* Oxford: Oxford University Press.

Carver, Martin. 1987. *Underneath English towns. Interpreting urban archaeology.* London: Batsford.

Castleford, John. 1992. "Archaeology, GIS, and the time dimension: an overview." In Lock, Gary and Moffet, Jonathan (eds.) *CAA91: computer applications and quantitative methods in archaeology.* BAR International Series **S577**, pp. 95-106.

Ceccarelli, Letizia and Niccolucci, Franco. 2003. "Modelling time through GIS technology: the ancient Prile lake (Tuscany, Italy)." In Doerr and Sarris 2003, pp. 133-138.

Chomicki, Jan. 1994. "Temporal query languages: a survey." In Gabbay, Dov M. and Ohlbach, Hans J. (eds.) *Lecture notes in artificial intelligence 827. Temporal Logic. First international conference, ICTL '94. Bonn, Germany, July 11-14, 1994. Proceedings.* London: Springer-Verlag, pp. 506-534.

Chrisman, Nicholas. 1997. *Exploring geographic information systems.* Chichester: John Wiley & Sons.

Claxton, J.B. 1995. "Future enhancements to GIS: implications for archaeological theory." In Lock and Stančič 1995, pp. 335-348.

Cook, Andrea, Wadsworth, Jeffrey, Southton, John R. and Van der Merwe, Nikolaas J. 2003. "AMS radiocarbon dating of rusty iron." *Journal of Archaeological Science,* **30**(1), pp. 95-101.

Daly, Patrick T. 2003. *Social practice and material culture: the use, discard, and deposition of ceramic material at two Iron Age hillforts in Oxfordshire.* Unpublished D.Phil. thesis. Oxford: University of Oxford.

Daly, Patrick and Evans, Thomas L. 2006a. "Afterword." In Evans and Daly 2006, pp. 253-255.

Daly, Patrick and Evans, Thomas L. 2006b. "Introduction. Archaeological theory and digital pasts." In Evans and Daly 2006, pp. 3-9.

Daly, Patrick and Lock, Gary. 1999. "Timing is everything: commentary on managing temporal variables in geographic information systems." In Barceló *et al.* 1999, pp. 287-293.

Davies, Martin. 1993. "The application of the Harris Matrix to the recording of standing structures." In Harris *et al.* 1993, pp. 167-189.

Dean, Jeffrey S. 1997. "Dendrochronology." In Taylor and Aitken 1997, pp. 31-64.

Delano Smith, Catherine. 1992. "The Annales for archaeology?" *Antiquity* **66**(251), pp.539-541.

Dietler, Michael. 1998. "A tale of three sites: the monumentalization of Celtic oppida and the politics of collective memory and identity." *World Archaeology,* **30**(1), pp. 72-89.

Doerr, Martin and Sarris, Apostolos (eds.) 2003. *CAA 2002: the digital heritage of archaeology.* Athens: Hellenic Ministry of Culture.

Dolukhanov, Pavel, Sokoloff, Dmitry and Shukurov, Anvar. 2001. "Radiocarbon chronology of upper Palaeolithic sites in eastern Europe at improved resolution." *Journal of Archaeological Science,* **28**(7), pp. 699-712.

Dragovich, D. 2000. "Rock engraving chronologies and accelerator mass spectrometry radiocarbon age of desert varnish." *Journal of Archaeological Science,* **27**(10), pp. 871-876.

Egenhofer, Max J. and Golledge, Reginald G. (eds.) 1998. *Spatial and temporal reasoning in geographic information systems.* Oxford: Oxford University Press.

Ericson, Jonathon E., Dersch, Oliver and Rauch, Friedel. 2004. "Quartz hydration dating." *Journal of Archaeological Science,* **31**(7), pp. 883-902.

Evans, Thomas L. 2006. "You, me and it. The application of simple quantitative techniques in the examination of gender, identity and social reproduction in the early to middle Iron Age of north-eastern France." In Evans and Daly 2006, pp. 61-93.

Evans, Thomas L. and Daly, Patrick (eds.) 2006. *Digital archaeology. Bridging method and theory.* Abingdon: Routledge.

Farid, Shahina. 2000. "The excavation process at Çatalhöyük." In Hodder 2000b, pp. 19-35.

Feathers, James K. 2002. "Luminescence dating in less than ideal conditions: case studies from Klasies River main site and Duinefontein, South Africa." *Journal of Archaeological Science,* **29**(2), pp. 177-194.

Fletcher, Roland. 1992. "Time perspectivism, *Annales*, and the potential of archaeology." In Knapp 1992a, pp. 35-49.

Frachetti, Michael. 2006. "Digital archaeology and the scalar structure of pastoral landscapes." In Evans and Daly 2006, pp. 128-147.

Frank, Andrew U. 1998. "Different types of 'times' in GIS." In Egenhofer and Golledge 1998, pp. 40-62.

Friedman, Jonathan. 1982. "Catastrophe and continuity in social evolution." In Renfrew, Colin, Rowlands, Michael and Segraves, Barbara (eds.) *Theory and explanation in archaeology.* London: Academic Press, pp. 175-196.

Gaffney, V., Stančič, Z. and Watson, H. 1995. "The impact of GIS on archaeology: a personal perspective." In Lock and Stančič 1995, pp. 211-229.

Gaffney, V. and Van Leusen, M. 1995. "Postscript - GIS, environmental determinism and archaeology: a parallel text." In Lock and Stančič 1995, pp. 367-382.

Gell, Alfred. 1992. *The anthropology of time. Cultural constructions of temporal maps and images.* Oxford: Berg.

Gillings, Mark. 1998. "Embracing uncertainty and challenging dualism in the GIS-based study of a palaeo-flood plain." *European Journal of Archaeology,* **1**(1), pp. 117-144.

Gillings, Mark and Pollard, Joshua. 1999. "Non-portable stone artifacts and contexts of meaning: the tale of Grey Wether." *World Archaeology,* **31**(2), pp. 179-193.

Gkiasta, Marina, Russell, Thembi, Shennan, Stephen and Steele, James. 2003. "Neolithic transition in Europe: the radiocarbon record revisited." *Antiquity,* **77**(1), pp. 45-62.

Going, C.J. 1992. "Economic 'long waves' in the Roman period? A reconnaissance of the Romano-British ceramic evidence." *Oxford Journal of Archaeology,* **11**(1), pp. 93-117.

Gosden, Chris and Lock, Gary. 1998. "Prehistoric histories." *World Archaeology,* **30**(1), pp. 2-12.

Gosden, Chris and Marshall, Yvonne. 1999. "The cultural biography of objects." *World Archaeology,* **31**(2), pp. 169-178.

Gosden, Christopher. 1994. *Social being and time.* Oxford: Blackwell.

Gose, Wulf A. 2000. "Palaeomagnetic studies of burned rocks." *Journal of Archaeological Science,* **27**(5), pp. 409-421.

Gräslund, Bo. 1987. *The birth of prehistoric chronology.* Cambridge: Cambridge University Press.

Green, Christopher. 2002. *C14View: an experiment in the creation of a temporal GIS for archaeology.* Unpublished M.Sc. dissertation. Leicester: University of Leicester.

Gregory, Lady. 1904. *Gods and fighting men.* 2nd edn. (1970) Gerrards Cross: Colin Smythe.

Hamilton, Caroline. 2000. "Faultlines: the construction of archaeological knowledge at Çatalhöyük." In Hodder 2000b, pp. 119-127.

Hammond, Norman. 1993. "Matrices and Maya archaeology." In Harris *et al.* 1993, pp. 139-152.

Harding, Jan. 2005. "Rethinking the great divide: long-term structural history and the temporality of the event." *Norwegian Archaeological Review*, **38**(2), pp. 88-101.

Harding, Jan, Thomas, Julian, Murray, Tim and Olivier, Laurent. 2006. "Comments on Jan Harding (2005): Rethinking the great divide: long-term structural history and the temporality of the event. Norwegian Archaeological Review 38, 88-101." *Norwegian Archaeological Review*, **39**(1), pp. 80-97.

Harris, Edward C. 1989. *Principles of archaeological stratigraphy.* 2nd edn. London: Academic Press.

Harris, Edward C., Brown III, Marley R. and Brown, Gregory J. (eds.) 1993. *Practices of archaeological stratigraphy.* London: Academic Press.

Harris, T. M. and Lock, G.R. 1995. "Toward an evaluation of GIS in European archaeology: the past, present and future of theory and applications." In Lock and Stančič 1995, pp. 349-365.

Hermon, Sorin and Niccolucci, Franco. 2003. "A fuzzy logic approach to typology in archaeological research." In Doerr and Sarris 2003, pp. 307-310.

Heuvelink, Gerard B.M. and Burrough, Peter A. 2002. "Guest editorial: developments in statistical approaches to spatial uncertainty and its propagation." *International Journal of Geographical Information Science,* **16**(2), pp. 111-113.

Higham, T.F.G. and Horn, P.L. 2000. "Seasonal dating using fish otoliths: results from the Shag River Mouth site, New Zealand." *Journal of Archaeological Science,* **27**(5), pp. 439-448.

Higham, Tom, Bronk Ramsey, Christopher and Owen, Clare (eds.) 2004. *Radiocarbon and archaeology: proceedings of the 4th Symposium, Oxford 2002.* Oxford: Oxford University School of Archaeology.

Hodder, Ian. 1993. "The narrative and rhetoric of material culture sequences." *World Archaeology,* **25**(2), pp. 268-282.

Hodder, Ian. 1996a. "Conclusions." In Hodder 1996b, pp. 359-366.

Hodder, Ian (ed.) 1996b. *On the surface: Çatalhöyük 1993-1995.* Cambridge: McDonald Institute for Archaeological Research (British Institute for Archaeology at Ankara).

Hodder, Ian. 1996c. "Re-opening Çatalhöyük." In Hodder 1996b, pp. 1-18.

Hodder, Ian. 2000a. "Developing a reflexive method in archaeology." In Hodder 2000b, pp. 3-14.

Hodder, Ian (ed.) 2000b. *Towards a reflexive method in archaeology: the example at Çatalhöyük.* Cambridge: McDonald Institute for Archaeological Research (British Institute for Archaeology at Ankara).

Hodder, Ian and Orton, Clive. 1976. *Spatial analysis in archaeology.* Cambridge: Cambridge University Press.

Holtorf, Cornelius. 1996. "Towards a chronology of megaliths: understanding monumental time and cultural memory." *Journal of European Archaeology,* **4**, pp. 119-152.

Holtorf, Cornelius. 1998. "The life-histories of megaliths in Mecklenburg-Vorpommern (Germany)." *World Archaeology,* **30**(1), pp. 23-38.

Holtorf, Cornelius. 2002. "Notes on the life history of a pot sherd." *Journal of Material Culture,* 7(1), pp. 49-71.

Hornsby, Kathleen and Egenhofer, Max J. 2000. "Identity-based change: a foundation for spatio-temporal knowledge representation." *International Journal of Geographical Information Science,* **14**(3), pp. 207-224.

Huggett, Jeremy. 2000. "Computers and archaeological culture change." In Lock, Gary and Brown, Kayt (eds.) *On the theory and practice of archaeological computing.* Oxford: Oxford University Committee for Archaeology, pp. 5-22.

Ingold, Tim. 1993. "The temporality of the landscape." *World Archaeology,* **25**(2), pp. 152-174.

Johnson, Ian. 1999. "Mapping the fourth dimension: the TimeMap project" Draft of paper published in the Proceedings of the 1997 computer applications in archaeology conference, Birmingham, UK, 1999, http://www.timemap.net/tm/documents/publications/1997_johnson_TimeMap_Project.pdf.

Johnson, Ian. 2002. "Contextualising archaeological information through interactive maps." *Internet Archaeology,* **12**.

Johnson, Ian. 2004a. "Aoristic analysis: seeds of a new approach to mapping archaeological distributions through time." In Magistrat de Stadt Wien, Referat Kulturelles Erbe and Stadtarchäologie Wien (eds.) *[Enter the past] The E-way into the four dimensions of cultural heritage. CAA2003. Computer applications and quantitative methods in archaeology, proceedings of the 31st conference, Vienna, Austria, April 2003.* BAR International Series **1227**, pp. 448-452.

Johnson, Ian. 2004b. "Putting time on the map. Using TimeMap for map animation and web delivery." *Geo Informatics,* **7**, pp. 26-29.

Johnson, Ian. 2005. "Indexing and delivery of historical maps using TimeMap." *National Library of Australia News,* **15**(4), http://www.nla.gov.au/pub/nlanews/2005/jan05/article2.html.

Johnson, Ian and Wilson, Andrew. 2002. "The *Time*Map kiosk: delivering historical images in a spatio-temporal context." In Burenhult 2002, pp. 71-78.

Johnson, Ian and Wilson, Andrew. 2003. "The TimeMap project: developing time-based GIS display for cultural data." *Journal of GIS in Archaeology,* **1**, pp. 123-135.

Jones, Christopher. 1997. *Geographical information systems and computer cartography.* Harlow: Longman.

Jones, Rick. 1991. "Archaeology, the *longue durée* and the limits of the Roman Empire." In Bintliff 1991a, pp. 93-107.

Joyce, Rosemary. 2002a. "Dialogues heard and unheard, seen and unseen." In Joyce 2002b, pp. 39-67.

Joyce, Rosemary (ed.) 2002b. *The languages of archaeology. Dialogue, narrative, and writing.* Oxford: Blackwell.

Joyce, Rosemary. 2002c. "The return of the first voice." In Joyce 2002b, pp. 133-144.

Joyce, Rosemary and Preucel, Robert. 2002. "Writing the field of archaeology." In Joyce 2002b, pp. 18-38.

Knapp, A.B. (ed.) 1992a. *Archaeology, Annales and ethnohistory.* Cambridge: Cambridge University Press.

Knapp, A.B. 1992b. "Archaeology and *Annales*: time, space and change." In Knapp 1992a, pp. 1-21.

Knapp, A.B. 1992c. "Independence and imperialism: politico-economic structures in the Bronze Age Levant." In Knapp 1992a, pp. 83-98.

Kobyliński, Zbigniew. 1993. "Polish medieval excavations and the Harris matrix: applications and developments." In Harris *et al.* 1993, pp. 57-67.

Kraak, Menno-Jan and Ormeling, Ferjan. 1996. *Cartography: visualization of spatial data.* Harlow: Prentice Hall.

Langran, Gail. 1992. *Time in geographic information systems.* London: Taylor & Francis.

Laurini, Robert and Thompson, Derek. 1992. *Fundamentals of spatial information systems.* London: Academic Press.

Leibhammer, Nessa. 2000. "Rendering realities." In Hodder 2000b, pp. 129-142.

Leone, Mark. 1978. "Time in American archaeology." In Redman, Charles, Langhorne, William, Berman, Mary J., Versaggi, Nina, Curtin, Edward and Wanser, Jeffery (eds.) *Social archaeology. Beyond subsistence and dating.* New York: Academic Press, pp. 25-36.

Lipschutz, S. and Schiller, J. 1998. *Introduction to probability and statistics.* London: McGraw-Hill.

Llobera, Marcus. 2006. "What you see is what you get? Visualscapes, visual genesis and hierarchy." In Evans and Daly 2006, pp. 148-167.

Lock, Gary and Stančič, Zoran (eds.) 1995. *Archaeology and geographic information systems.* London: Taylor & Francis.

Lock, Gary. 2003. *Using computers in archaeology.* London: Routledge.

Lock, Gary. 2006. "Computers, learning and teaching in archaeology. Life past and present on the screen." In Evans and Daly 2006, pp. 226-235.

Lock, Gary and Daly, Patrick. 1999. "Looking at change, continuity and time in GIS: an example from the Sangro Valley, Italy." In Barceló *et al.* 1999, pp. 259-263.

Lock, Gary and Harris, Trevor. 2000. *Analysing change through time within a cultural landscape: conceptual and functional limitations of a GIS approach.* http://www.arkeologi.uu.se/afr/projects/BOOK/lockframe.htm.

Lopiparo, Jeanne. 2002. "A second voice: crafting cosmos." In Joyce 2002b, pp. 68-89.

Lucas, Gavin. 1997. "Forgetting the past." *Anthropology Today,* **13**(1), pp. 8-14.

Lucas, Gavin. 2001. *Critical approaches to fieldwork. Contemporary and historical archaeological practice.* London: Routledge.

Lucas, Gavin. 2005. *The archaeology of time.* Abingdon: Routledge.

Mannino, Marcello A., Baruch, F.S. and Thomas, Kenneth D. 2003. "Sampling shells for seasonality: oxygen isotope analysis on shell carbonates of the inter-tidal gastropod *monodonta lineata* (da Costa) from populations across its modern range and from a Mesolithic site in southern Britain." *Journal of Archaeological Science,* **30**(6), pp. 667-679.

Marble, D.F. 1990. "The potential methodological impact of geographic information systems on the social sciences." In Allen, K.M.S., Green, S.W., and Zubrow, E.W. (eds.) *Interpreting space: GIS and archaeology.* London: Taylor & Francis.

Marsh, Geoff. 1981. "London's samian supply and its relationship to the development of the Gallic samian industry." In Anderson, A.C. and Anderson, A.S. *Roman pottery research in Britain and north-west Europe.* BAR International Series **123**(i), pp. 173-238.

Marshall, P., Hamilton, W. D., Van der Plicht, J., Bronk Ramsey, C., Cook, G. and Goslar, T.. 2009a. "Scientific dating." In Beamish 2009, pp. 62-81.

Marshall, P., Hamilton, W. D., Beamish, M., Woodward, A., Van der Plicht, J., Bronk Ramsey, C., Cook, G. and Goslar, T.. 2009b. "Scientific dating." In Beamish, M. *Willington, South Derbyshire. ALSF Project Number 2517.* http://ads.ahds.ac.uk/catalogue/resources.html?willington_eh_2008

Matthews, Wendy, French, Charles, Lawrence, Timothy and Cutler, David. 1996. "Multiple surfaces: the micromorphology." In Hodder 1996b, pp. 301-342.

McGlade, James. 1987. "Chronos and the Oracle: some thoughts on time, timescales and simulation." *Archaeological Review from Cambridge,* **6**(1), pp. 21-31.

McGlade, James. 1997. "The limits of social control: coherence and chaos in a prestige goods economy." In Van der Leeuw and McGlade 1997b, pp. 298-330.

McGlade, James. 1999. "The times of history: archaeology, narrative and non-linear causality." In Murray 1999b, pp. 139-163.

McGlade, James and Van der Leeuw, Sander. 1997. "Introduction: archaeology and non-linear dynamics - new approaches to long term change." In Van der Leeuw and McGlade 1997b, pp. 1-31.

McGuire, Randall. 1992. *A Marxist archaeology.* San Diego: Academic Press.

Mellor, D.H. 1981. *Real time.* Cambridge: Cambridge University Press.

Miller, P. 1995. "How to look good and influence people: thoughts on the design and interpretation of an archaeological GIS." In Lock and Stančič 1995, pp. 319-333.

Millett, Martin. 1987. "Boudicca, the first Colchester potters' shop, and the dating of Neronian Samian." *Britannia,* **18**, pp. 93-123.

Mithen, Stephen J. 1997. "Simulating mammoth hunting and extinctions: implications for North America." In Van der Leeuw and McGlade 1997b. London: Routledge, pp. 176-215.

Mizoguchi, Koji. 1993. "Time in the reproduction of mortuary practices." *World Archaeology,* **25**(2), pp. 223-235.

Moore, Alan and Gibbons, Dave. 1986. *Watchmen.* London: Titan Books.

Moore, Henrietta. 1995. "The problem of origins: poststructuralism and beyond." In Hodder, Ian, Shanks, Michael, Alexandri, Alexandra, Buchli, Victor, Carman, John, Last, Jonathan and Lucas, Gavin (eds.) *Interpreting archaeology. Finding meaning in the past.* London: Routledge, pp. 51-53.

Moreland, John F. 1992. "Restoring the dialectic: settlement patterns and documents in medieval central Italy." In Knapp 1992a, pp. 112-129.

Murray, Tim. 1999a. "A return to the Pompeii premise." In Murray 1999b, pp. 8-27.

Murray, Tim (ed.) 1999b. *Time and archaeology.* London: Routledge.

Orton, Clive. 1980. *Mathematics in archaeology.* Cambridge: Cambridge University Press.

O'Sullivan, David and Unwin, David J. 2003. *Geographic information analysis.* Hoboken: John Wiley & Sons.

Palmer, Carol and Daly, Patrick. 2006. "Jouma's tent. Bedouin and digital archaeology." In Evans and Daly 2006, pp. 97-127.

Pettitt, P.B., Davies, W., Gamble, C.S. and Richards, M.B. 2003. "Palaeolithic radiocarbon chronology: quantifying our confidence beyond two half-lives." *Journal of Archaeological Science,* **30**(12), pp. 1685-1693.

Peuquet, Donna J. 1994. "It's about time: a conceptual framework for the representation of temporal dynamics in geographic information systems." *Annals of the Association of American Geographers,* **84**(3), pp. 441-461.

Peuquet, Donna J. 2001. "Making space for time: issues in space-time data representation." *Geoinformatica,* **5**(1), pp. 11-32.

Pluciennik, Mark. 1999. "Archaeological narratives and other ways of telling." *Current Anthropology,* **40**, pp. 653-678.

Premo, L.S. 2004. "Local spatial autocorrelation statistics quantify multi-scale patterns in distributional data: an example from the Maya lowlands." *Journal of Archaeological Science,* **31**(7), pp. 855-866.

Ramenofsky, Ann. 1998. "The illusion of time." In Ramenofsky, Ann and Steffen, Anastasia (eds.) *Unit issues in archaeology. Measuring time, space, and material.* Salt Lake City: University of Utah Press, pp. 74-84.

Raper, Jonathan. 2000. *Multidimensional geographic information science.* London: Taylor & Francis.

Ratcliffe, Jerry H. 2000. "Aoristic analysis: the spatial interpretation of unspecific temporal events." *International Journal of Geographical Information Science,* **14**(7), pp. 669-679.

Ratcliffe, Jerry H. 2002. "Aoristic signatures and the spatio-temporal analysis of high volume crime patterns." *Journal of Quantitative Criminology,* **18**(1), pp. 23-43.

Reimer, P.J., Baillie, M.G.L., Bard, E., Bayliss, A., Beck, J.W., Bertrand, C., Blackwell, P.G., Buck, C.E., Burr, G., Cutler, K.B., Damon, P.E., Edwards, R.L., Fairbanks, R.G., Friedrich, M., Guilderson, T.P., Hughen, K.A., Kromer, B., McCormac, F.G., Manning, S., Bronk Ramsey, C., Reimer, R.W., Remmele, S., Southon, J.R., Stuiver, M., Talamo, S., Taylor, F.W., Van der Plicht, J. and Weyhenmeyer, C.E. 2004. *Radiocarbon,* **46**, pp. 1029-1058.

Richter, D., Waiblinger, J., Rick, W.J. and Wagner, G.A. 2000. "Thermoluminescence, electron spin resonance and ^{14}C-dating of the late middle and early upper Palaeolithic site of Geißenklösterle Cave in southern Germany." *Journal of Archaeological Science,* **27**(1), pp. 71-89.

Russell, Thembi. 2004. *The spatial analysis of radiocarbon databases. The spread of the first farmers in Europe and of the fat-tailed sheep in southern Africa.* BAR International Series **1294**.

Russell, Thembi and Steele, James. 2002. "Visualising the Neolithic transition in Europe." In Burenhult 2002, pp. 511-516.

Schiffer, Michael B. 1987. *Formation processes of the archaeological record.* Albuquerque: University of New Mexico Press.

Scott, E.M., Bryant, C., Carmi, I., Cook, G., Gulliksen, S., Harkness, D., Heinemeier, J., McGee, E., Naysmith, P., Possnert, G., Van der Plicht, H. and Van Stroydonck, M. 2004. "Precision and accuracy in applied ^{14}C dating: some findings from the fourth International Radiocarbon Inter-comparison." *Journal of Archaeological Science,* **31**(9), pp. 1209-1213.

Shanks, Michael and Tilley, Christopher. 1987a. "Abstract and substantial time." *Archaeological Review from Cambridge,* **6**(1), pp. 32-41.

Shanks, Michael and Tilley, Christopher. 1987b. *Social theory and archaeology.* Cambridge: Polity.

Sherratt, Andrew. 1992. "What can archaeologists learn from Annalistes?" In Knapp 1992a, pp. 135-142.

Šmejda, Ladislav. 2004. "Potential of GIS for analysis of funerary areas: prehistoric cemetery at Holešov, dist. Kroměříž, Czech Republic." In Šmejda, Ladislav and Turek, Jan (eds.) *Spatial analysis of funerary areas.* Plzeň: University of West Bohemia, Department of Archaeology, pp. 57-68.

Smith, Michael. 1992a. "Braudel's temporal rhythms and chronology theory in archaeology." In Knapp 1992a, pp. 23-34.

Smith, Michael E. 1992b. "Rhythms of change in Postclassic central Mexico: archaeology, ethnohistory, and the Braudelian model." In Knapp 1992a, pp. 51-74.

Spence, Craig. 1993. "Recording the archaeology of London: the development and implementation of the DUA recording system." In Harris *et al.* 1993, pp. 23-46.

Squair, Robert. 1994. "Time and the privilege of retrospect." In Mackenzie, Iain (ed.) *Archaeological theory: progress or posture?* Aldershot: Avebury, pp. 92-113.

Stead, Stephen D. 1998. "Temporal dynamics and geographic information systems." In Egenhofer, Max J. and Golledge, Reginald G. (eds.) *Spatial and temporal reasoning in geographic information systems.* Oxford: Oxford University Press, pp. 214-219.

Steiniger, Stefan and Bocher, Erwan. Forthcoming. "An overview of current free and open source desktop GIS developments." *International Journal of Geographic Information Science.*

Sternberg, Robert S. 1997. "Archaeomagnetic dating." In Taylor and Aitken 1997, pp. 323-356.

Taylor, Jeremy. 1996. *Iron Age and Roman landscapes in the East Midlands: a case study in integrated survey.* Unpublished Ph.D. thesis. Durham: University of Durham.

Taylor, R.E. 1997. "Radiocarbon dating." In Taylor and Aitken 1997, pp. 65-96.

Taylor, R.E. and Aitken, Martin J. (eds.) 1997. *Chronometric dating in archaeology.* London: Plenum Press.

Thomas, Julian. 1996. *Time, culture and identity. An interpretive archaeology.* London: Routledge.

Thomas, Julian. 2004. *Archaeology and modernity.* London: Routledge.

Thomas, Julian (ed.) 2007. *Place and memory: excavations at the Pict's Knowe, Holywood and Holm Farm, Dumfries and Galloway, 1994-8.* Oxford: Oxbow.

Tobler, W.R. 1970. "A computer movie simulating urban growth in the Detroit region." *Economic Geography* **46**(2), pp. 234-240.

Trément, Frédéric. 2000. "Prospection et chronologie: de la quantification du temps au modèle de peuplement. Méthodes appliquées au secteur des étangs de Saint-Blaise (Bouches-du-Rhône, France)." In Francovich, Riccardo and Patterson, Helen (eds.) *Extracting meaning from ploughsoil assemblages.* Oxford: Oxbow, pp. 77-91.

Trigger, Bruce. 1978. *Time and traditions. Essays in archaeological interpretation.* Edinburgh: Edinburgh University Press.

Tringham, Ruth and Stevanovic, Mirjana. 2000. "Different excavation styles create different windows into Çatalhöyük." In Hodder 2000b, pp. 111-118.

Van der Leeuw, Sander and McGlade, James. 1997a. "Structural change and bifurcation in urban evolution: a non-linear dynamical perspective." In Van der Leeuw and McGlade 1997b, pp. 331-372.

Van der Leeuw, Sander and McGlade, James (eds.) 1997b. *Time, process and structured transformation in archaeology.* London: Routledge.

Van der Plicht, J., Van der Sanden, W.A.B., Aerts, A.T. and Streurman, H.J. 2004. "Dating bog bodies by means of ^{14}C-AMS." *Journal of Archaeological Science,* **31**(4), pp. 471-491.

Van Dyke, Ruth and Alcock, Susan (eds.) 2003. *Archaeologies of memory.* Oxford: Blackwell.

Vellanoweth, René 2001. "AMS radiocarbon dating and shell bead chronologies: middle Holocene trade and interaction in western North America." *Journal of Archaeological Science,* **28**(9), pp. 941-950.

Weber, Christopher R. 1998. "The representation of spatio-temporal variation in GIS and cartographic displays: the case for sonification and auditory data representation." In Egenhofer and Golledge 1998, pp. 74-84.

Wheatley, David. 1996. "Between the lines: the role of GIS-based predictive modelling in the interpretation of extensive survey data." *Analecta Praehistorica Leidensia,* **28**(II), pp. 275-292.

Wheatley, David and Gillings, Mark. 2002. *Spatial technology and archaeology. The archaeological applications of GIS.* London: Taylor & Francis.

Whittle, Alastair, Barclay, Alistair, Bayliss, Alex, McFadyen, Leslie, Schulting, Rick and Wysocki, Michael. 2007. "Building for the dead: events, processes and changing worldviews from the thirty-eighth to the thirty-fourth centuries cal. BC in southern Britain." *Cambridge Archaeological Journal,* **17**(1), pp. 123-147.

Williams, Howard. 1998. "Monuments and the past in early Anglo-Saxon England." *World Archaeology,* **30**(1), pp. 90-108.

Wilson, Andrew. 2001. "Sydney TimeMap: integrating historical resources using GIS." *History and Computing,* **13**(1), pp. 45-68.

Worboys, Michael and Duckham, Matt. 2004. *GIS. A computing perspective.* 2nd edn. London: CRC Press.

Yoneda, Minoru, Tanaka, Atsushi, Shibata, Yasuyuki, Morita, Masatoshi, Uzawa, Kazuhiro, Hirota, Masashi and Uchida, Masao. 2002. "Radiocarbon marine reservoir effect in human remains from the Kitakogane site, Hokkaido, Japan." *Journal of Archaeological Science,* **29**(5), pp. 529-536.

Zimmermann, H. 2001. *Fuzzy set theory - and its applications.* London: Kluwer Academic.

Zubrow, Ezra. 1997. "Clusters of death, pockets of survival: dynamic modelling and GIS." In Van der Leeuw and McGlade 1997b, pp. 216-253.

Zubrow, Ezra B.W. 2006. "Digital archaeology. A historical context." In Evans and Daly 2006, pp. 10-31.

www.ingramcontent.com/pod-product-compliance
Lightning Source LLC
LaVergne TN
LVHW071216210926
848482LV00015B/820